ORGANIC CHEMISTRY 2 PRIMER 2021

BY RHETT C. SMITH, TANIA HOUJEIRY, AND ANDREW G. TENNYS

Marketed by Proton Guru

Find additional online resources and guides at protonguru.com.

There is a lot of online video content to accompany this book at the Proton Guru YouTube Channel! Just go to YouTube and search "Proton Guru Channel" to easily find our content.

**Correlating these reactions with your course:** The homepage at protonguru.com provides citations to popular textbooks for further reading on each reaction in this book, so that you can follow along using this book in any course using one of these texts.

**Instructors:** Free PowerPoint lecture slides to accompany this text can be obtained by emailing IQ@protonguru.com from your accredited institution email account. The homepage at protonguru.com provides a link to citations to popular textbooks for further reading on each Lesson topic in this primer.



Executive Editor: Rhett C. Smith, Ph.D. You can reach him through our office at: IQ@protonguru.com



Printed in the United States of America

10 9 8 7 6 5 4 3 2 1

ISBN 978-0-9991672-5-0 (IQ-Proton Guru)

# Organic Chemistry 2 Primer 2021

Rhett C. Smith, Ph.D.
Tania Houjeiry, Ph.D.
Andrew G. Tennyson, Ph.D.

# Table of Contents

# Purpose of the Primer

This brief, plain language Primer is meant to provide the basic principles and core concepts of organic chemistry. It is meant to be an excellent *primer* to read *before* lecture so that once you get to lecture you will already have some knowledge of what will be discussed. It is also meant to be an excellent pre-exam review. Especially in cases where you are preparing for a final exam – which may be cumulative and cover material that you have not specifically studied for weeks or months – it should be useful as a more compact review than will be offered by a text book or other study guides.

When used with the supporting online videos (**go to YouTube and search "Proton Guru Channel"**) and practice problems (like those found in "Organic Chemistry 2 Practice Problems with Solutions" by Rhett C. Smith), **this primer material can be used to replace a text book**. Video content offers additional explanations through worked problems.

# Online Content

There are many resources to support the material covered in this book. Many of these resources are found on protonguru.com.

**The Proton Guru website site includes:**

- Self-assessment quizzes, links to video content including solved practice problem videos and lecture videos.

Proton Guru also has a reaction and practice problem guide that can be used with this Primer. It is "Organic Chemistry 2 Practice Problems with Solutions" by Rhett C. Smith and can be found on Amazon.com.

**BONUS Lessons from PART II. Substitution, Elimination and Oxidation**

These lessons are also found in Volume 1, "Organic Chemistry 1 Primer" by Smith, Tennyson and Houjeiry. They are repeated here because some courses cover this material in the second semester course.

Lesson II.11 Reactions of Alcohols I: Substitution and Elimination

Lesson II.12 Oxidation and Reduction: Definitions

Lesson II.13 Reactions of Alcohols II: Oxidation

Lesson II.14 Acid Cleavage of Ethers

Lesson II.15 Ring-Opening of Epoxides: Steric versus Electronic Effects on Product Distribution

## Lesson II.11. Substitution and Elimination of Alcohols

*Lesson II.11.1. Activating a Poor Leaving Group with an Acid*

In our previous studies on substitution and elimination reactions, we noted that the presence of a good leaving group on the substrate was required for reaction to occur. If we attempt any of these reactions on an unmodified alcohol, our efforts will fail because the leaving group ($HO^-$) would be a strong base, and therefore a very poor leaving group. However, there are several ways to activate the OH unit of an alcohol to convert it into a good leaving group. Once this activation has taken place, the substrate is viable for both substitution and elimination reactions.

The first way to activate an alcohol is to protonate the OH group with a strong acid:

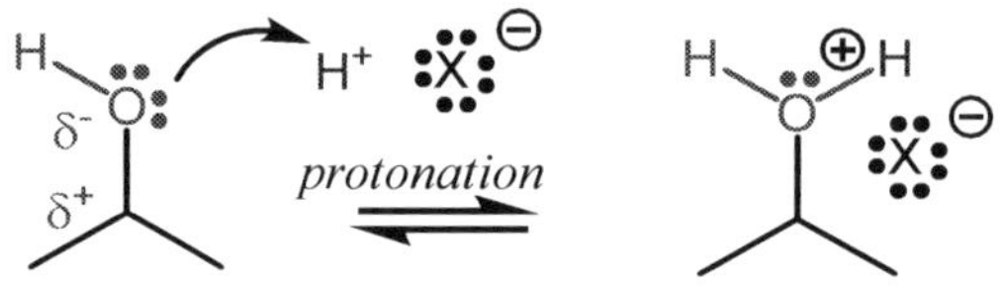

H $\delta^-$ $\delta^+$ $H^+$ X *protonation* H H X

(X = Cl, Br, I or $HSO_4$) **Oxonium Intermediate**

Once protonated, the resultant **oxonium intermediate** has a good leaving group ($H_2O$). This makes it a viable substrate for both substitution and elimination reactions.

If a reasonably strong nucleophile is present ($Cl^-$, $Br^-$ or $I^-$), a substitution pathway will dominate, with $H_2O$ as the leaving group from the oxonium intermediate. **The $S_N1$ pathway is favored for 3° alcohols** because this case leads to the most stable carbocation intermediate. If the $S_N1$ route is not viable (methyl or 1° alcohols), then the $S_N2$ product will be the major product. For secondary alcohols, conditions can be tuned to favor one or the other pathway. All of the guiding principles for $S_N2$ and $S_N1$ reactions still hold, as they were described in Lessons II.3–II.5, for protonated alcohols.

Example II.11.1

Provide a reasonable mechanistic pathway leading to the major product of the reaction shown:

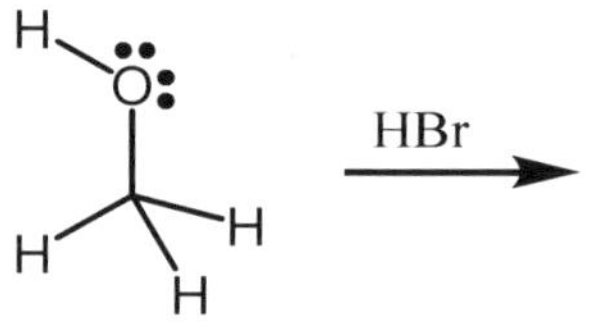

H O H H H HBr →

Solution II.11.1

The first step is protonation of the OH group to form an oxonium intermediate. This is a methyl alcohol, so the 2nd step, substitution, is an $S_N2$ reaction on the activated substrate:

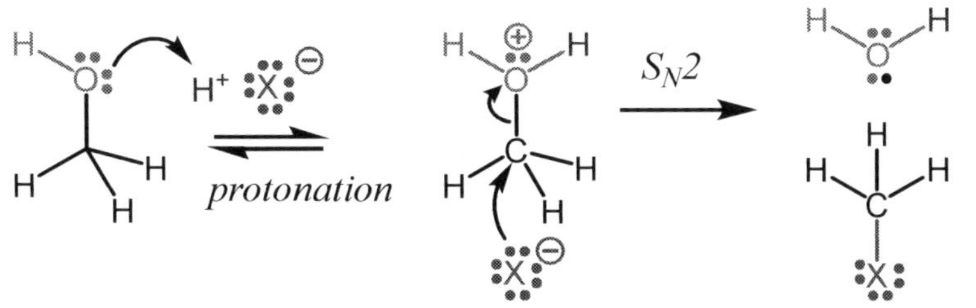

If the conjugate base of the acid used to protonate the alcohol is bulky (as is $HSO_3^-$, the conjugate base of $H_2SO_4$), then elimination will dominate. These reactions proceed by an E1 mechanism because E2 requires a strong base, which, of course, could not be present when we have added a strong acid! However, we know that the E1 reaction involves a carbocation intermediate. So, for primary substrates the E1 pathway will not be very effective. However, if we convert the alcohol to a good leaving group by some other route (as in Lesson II.11.3), it can then be reacted with a base to form an alkene via the E2 mechanism.

Example II.11.2

Provide a reasonable mechanistic pathway leading to the major product of the reaction shown:

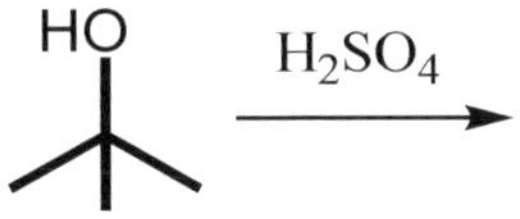

Solution II.11.2

The first step is protonation of the OH group to form the oxonium intermediate. There is no reasonable nucleophile, so an elimination reaction via the E1 pathway follows:

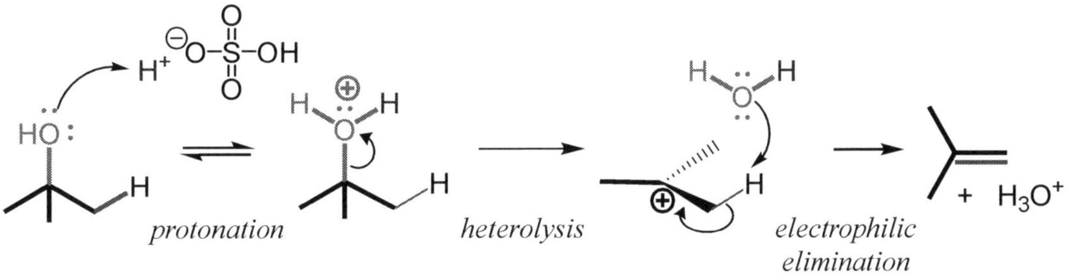

*Lesson II.11.2. Activating a Poor Leaving Group with $PBr_3$ or $SOCl_2$*

There are ways to activate the OH group of an alcohol for substitution/elimination reactions other than using a strong acid to protonate it. The two alternatives we will cover involve using thionyl chloride ($SOCl_2$) or phosphorus tribromide ($PBr_3$). The reaction of an alcohol with $PBr_3$ proceeds as follows:

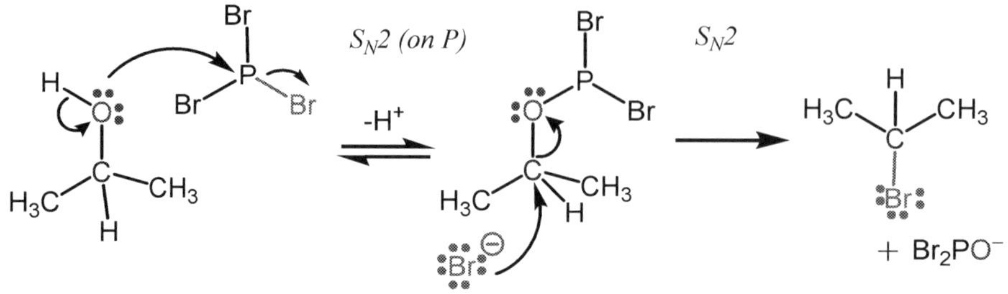

The reaction of an alcohol with $SOCl_2$ proceeds as follows:

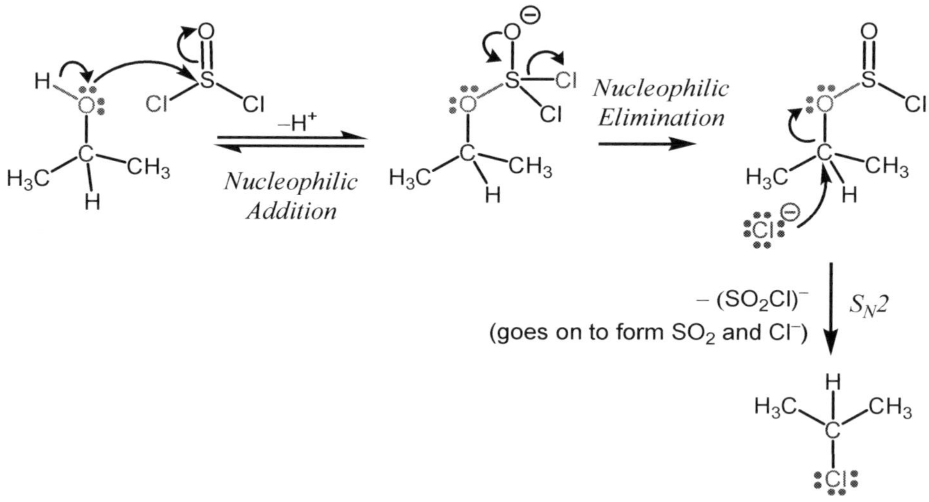

After the first activation step, the displacement of the $[OPBr_2]^-$ or $[OS(O)Cl]^-$ leaving group is an $S_N2$ reaction. Everything we learned about the $S_N2$ reaction in Lessons II.3 and II.5 still apply here, so a 3° alcohol still will not undergo an $S_N2$ reaction, even if we activate it with $PBr_3$ or $SOCl_2$, for example and if the site that is attacked s chiral, its configuration will be inverted.

*Lesson II.11.3. Changing an Alcohol into a Sulfonate Ester*

The substitution reactions of alcohols we have examined all result in net substitution of an OH group for a halide. If we want to substitute an OH group with something else, a different activation pathway will be needed. One convenient approach is to convert the OH into a **sulfonate ester**, which proceeds via the following mechanism:

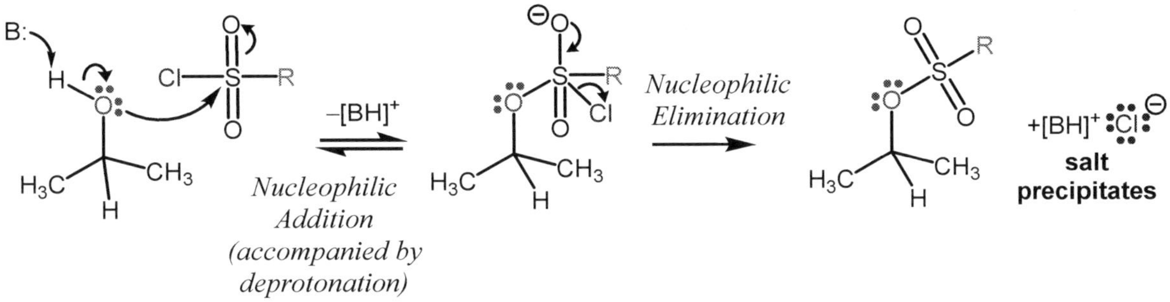

−[BH]$^+$

*Nucleophilic Addition (accompanied by deprotonation)*

*Nucleophilic Elimination*

+[BH]$^+$ $Cl^-$ **salt precipitates**

After the sulfonate ester unit has been formed, the compound is isolated. The sulfonate anion is an excellent leaving group due to its high resonance stabilization. Sulfonate anions are the best leaving groups that we will learn in this book, and they are so important that a few of them have been given common names and abbreviations (which you, the student, must memorize):

| *When R =* | *Anion Name* | *Abbreviation* |
|---|---|---|
|  p-tolyl ($C_6H_4CH_3$) | Tosylate | $TsO^-$ |
| $CH_3$ | Mesylate | $MsO^-$ |
| $CF_3$ | Triflate | $TfO^-$ |

Example II.11.3

Provide the product for each step leading to the major product of the reaction shown:

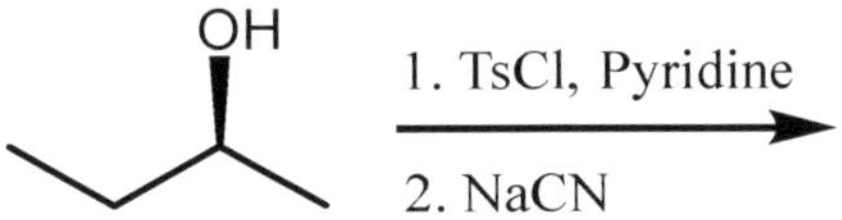

1. TsCl, Pyridine
2. NaCN

Solution II.11.3

The first step simply replaces the H on O with Ts, to form the tosyl group (OTs). Note that the stereochemistry of the C is unchanged because the reaction takes place at the O:

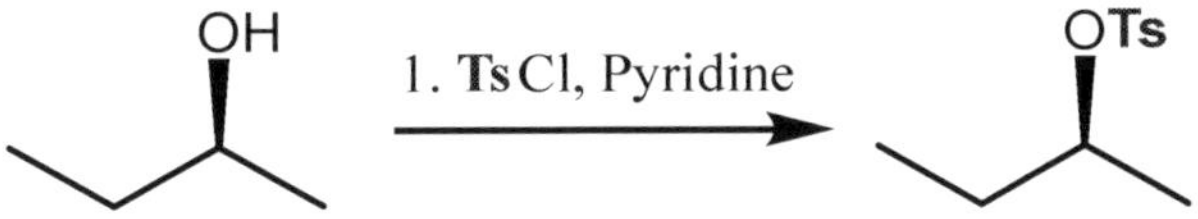

1. **Ts**Cl, Pyridine

The tosyl group is a very good leaving group and will be readily displaced by the good nucleophile $NC^-$ via an $S_N2$ pathway with inversion of configuration:

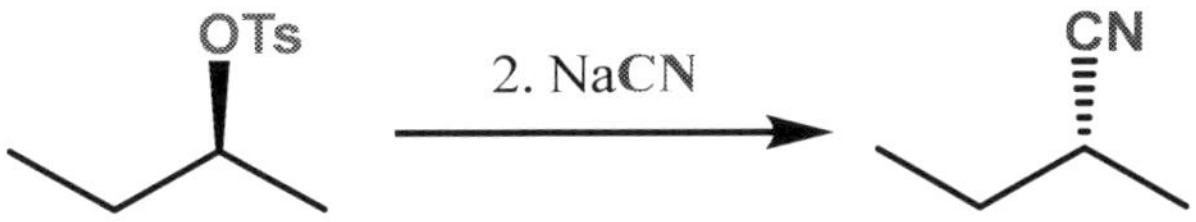

2. Na**CN**

## Lesson II.12. Oxidation and Reduction: Definitions

In organic chemistry, we use a simplified definition of oxidation and reduction, with a focus on the carbon atoms in a structure. In the simplest definition, oxidation is defined as a reaction leading to more C–O bonds and/or fewer C–H bonds, whereas reduction is defined as a reaction leading to fewer C–O bonds and/or more C–H bonds. For a broader definition, we would substitute "C–O bonds" with "C–EN" bonds, where "EN" is any element more electronegative than C.

Example II.12.1

Label each of the following reactions as being an oxidation, a reduction or neither:

A) $\xrightarrow{H_2,\ Pd}$

B) O $\xrightarrow[2.\ H_3O^+]{1.\ NaBH_4}$ OH

C) OH $\xrightarrow{H_2CrO_4}$ O, OH

D) $\xrightarrow{Br_2}$ Br, Br

Solution II.12.1

Reaction A: has more C–H bonds in the product than in the reactant, so this is a reduction. Reaction B has fewer C–O bonds in the product than in the reactant, so this is a reduction. Reaction C has more C–O bonds in the product than in the reactant, so this is an oxidation. Reaction D has more C–EN bonds in the product than in the reactant, so this is an oxidation.

## Lesson II.13. Oxidation Reactions of Alcohols

*Lesson II.13.1. Oxidation of Alcohols Using Chromium Reagents*

The carbon in an alcohol to which the OH group is attached (called the "carbinol" carbon) can be oxidized by several chromium reagents, typically with added acid. Those that we will cover in this book are: $H^+/CrO_4^{2-}$, $H^+/Cr_2O_7^{2-}$, $CrO_3/H_2SO_4$ (reagents for what is called the "Jones Oxidation"), and pyridinium chlorochromate (PCC). For PCC the proper solvent to assure the reactivity described here is $CH_2Cl_2$, so you will often see this written in the reaction conditions around the arrow as well.

PCC is a reagent that can replace one C–H bond on the carbinol with another bond to the O of the OH group (which must be accompanied by loss of H from the OH group). This leads to the formation of an aldehyde (from 1° alcohol) or ketone (from 2° alcohol):

OH, H3C, H, H — PCC / $CH_2Cl_2$ → O, H3C, H — aldehyde

OH, H3C, R', H — PCC / $CH_2Cl_2$ → O, H3C, R' — ketone

All of the chromium reagents listed above, other than PCC, are more powerful oxidants. They are sufficiently strong oxidants to replace *all* of the C–H bonds on the carbinol with bonds to O. This will lead to the formation of ketones (from 2° alcohol) or carboxylic acids (from 1° alcohol):

OH, H3C, H, H — $H^+/CrO_4^{2-}$, $H^+/Cr_2O_7$, or $CrO_3/H_2SO_4$ → O, H3C, OH — carboxylic acid

OH, H3C, R', H — $H^+/CrO_4^{2-}$, $H^+/Cr_2O_7$, or $CrO_3/H_2SO_4$ → O, H3C, R' — ketone

Note that, in the case of 2° alcohols, reaction with PCC or the more powerful oxidizing agents both afford ketones because there is only one H on the carbinol that can be changed to a C–O bond. However, when the reactant is a 1° alcohol, PCC will yield different products than Jones Oxidation. Note that these oxidation reactions can only remove H atoms on the same carbon as the –OH group; the adjacent groups, like the $-CH_3$ groups in the examples above, are not changed at all by these oxidation reactions.

Example II.13.1

Provide the major product for each of the following reactions:

A) OH; PCC, $CH_2Cl_2$ →

B) OH; Jones Oxidation →

C) OH; $H_2CrO_4$ →

D) OH; $Na_2CrO_7$, $H_2SO_4$ →

Solution II.13.1

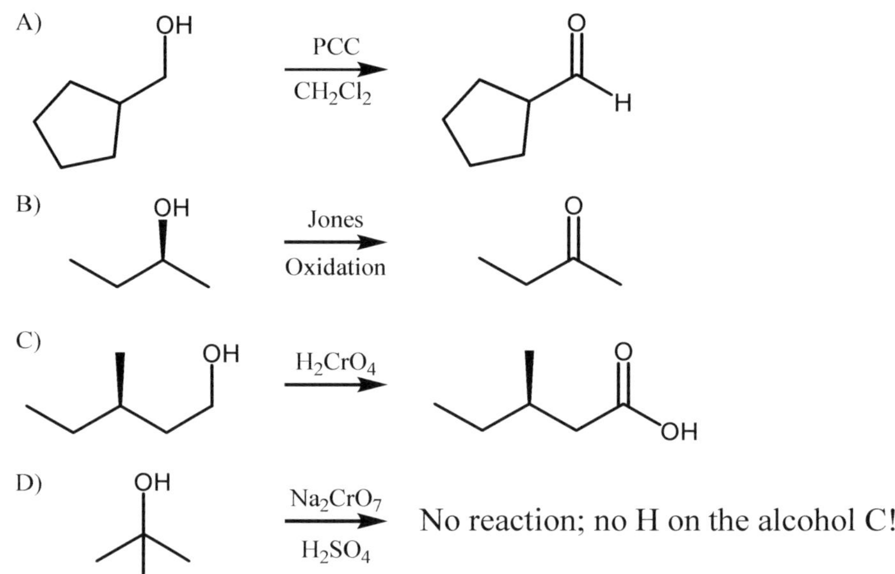

## Lesson II.14. Nomenclature and Acid Cleavage of Ethers

### *Lesson II.14.1. Naming Ethers*

When we name ethers, we divide them into two types: **symmetrical** (the two R groups are the same, as in $CH_3CH_2OCH_2CH_3$), and **unsymmetrical** (the two R groups are different, as in $CH_3OC_2H_5$).

We can effectively name ethers with either **common names** or **systematic names**. Common names are formulated by first naming each R group (the R and $R^1$ groups in the diagram below):

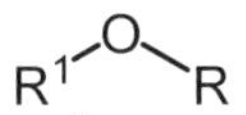

$R^1$–O–R
Alkyl ($R^1$) alkyl ether

The two alkyl groups are then listed alphabetically, followed by "ether". If the ether is symmetrical, it is simply named "di*alkyl* ether." These examples illustrate the common naming method:

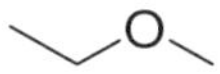

*ethyl methyl ether* 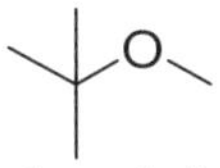 *t-butyl methyl ether* 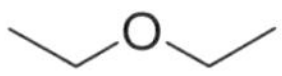 *diethyl ether*

IUPAC systematic names are often used for ethers as well. In this approach, the longer alkyl group is named as the parent chain and the O with the shorter alkyl group is named as an **alkoxy** substituent (methoxy, ethoxy, or alkyloxy for longer C chains). These two examples illustrate the systematic naming method:

*1-ethoxybutane* 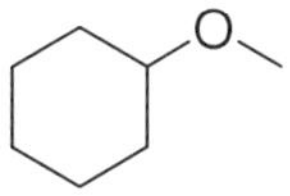 *methoxycyclohexane*

### *Lesson II.14.1. Substitution Reactions of Ethers*

In Lesson II.11, we saw that HX (X = Cl, Br or I) simultaneously provides $H^+$, which activates an alcohol OH group, and $X^-$, the nucleophile which displaces water from activated alcohols. These same HX acids can undergo similar reactions with ethers. Acid cleavage of ethers is an important reaction because ethers are relatively inert under most of the reaction conditions we have studied. As for alcohols **if $S_N1$ is possible in the ether, $S_N1$ will occur faster than $S_N2$**. This means that the major product will come from an $S_N1$ pathway for nucleophilic attack on 2° or 3° ether carbons, and only by $S_N2$ on methyl or 1° ether carbons.

The fact that the $S_N1$ pathway outpaces the $S_N2$ pathway for substitution has important ramifications, especially if the ether is not symmetrically substituted. Consider the reaction of ethyl isopropylether with HI. We have to choose which of the two carbon atoms on the O will lose the leaving group. In this case, the secondary side will be faster for the $S_N1$ pathway ($S_N1$ fails on primary sites), so the secondary C loses the leaving group to produce the most stable carbocation available:

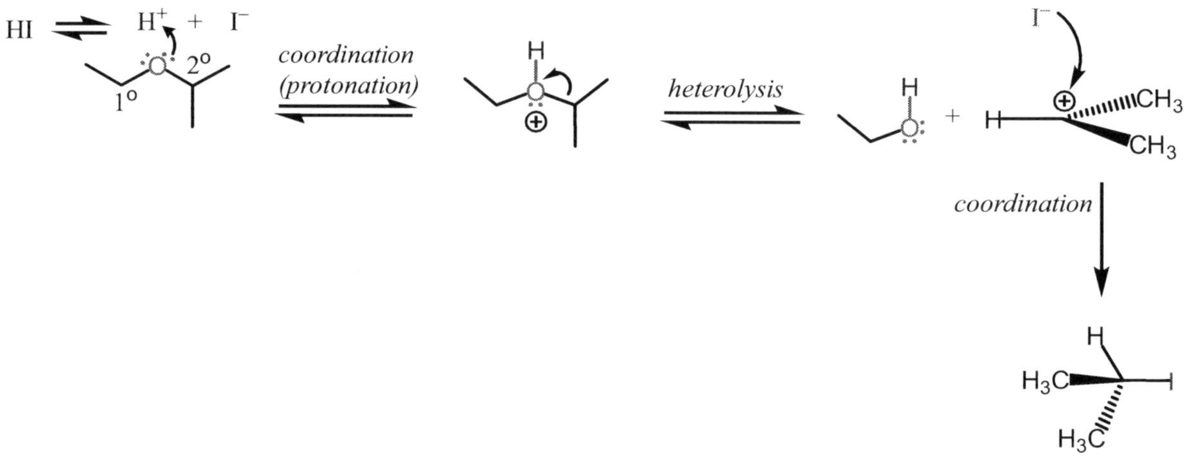

HI ⇌ $H^+$ + $I^-$; 1°, 2°; *coordination (protonation)*; *heterolysis*; *coordination*

The net result is production of one equivalent of ethanol and one equivalent of 2-iodopropane as the major products. Once we have identified whether an $S_N1$ or $S_N2$ pathway is going to take place, we can use what we learned in Lessons II.3–II.5 to determine the major products of the reaction.

Example II.14.1

Provide the major product for each of the following reactions:

A) HBr

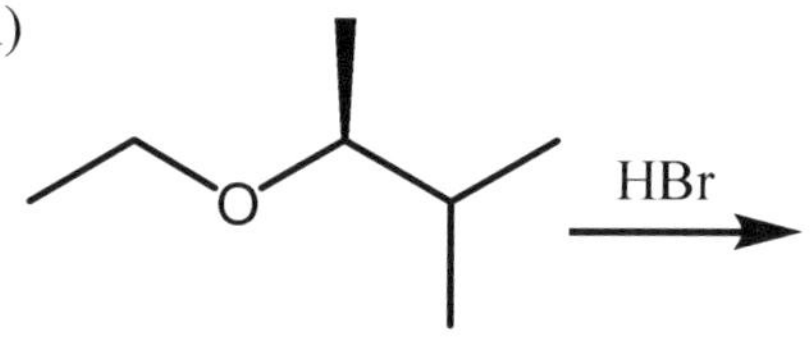

B) HBr

Solution II.14.1

In reaction A, the left-hand ether carbon is primary and the right-hand ether carbon is secondary. The secondary site can undergo $S_N1$ reaction, so this will be the major pathway. The secondary carbocation will form (after protonation and heterolysis):

A) OH

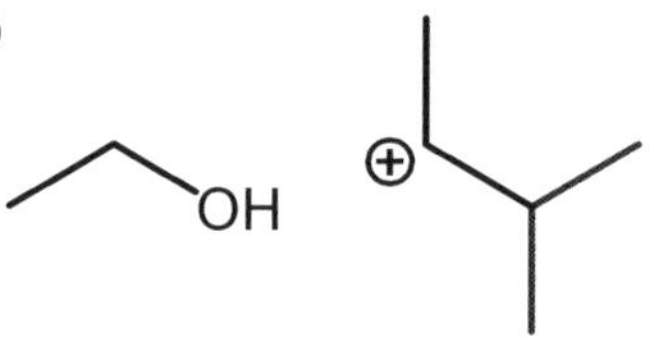

The secondary carbocation will rearrange to form the tertiary carbocation:

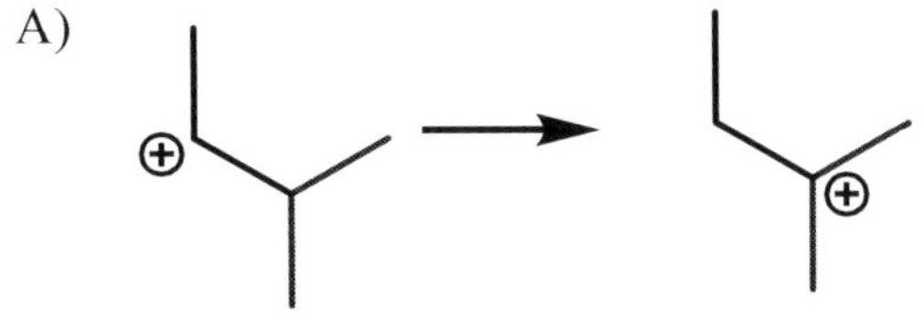

A)

The tertiary carbocation will then undergo coordination by the bromide. The final net reaction for reaction A is thus:

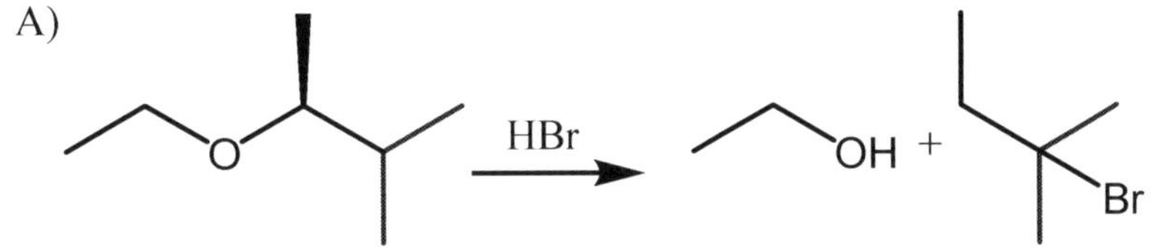

A)

HBr

OH +

Br

In reaction B, the left-hand ether carbon is primary and the right-hand ether carbon is methyl. Neither site can undergo $S_N1$ reaction, so $S_N2$ will be the major pathway. For $S_N2$, we know nucleophilic attack will occur more rapidly at the less sterically hindered site, so the major product comes from nucleophilic attack on the methyl site:

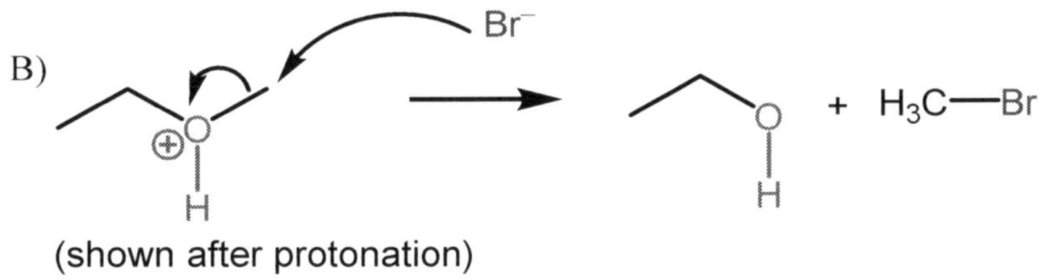

B)

$Br^-$

O

H

(shown after protonation)

O

H

+ $H_3C$—Br

One additional note about this reaction is that we can use more than one equivalent of the acid and drive the reaction further. We learned in Lesson II.11 that ROH reacts with HX to form RX and water. If we perform the reaction in Example II.14.1A with 2 or more equivalents of HBr, the first equivalent of HBr will react as shown in Solution II.14.1A, but the second equivalent of HBr will convert EtOH to ethyl bromide, so that the overall products will be as shown:

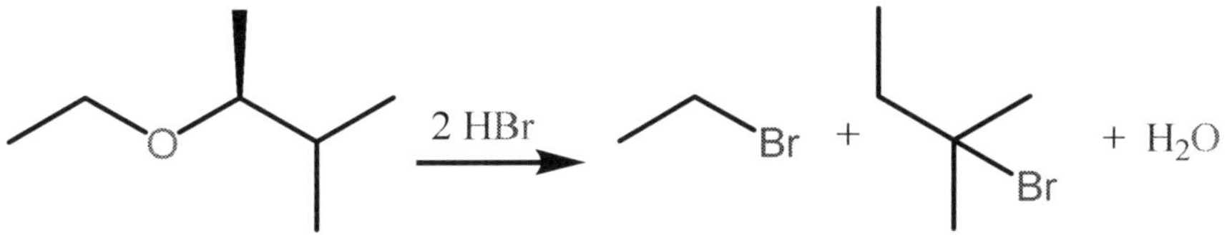

2 HBr

Br +

Br

+ $H_2O$

## Lesson II.15. Ring-Opening of Epoxides: Steric/Electronic Effects on Major Product

*Lesson II.15.1. Epoxides are More Reactive than Other Ethers*

Epoxides are ethers in which the O belongs to a three-membered ring:

In Lesson I.16, we learned that a three-membered ring has the most ring strain of any ring size. The significant ring strain in the epoxide makes it much less stable (more reactive) than acyclic ethers. The enhanced reactivity of epoxides means that the epoxide can react with good nucleophiles to open the epoxide ring, even if the epoxide O is not protonated:

Lots of ring strain δ+ δ+ Nu⁻ → ⊖O … Nu No ring strain

Note that, although ring strain is alleviated by ring opening, the anion produced is still a strong base and is thus relatively unstable. Because the anion produced by ring opening is a strong base, not every nucleophile will react with the neutral epoxide. **The nucleophile must be a relatively unstable anion (a strong base)**, such that the reactant anion has similar stability to the product anion. The net reaction is spontaneous due to relief of ring strain. It is also important to remember that, after nucleophilic attack under basic conditions, the resulting anion must be protonated to obtain the neutral alcohol product.

If the epoxide is reacted with a nucleophile under acidic conditions, however, even a poor nucleophile will react with the epoxide-derived oxonium intermediate, because the leaving group is a neutral OH group and quite stable:

$H^+$ coordination (protonation) H ⊕ δ+ δ+ Nu⁻ $S_N2$ → Nu OH

*Lesson II.15.2. Regioselectivity in Ring-Opening of Epoxides*

We must determine which side of an asymmetrically-substituted epoxide will be attacked by the nucleophile in a ring-opening reaction. Regardless of conditions, if one side of the epoxide is primary and one is secondary, the less-substituted side is attacked because it is less hindered; the sterics win out in this case, as seen for the following example under acidic conditions:

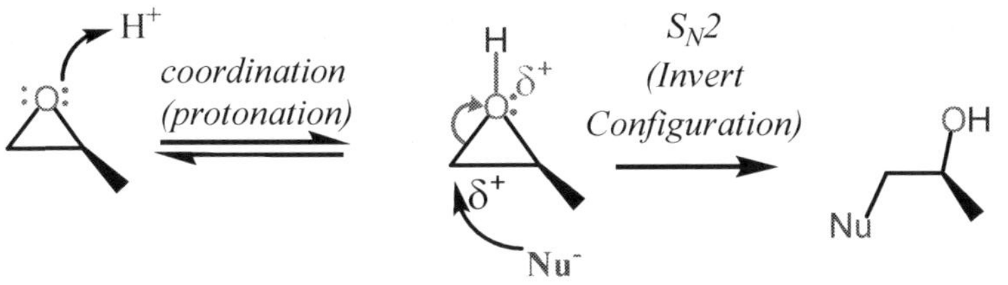

… and for this example, under basic conditions (followed by protonation):

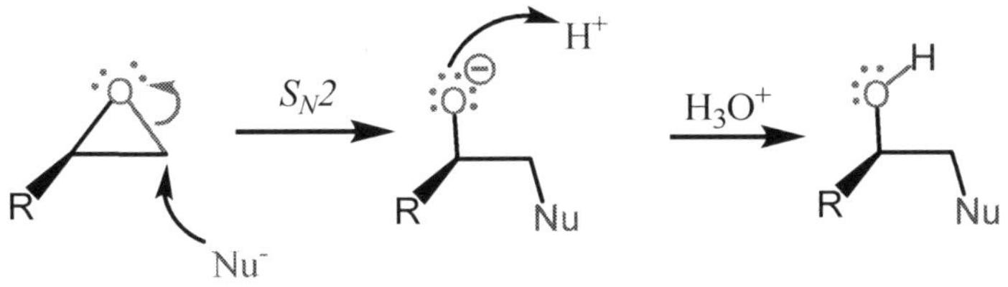

**The only time that the more substituted side is attacked is if one side is tertiary and the reaction is under acidic conditions.** The observed attack by the nucleophile on the tertiary site may seem to contradict the general trend that $S_N2$ reactions are faster at less sterically-hindered sites. In the case of an oxonium cation, however, there is so much more partial positive charge on the more substituted carbon that the attraction of the nucleophile's electrons to this site is much stronger than it would be if the nucleophile were attaching a neutral species. All epoxide ring opening reactions undergo an $S_N2$-like ring opening, with inversion of configuration:

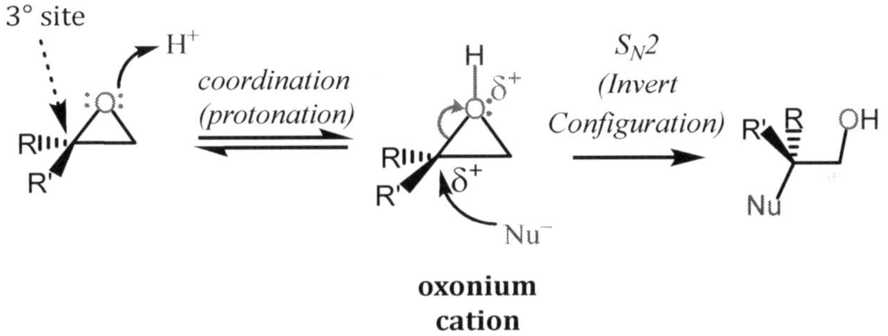

**oxonium cation**

**Part IV: Properties and Reactions of Conjugated and Aromatic Molecules**

## Lesson IV.1. Pi Conjugation Stabilizes a Molecule

*Lesson IV.1 π-Bond Conjugation is Stabilizing and π-Bond Cumulation is Destabilizi*

In Lessons III.1-10 (in volume 1 of this two-primer set), we considered a variety of for individual C=C double bonds in isolated alkene units. Next, we will consider the st molecules that have *more than one* C=C bond and then examine some reactions of dienes. Remember that alkene π-bonds are made up of electrons in carbon *p*-orbitals. In 1,4-pentadiene, the two C=C bonds are separated by an $sp^3$-hybridized carbon, so the *p*-orbitals on carbons 1 and 2 cannot overlap the *p*-orbitals on carbons 4 and 5, so each of the π bonds is **isolated**:

Each of the isolated π-bonds in 1,4-pentadiene react in the same way that a π-bond reacts in a monoalkene.

In *trans*-1,3-pentadiene there are no $sp^3$-hybridized carbons separating the two C=C bonds, so all four of the *p*-orbitals involved in making π bonds can overlap with each other, and we classify this type of diene as **conjugated**:

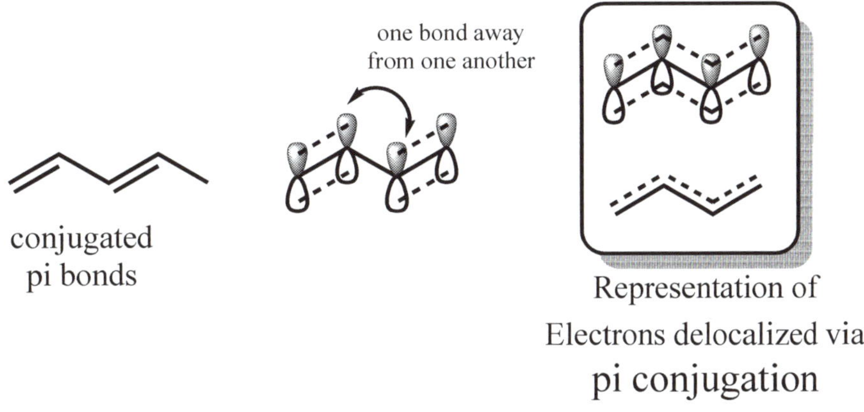


Representation of
Electrons delocalized via
pi conjugation

When two C=C bonds are conjugated, each of the 4 π-electrons can spread out over 4 positions, this delocalization is why **conjugation significantly stabilizes π-bonds**. We measure alkene stabilities using heats of hydrogenation ($\Delta H_h$). For 1-pentene and *trans*-2-pentene, the $\Delta H_h$ values are –30.1 and –28.6 kcal/mol, respectively. At first glance, we might expect the heat of hydrogenation for *trans*-1,3-pentadiene to be simply the sum of the values for 1-pentene and *trans*-2-pentene (–58.7 kcal/mol, because it has one monosubstituted alkene and one *trans* alkene). The experimentally observed $\Delta H_h$ for

-pentadiene is –54.1 kcal/mol, which demonstrates that **conjugation of the two C=C bonds es an extra 4.6 kcal/mol of stabilization:**

| | $\Delta H_h$ (kcal/mol) |
|---|---|
| 1-pentene | –30.1 |
| *trans*-2-pentene | –28.6 |
| 1,4-pentadiene | –60.8 |
| *trans*-1,3-pentadiene | –54.1 |
| 1,2-pentadiene | –69.8 |

When two C=C bonds begin at the same carbon, we get a C=C=C unit, in which the C=C bonds are classified as **cumulated**. Note that the central carbon is *sp*-hybridized, so the *p*-orbitals used to make the C=C bond to the left carbon is orthogonal to those used to make the C=C bond to the right, and thus the two π-bonds cannot overlap. This is illustrated for $H_2C{=}C{=}CH_2$ here:

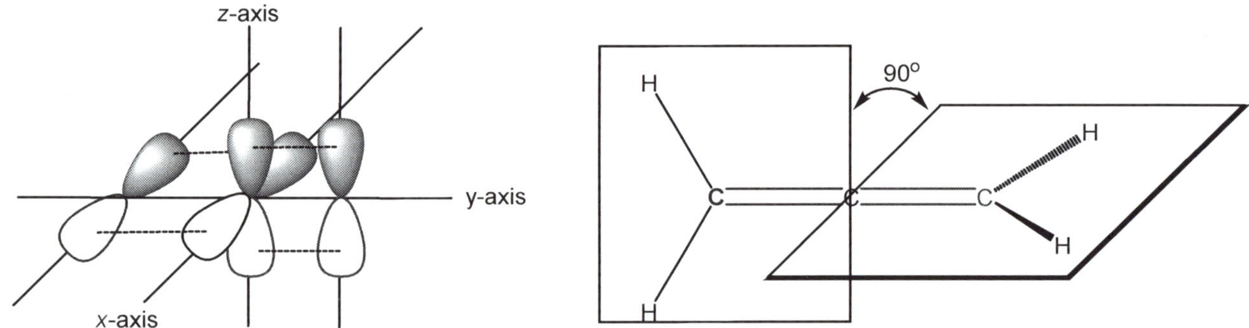


The diagram on the left shows the *p*-orbitals used to make the π-bonds ($p_x$ for the left π bond and $p_y$ for the right π-bond). The diagram on the right shows the geometry in a typical structural drawing.

When we measure the $\Delta H_h$ for such a C=C=C unit in 1,2-pentadiene, it is significantly higher than $\Delta H_h$ for 1,4-pentadiene (–69.8 vs. –60.8 kcal/mol, respectively), which indicates that **cumulated π-bonds are significantly less stable than isolated π-bonds**. This phenomenon is due to the fact that an *sp*-hybridized carbon is much more electronegative than an $sp^2$-hybridized carbon, so a C=C bond between an *sp*-hybridized carbon and an $sp^2$-hybridized carbon is more electron-deficient than a C=C bond between two $sp^2$-hybridized carbons. Remember that alkene units are stabilized by inductive electron-donating groups like alkyl groups ($sp^3$ hybridized C atoms). The differences in stability of cumulated, isolated and conjugated alkenes means that they have different reactivity. In our next couple of lessons, we will look at special reactivity of conjugated dienes.

Example IV.1.1

Which of the following alkenes would have the higher heat of hydrogenation?

vs

Solution IV.1.1

In the molecule on the left (1,4-cyclohexadiene), the two π-bonds are isolated, but in the molecule on the right (1,3-cyclohexadiene), the two π-bonds are conjugated. When π-bonds are isolated from each other by one or more $sp^3$-hybridized carbons, they cannot interact with each other and thus cannot be stabilized or destabilized by each other. However, when π-bonds are conjugated with each other, the π-electrons are able to delocalize over a greater number of positions, which increases stability. As a result, 1,3-cyclohexadiene will be more stable than 1,4-cyclohexadiene, which means that 1,3-cyclohexadiene will release less energy upon hydrogenation of its C=C bonds. Thus, a higher heat of hydrogenation will be observed with 1,4-cyclohexadiene.

isolated π-bonds

conjugated π-bonds

## Lesson IV.2. Addition Reactions of Conjugated Dienes

*Lesson IV.2.1 Introduction to Addition Reactions of Conjugated Dienes*

We saw in Lesson III.5 (OC1 Primer) that reaction of one equivalent of HX (X = Cl, Br or I) to a C=C bond will produce the more substituted alkyl halide as the major product (Markovnikov's Rule). However, two products are possible when one equivalent of HX reacts with a 1,3-diene:

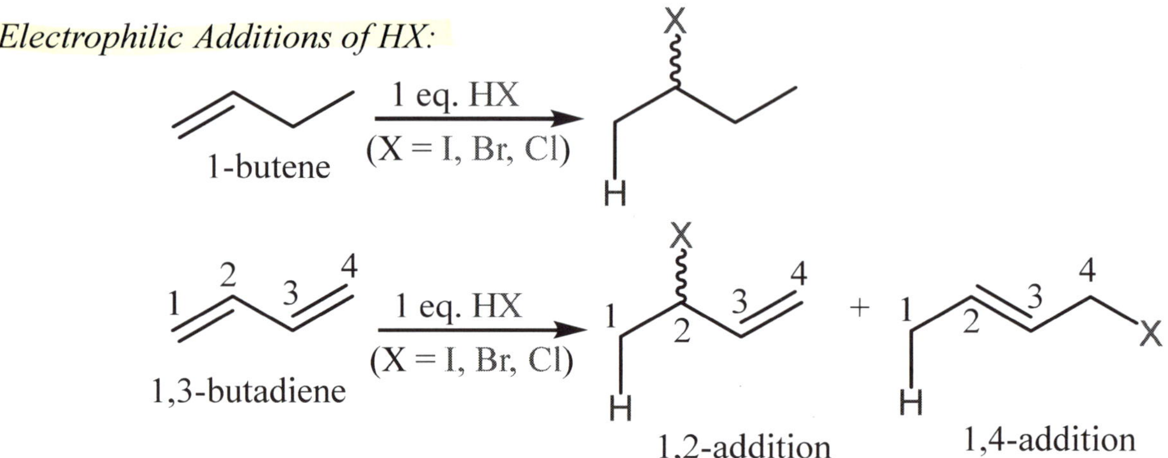


The **1,2-product** (also called the **direct addition** product) and **1,4-product** (also called the **conjugate addition** product) are possible because the carbocation formed by electrophilic addition of the proton to the C1 end of the C=C bond leads to a resonance-stabilized carbocation having δ+ character on both C2 and C4

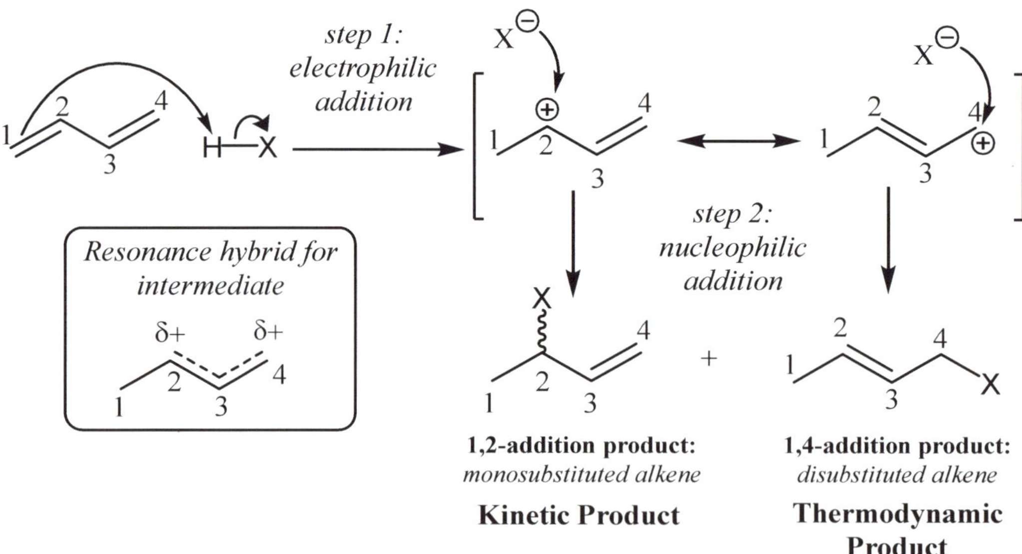


The major product isolated from these reactions depends on the conditions of the reaction.

*Lesson IV.2.2. Determining the Major Product*

Both 1,2- and 1,4-addition products have the first mechanistic step: electrophilic addition of the proton to one of the C=C bonds. The second step of the 1,2-addition is coordination of the bromide to C2, whereas the second step of 1,4-addition is coordination of bromide to C4. In the case of addition of HX to 1,3-butadiene, the product of 1,2-addition is a monosubstituted alkene, whereas the product of 1,4-addition is a disubstituted alkene. The 1,4-addition produces the more stable alkene (the **thermodynamic product**). The 1,2-addition is faster (lower energy of activation, $E_{a1,2}$ in the figure below), however, because the bromide is closer to C2 when it is produced in the electrophilic addition step. This is known as a **proximity effect**, and the result of this is that the 1,2-addition product is formed faster – it is the **kinetic product**. The qualitative reaction coordinate diagram below illustrates the energetics of these processes:

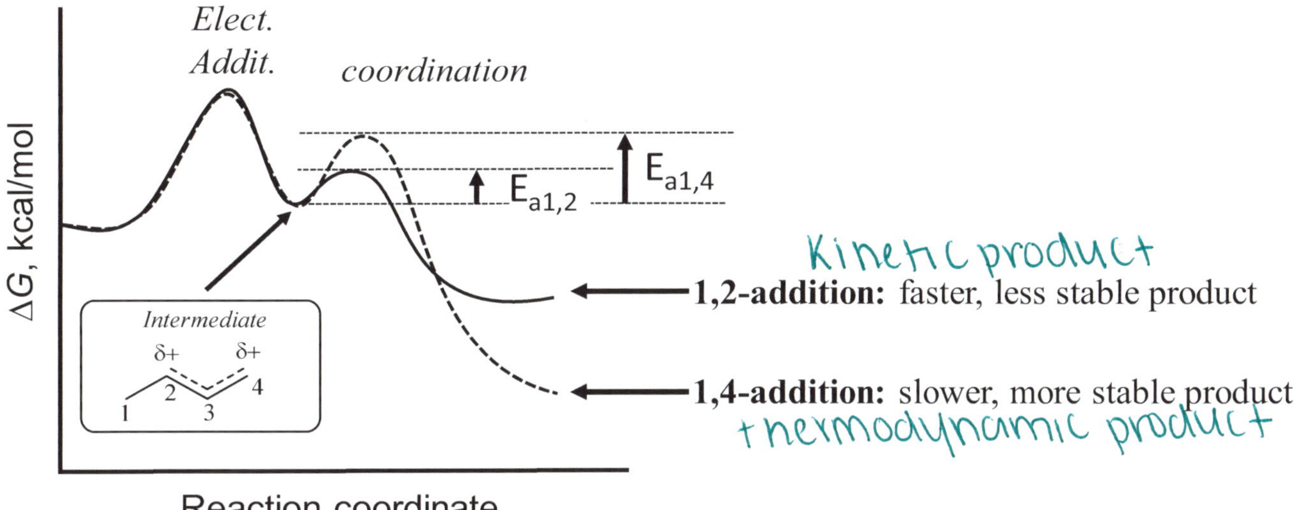


The 1,2-addition is faster, but the 1,4-addition produces the more stable product in this case. So, how to we determine which is the major product? A fundamental property of chemical reactions is that **the major product of irreversible reactions is the kinetic product** (the one formed at a higher rate). For a reaction **at equilibrium (reversible), the major product is the thermodynamic product** (the more stable product). We can control which product is the major product by controlling the reaction conditions. **At low temperature (below ~20 °C for these reactions) the reaction is irreversible** because there is not enough energy to overcome the larger energy barrier for the reverse reaction, so the kinetic product is the major product, and the reaction is said to be under **kinetic control**.

**At higher temperature (above ~50 °C for these reactions), the reaction is reversible** because there is enough energy to overcome the energy barrier to the reverse reaction, so the thermodynamic product is the major product, and the reaction is said to be under **thermodynamic control**.

**The 1,2-product is always the kinetic product** due to the proximity effect. The thermodynamic product is always the more stable alkene, which may be either the 1,2- or 1,4-addition product depending on the substitution pattern. For example, consider addition of HBr to 2,5-dimethyl-2,4-hexadiene:

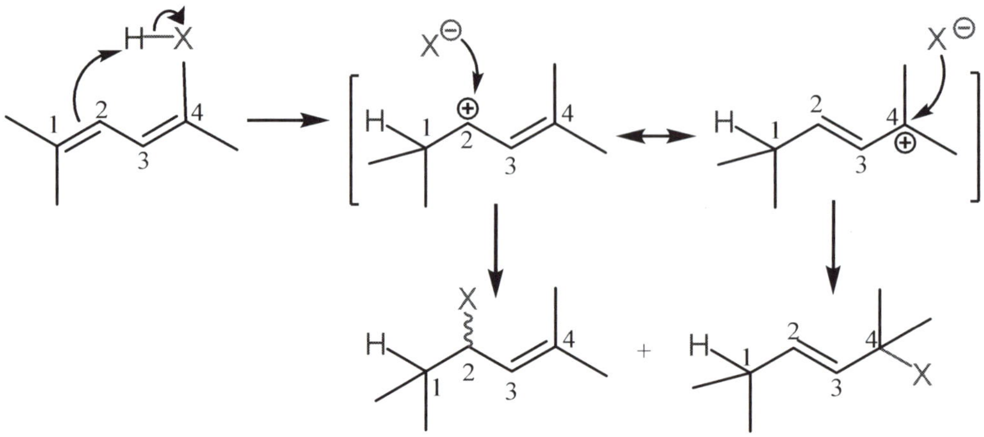

**1,2-addition product:**
*trisubstituted alkene*
**Kinetic product**
**and**
**Thermodynamic product**

**1,4-addition product:**
*disubstituted alkene*

In this case, the 1,2-addition product is *both* the kinetic product *and* the thermodynamic product.

To approach these problems effectively, follow this procedure:

1. Number the π-conjugated part 1–4.
2. Add $H^+$ to carbon #1 of the conjugated segment
3. Draw both resonance contributors for the cation you get
4. Place the $X^-$ on each cation to get your final products
5. The 1,2-product is the kinetic product
6. The most substituted alkene is the thermodynamic product

Conjugated dienes can also be halogenated using $Br_2$. The mechanism for this reaction is similar to what we observed for reaction with HBr, and the principles for determining kinetic/thermodynamic products of halogenation mirror what we have discussed for reaction of dienes with HX:

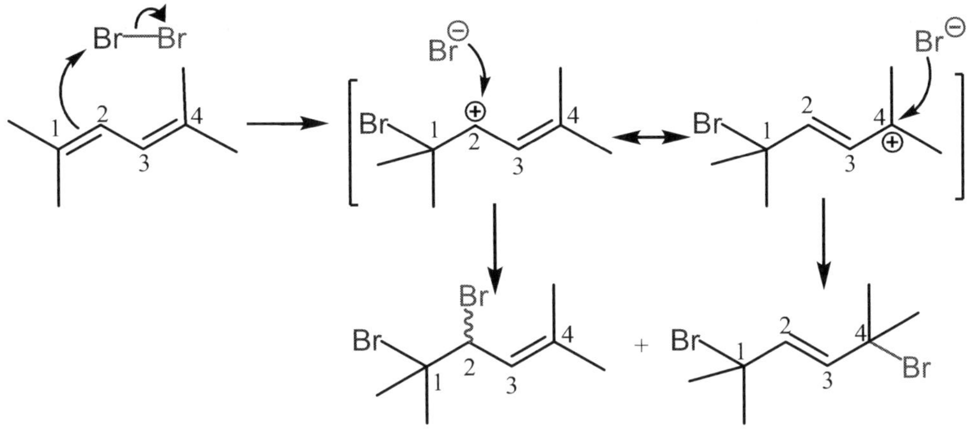

Br—Br

Br⊖ Br⊖

Br 1 2 ⊕ 3 4 ↔ Br 1 2 3 4 ⊕

Br Br 1 2 3 4 + Br 1 2 3 4 Br

**1,2-addition product:** **1,4-addition product:**

Example IV.2.1

Provide the major product for each reaction. Label each reaction as favoring either the kinetic or thermodynamic product.

+ HBr 60 °C / 0 °C

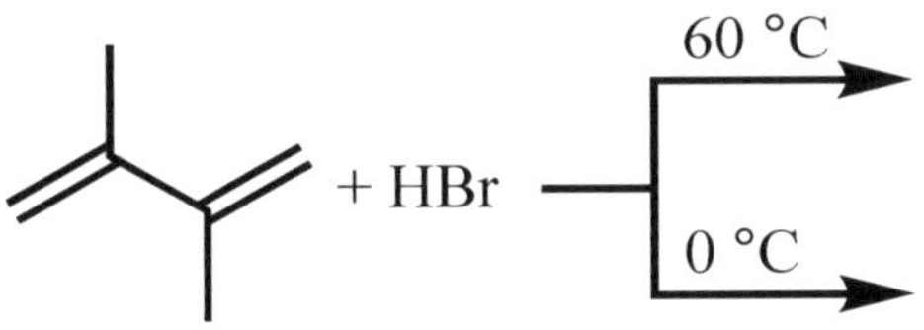

Solution IV.2.1

First, we consider what the 1,2- and 1,4-addition products would be:

*1,2-addition* *1,4-addition*

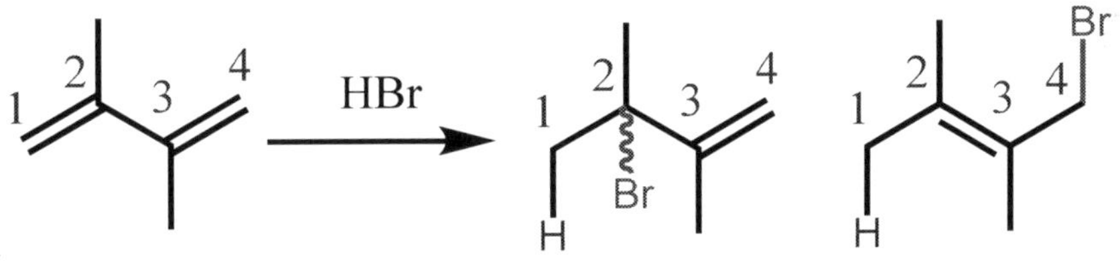

1 2 3 4 HBr → 1 2 3 4 H Br 1 2 3 4 Br H

a

At 60 °C, the thermodynamic product is favored. This will be the most stable alkene. In this case, that is the 1,4-addition product, a tetrasubstituted alkene.

At 0 °C, the kinetic product is favored. This will always be the 1,2-addition product, due to the proximity effect.

## Lesson IV.3. Diels-Alder Reaction

### *IV.3.1 Introduction to the Diels-Alder Reaction*

The Diels-Alder reaction is a [4 + 2] **cycloaddition reaction** of a diene and an alkene to form a cyclohexene ring. The reaction takes place between a diene and a monoalkene. In this particular reaction, we refer to the monoalkene as a **dienophile** (meaning "diene-loving"). The scheme below shows the general mechanism of the reaction:

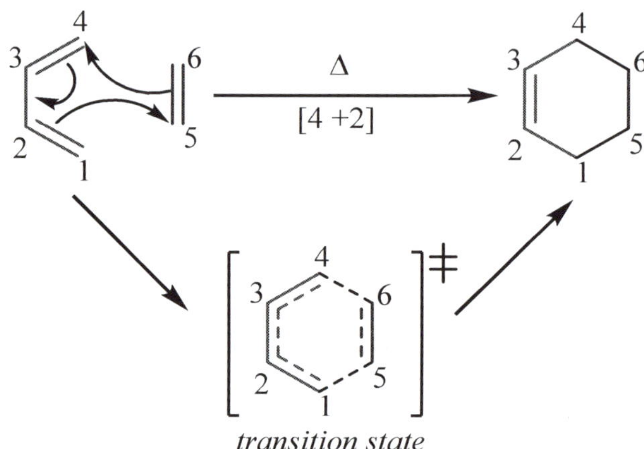

*transition state*

The Diels-Alder reaction is an example of a **pericyclic reaction**: a concerted, cyclic flow of π electrons within a cyclic transition state structure. The driving force for this reaction is that two of the π-bonds in the reactants become σ-bonds in the product. Because σ-bonds are stronger than π-bonds, this makes the reaction thermodynamically favorable.

The diene of the Diels-Alder reaction has to be a s-*cis* conformation for the cycloaddition to take place in a concerted mechanism. Even if the diene in the starting material is in the thermodynamically more stable configuration (s-*trans*), when the reaction is heated, the heat will promote the rotation of the C—C σ-bond, allowing access to the required s-*cis* conformation.

*s-trans* *s-cis*

The Diels-Alder reaction proceeds more rapidly when the dienophile has an electron-withdrawing group (an electronegative atom, or an atom with a δ+ on it) attached to the C=C carbons in the monoalkene. The presence of an electron-withdrawing group such as carbonyl groups (C=O), nitro groups ($-NO_2$) or nitrile groups (–C≡N) helps the dienophile pull electrons from the diene to get the reaction started.

Another important point about the Diels-Alder reaction is that because the mechanism is concerted, there is no chance for rearrangement of what direction substituents are pointing on the reactants. The result is that the **Diels-Alder reaction is stereoselective:** a *cis*-alkene (dienophile) will produce a *cis*-cyclohexene and a *trans*-alkene will produce a *trans*-cyclohexene:

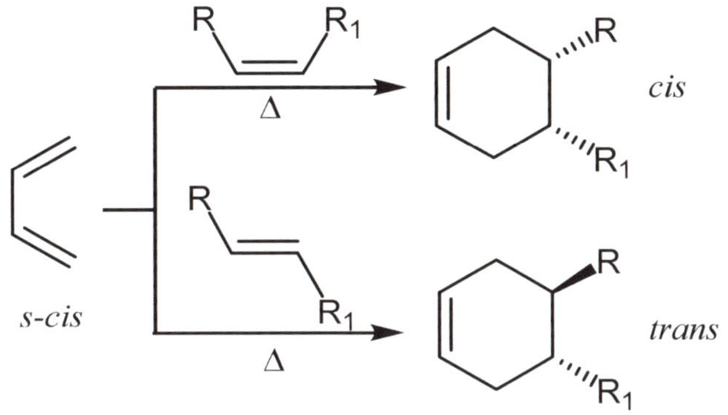

R R1 Δ cis
s-cis R R1 Δ R trans R1

Example VI.3.1

Give the final product of the following Diels-Alder cycloaddition

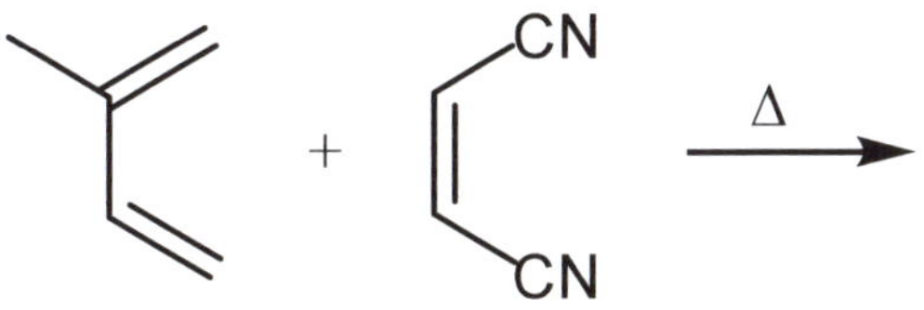

CN + Δ CN

Solution VI.3.1

A good starting point is to number the diene and dienophile carbons so we can keep track of where they go in the product:

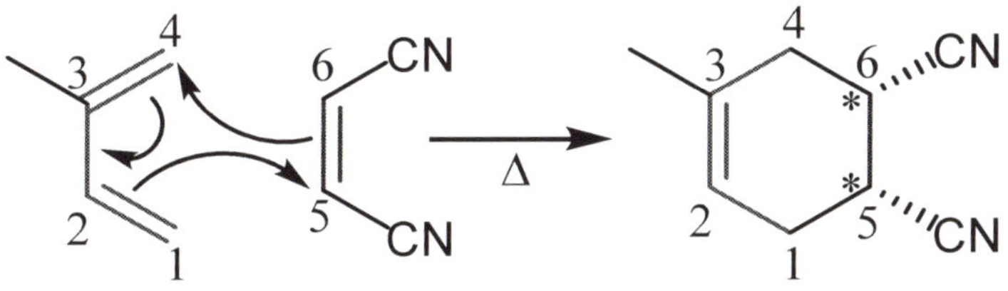

4 3 6 CN 2 1 5 CN Δ 4 3 6 CN * * 2 1 5 CN + enantiomer

We can then move the electrons like we did for the general reaction. We have to be certain that we keep the substituents on carbons 5 and 6 *cis*- to one another like they were in the starting material. It is also important to recognize that the product is chiral (two chiral centers marked by asterisks), but that the starting materials were achiral. We know that this situation results in a racemic mixture of two enantiomers.

*IV.3.2 Cyclic Dienes: Endo- and Exo- Products*

Cyclic dienes make good substrates for the Diels-Alder reaction because they may be held in the needed s-*cis* conformation. When you try to figure out the product of a Diels-Alder reaction involving a cyclic diene, remember that only the doubly-bound carbon atoms are involved in the movement of electrons. The "bridge" (carbons labelled B1 and B2) linking carbons 1 and 4 in the starting material still bridge carbons 1 and 4 in the product:

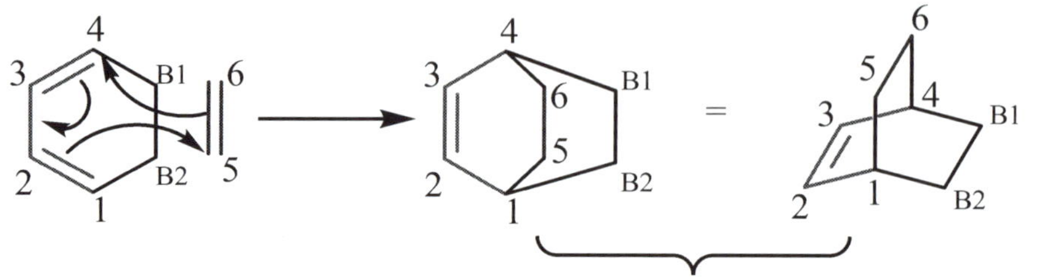

these two are the same structure drawn two different ways

When a cyclic diene and a substituted alkene are used in a Diels-Alder reaction, there can be two product choices: one with the substituent "R" pointing in the same direction as the product π-bond (down in the drawing below), and one with "R" pointing in the same direction as the product π-bond (up in the drawing below):

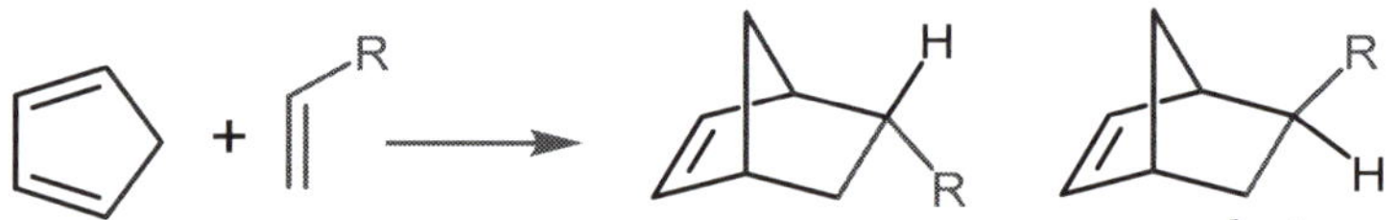

***endo product***
(major product)

***exo product***

When the R group points towards the same direction as the π-bond, this is called the **endo product**. In the **exo product**, substituent R points in the opposite direction of the π-bond. Generally, **the endo product is the major product**. Also notice that the products shown above are chiral, so they the major product is a racemic mixture of the two enantiomers of the endo product, as the example below illustrates.

---

Example VI.3.2

Give the major product of the following Diels-Alder cycloaddition

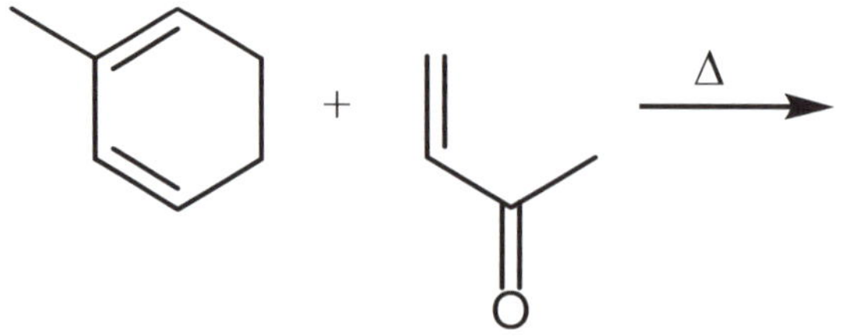

Solution VI.3.1

First, highlight the reactive pieces (the diene and dienophile carbons) by numbering them:

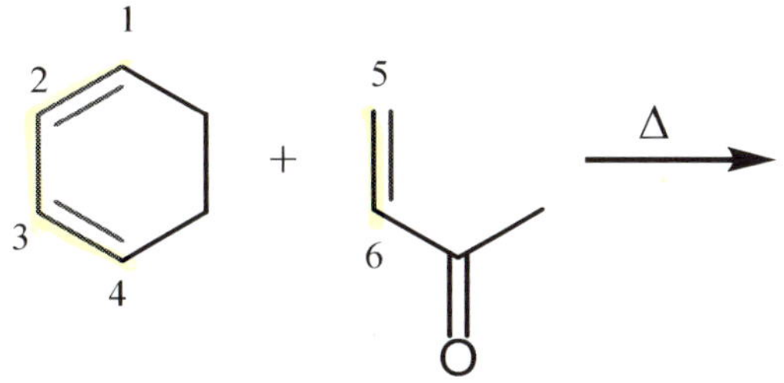

Use these carbons to form the cyclohexene ring:

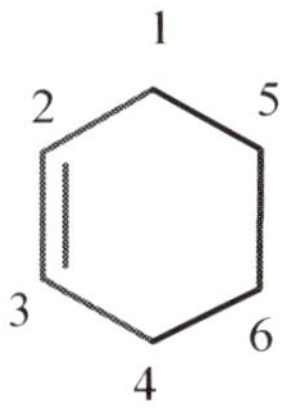

Add the bridging C that link C1 to C4 in the starting material, then redraw it in a more realistic geometry:

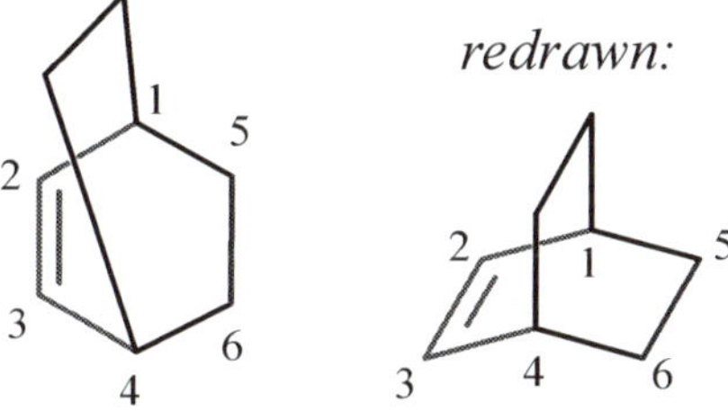

Finally, add the ketone to C6, making sure to draw it pointing the same way as the C=C (here in the downward direction), to show the *endo* product. It is also a racemic mixture of *R*- and *S*-isomers at the labelled chiral carbon:

endo-*product*
*(racemic)*

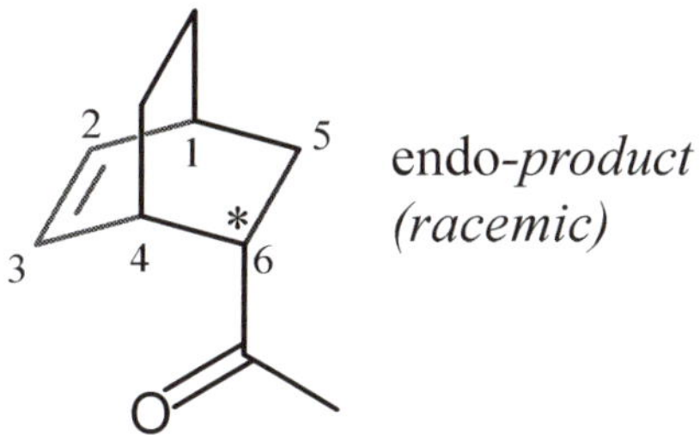

## Lesson IV.4. Aromaticity: A Highly Stabilizing Effect

### *Lesson IV.4.1 Aromaticity is a Special Type of Resonance Stabilization*

We know that resonance delocalization provides additional stabilization to a molecule. In Lesson IV.1, we saw that π-conjugation provides another means of stabilization via electron delocalization. In this lesson, we study **aromaticity**, a special type of resonance delocalization that can only occur in a cyclic π system. A molecule that exhibits aromaticity is said to be **aromatic**. The canonical aromatic compound is benzene ($C_6H_6$):

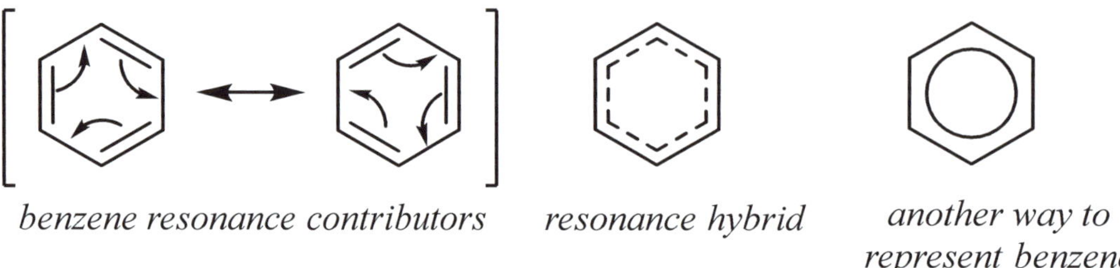

*benzene resonance contributors* *resonance hybrid* *another way to represent benzene*

As illustrated in the graphic above, benzene has two resonance contributors that are identical in energy. The resonance hybrid more accurately depicts the facts that (1) the electrons are delocalized evenly around the ring and (2) rather than alternating C–C single and double bonds, all 6 C–C bonds in benzene instead have bond orders of 1.5. To emphasize the uniformity of the electron density and bonding throughout the ring, benzene is sometimes drawn with a circle in the middle of the ring. Despite having partial C=C character, the C–C bonds in aromatic systems are so much more stable than in alkenes that they will not undergo typical alkene reactions (see Lessons III.2–10 in OC1 Primer).

Example IV.4.1

Provide the product of this reaction:

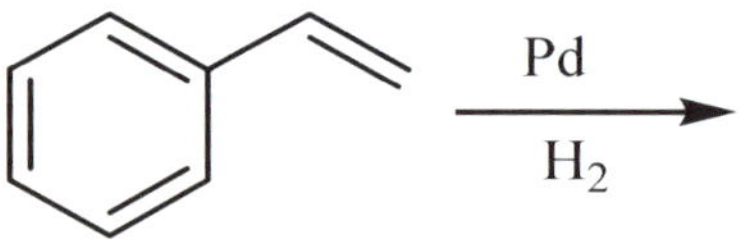

Pd

$H_2$

Solution IV.4.1

These conditions are used for the hydrogenation of alkenes. Hydrogenating the C=C bonds in the benzene would disrupt aromaticity, which is too energetically costly, so only the non-aromatic C=C bond undergoes hydrogenation to afford the product:

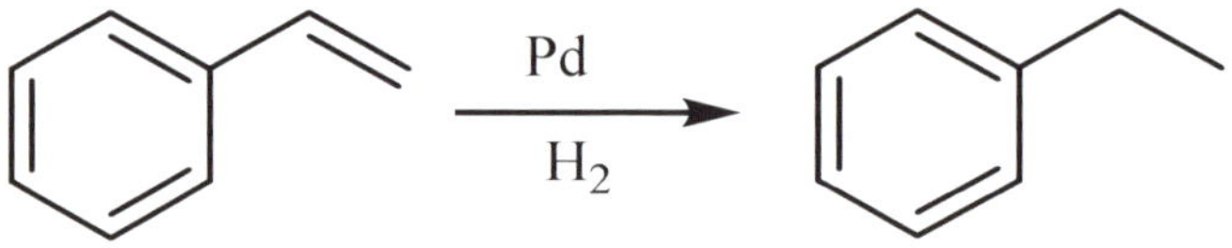

Pd

$H_2$

Benzene is not the only molecule that exhibits the special type of stability that is called aromaticity. An aromatic molecule may have atoms other than C in it, or may even be ionic. There are certain criteria for a molecule to exhibit aromaticity:

1. The molecule must be cyclic and planar
2. The π-electrons must be delocalized around the whole ring. This requires each atom to have either a lone pair, a π-bond, or an empty *p*-orbital
3. The number of electrons that are delocalized around the ring must be $4n+2$ (where $n$ is an integer). This last criterion is called **Hückel's rule**.

A molecule that meets criteria 1 and 2 for aromaticity, but instead has $4n$ electrons, is **antiaromatic**. A molecule that is neither aromatic nor antiaromatic is simply referred to as **nonaromatic**.

Example IV.4.2

Identify each of the following as being aromatic (A), antiaromatic (AA) or nonaromatic (NA):

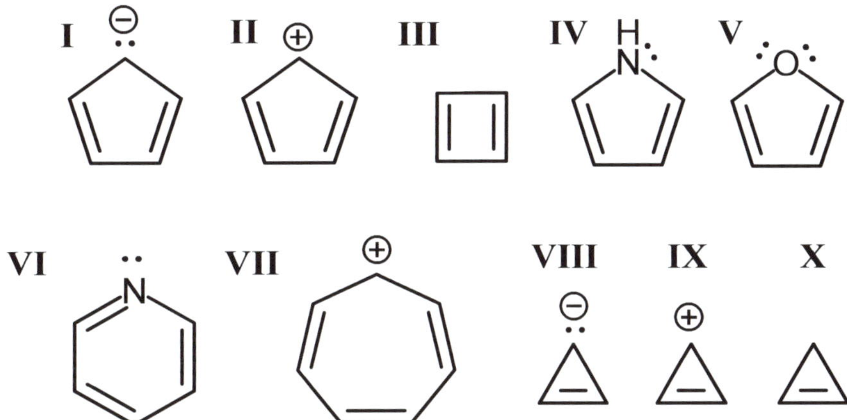


Solution IV.4.2

We must examine each molecule to see what classification is correct. They are all cyclic and, although there are some monocyclic π-conjugated molecules that are not planar (ring size 8 or greater where the number of π electrons $4n+2$), these are also planar. Let us see whether each of these molecules has resonance delocalization around the ring and then check for how many electrons are delocalized:

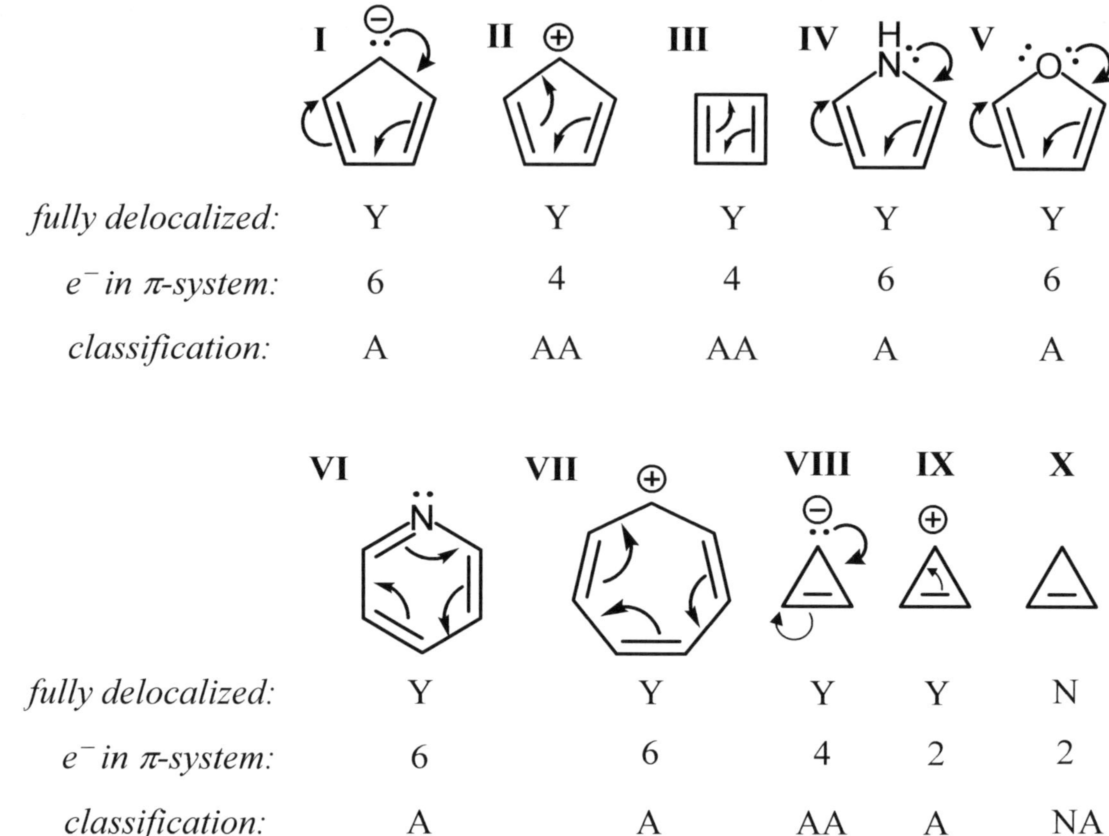


| | I | II | III | IV | V |
|---|---|---|---|---|---|
| *fully delocalized:* | Y | Y | Y | Y | Y |
| *e⁻ in π-system:* | 6 | 4 | 4 | 6 | 6 |
| *classification:* | A | AA | AA | A | A |

| | VI | VII | VIII | IX | X |
|---|---|---|---|---|---|
| *fully delocalized:* | Y | Y | Y | Y | N |
| *e⁻ in π-system:* | 6 | 6 | 4 | 2 | 2 |
| *classification:* | A | A | AA | A | NA |

A few points are worth noting. First, delocalization around the entire ring can only occur if each carbon has an unhybridized *p*-orbital (empty or in a π bond) or a lone pair (remember that resonance structures involve the movement of π-bonds and lone pairs). Each molecule except **X** satisfies this criterion.

Another important point has to do with the lone pairs on the O and N atoms of IV, V and IV. In VI, the lone pair on N does not participate in the resonance delocalization. This is because the N in VI already has a π bond to it; it does not need to use its lone pair to participate in resonance. In fact, the lone pair on the N in VI is in an $sp^2$-hybridized orbital at 90° angle to the π system. We only count the electrons in the π system (the number delocalized), so the lone pair electrons are not counted in our check for Hückel's rule. In contrast, the N in IV does not have a π bond as it is drawn, so its lone pair is in an orbital that can participate in the π system, so these electrons are counted. In V, notice that there are two lone pairs on the O atom. Only one lone pair can be delocalized into the π-system, whereas the other is at 90° to the π-system, so only one lone pair is counted in our assessment of the number of electrons in the π system.

## Lesson IV.5. Aromaticity Effects on Acidity and Basicity

### *IV.5.1 Aromaticity Effects on Acidity*

As we saw in Lesson IV.4, some anions are aromatic. Aromatic anions are significantly more stable than anions that are not aromatic (e.g., $H_3C^-$). Recall that more stabilized anions are weaker bases, and consequently they are the conjugate bases derived from stronger acids. With these facts in mind, it is unsurprising that hydrocarbons which yield aromatic anions upon deprotonation are significantly stronger acids than hydrocarbons which do not. Consider the following p$K_a$ values:

p$K_a$ > 50 — Typical Carbanion

p$K_a$ = 16 — Aromatic Carbanion

As you can see, cyclopentadiene has a much lower p$K_a$ than does cyclopentane. The remarkable stability of its conjugate base, as a result of its aromatic nature, makes cyclopentadiene over $10^{34}$ times more acidic than cyclopentane! We must be careful to consider whether an anion is aromatic or not when we are asked a question that requires us to consider the thermodynamic favorability of a reaction.

### *IV.5.2 Aromaticity Effects on Basicity*

Recall that a Lewis base is an electron pair donor. The ability of a compound to donate its lone pair and act as a Lewis base can be influenced significantly by whether or not a lone pair is tied up in resonance or in an aromatic system. In cases where the lone pair is involved in aromaticity, pulling that lone pair away to do a Lewis acid–base reaction would thus remove aromatic stabilization. We know that aromaticity provides a large amount of stabilization, so such an acid–base reaction would be thermodynamically unfavorable. Consider the basicity of pyridine versus that of pyrrole:

Pyrrole:
Aromatic

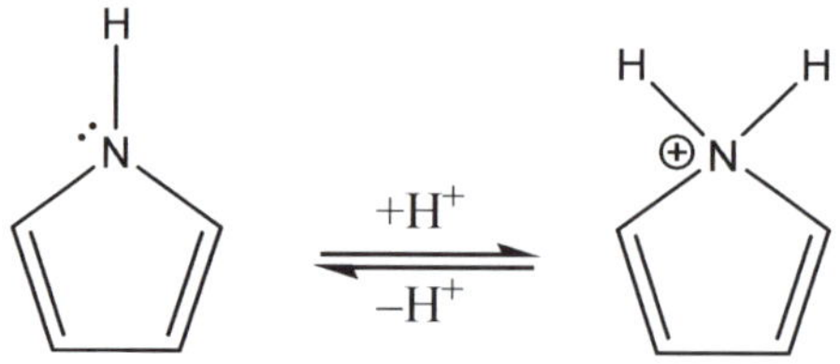

$pK_b = 14$

$+H^+$ / $-H^+$

Not Aromatic;
much stability lost!

Pyridine:
Aromatic

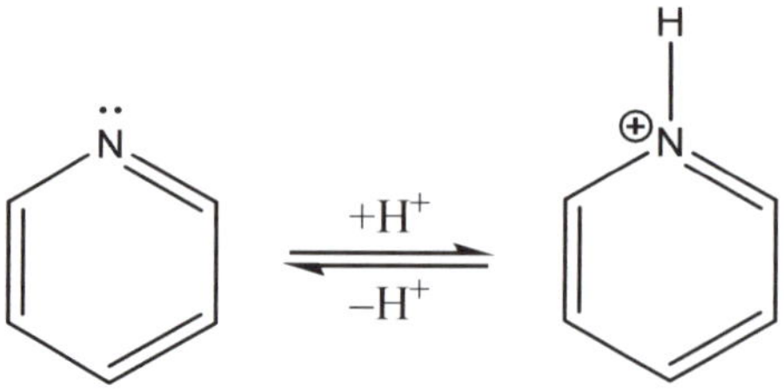

$pK_b = 9$

$+H^+$ / $-H^+$

Still Aromatic!

Recall from general chemistry that a lower $pK_b$ corresponds to a stronger base. The higher $pK_b$ value of pyrrole (14) versus that of pyridine (9) thus indicates that pyridine is about 100,000 times more basic than is pyrrole. This is because the lone pair on the N in pyrrole is delocalized as a key component of the aromatic system. If the lone pair instead donates to an acid, this aromaticity is disrupted. In contrast, the lone pair on the N in pyridine is not delocalized (only the π-bonding electrons are delocalized), so the product of protonation is still aromatic, and thus protonation of pyridine is a much more favorable reaction than is protonation of pyrrole. The general concept to take away from this Lesson is that **reactions that create aromaticity in products tend to be thermodynamically favorable, whereas reactions that disrupt the aromaticity of a starting material tend to be thermodynamically unfavorable.**

Example IV.5.1

A) Which is the most basic nitrogen atom in this molecule?

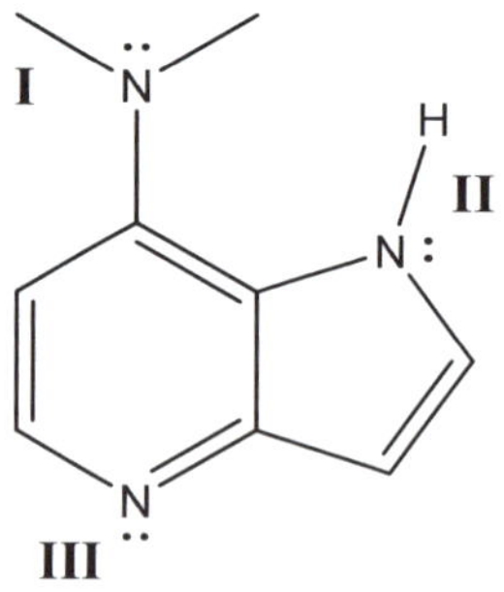

I

II

III

B) Which will undergo heterolysis of the C–Br bond (to form a carbocation and bromide) most rapidly?

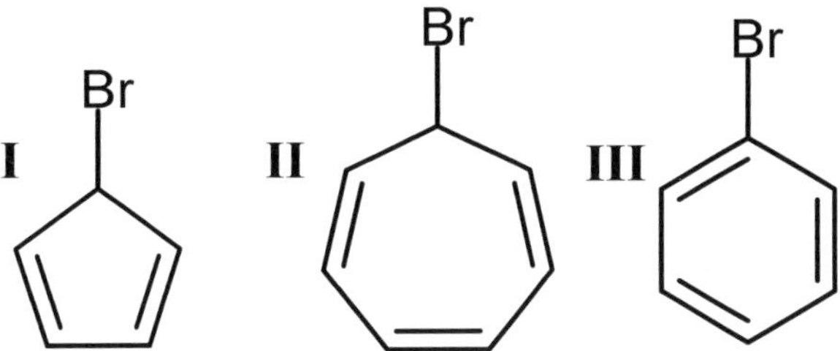


Solution IV.5.1

A) We must consider whether each lone pair participates in resonance delocalization or not. We first check the lone pair on nitrogen **I** and see that it can participate in resonance:

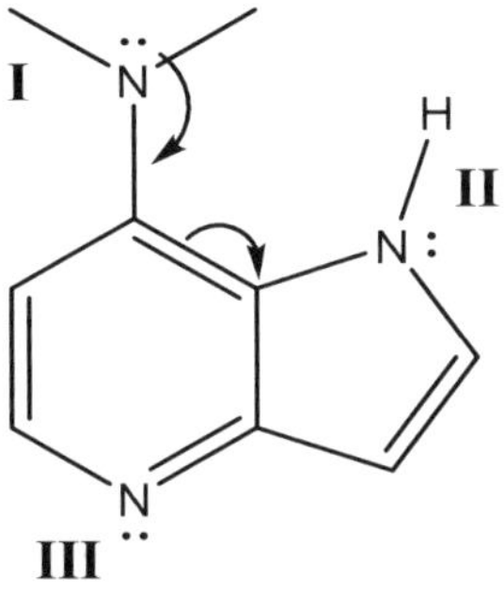

The lone pair on nitrogen **I** is engaged in resonance delocalization. To take that lone pair out of the delocalized system and donate it to an acid instead would cost the resonance delocalization energy. This will not be a very basic site. Next, we check nitrogen **II**:

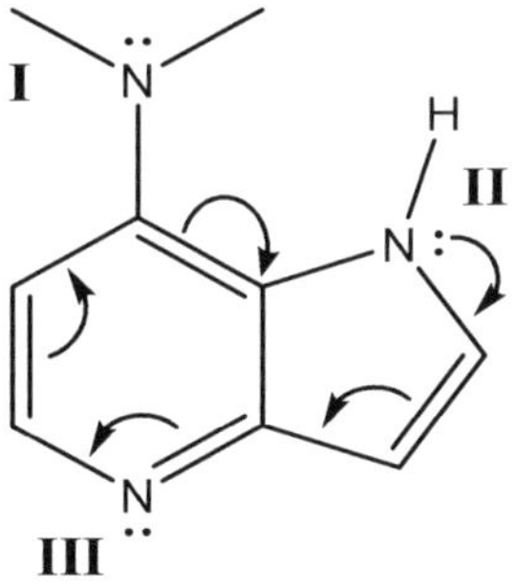

Not only is the lone pair on nitrogen **II** engaged in delocalization, but it is required to make the molecule aromatic. To remove this lone pair would require significant energy to break up the aromaticity. This is an even less basic nitrogen than is nitrogen **I**. The lone pair on nitrogen **III**, however, is not needed for resonance delocalization or for aromaticity. Nitrogen **III** will be by far the most basic N in the molecule.

The carbocations produced from C–Br heterolysis in **I–III** are:

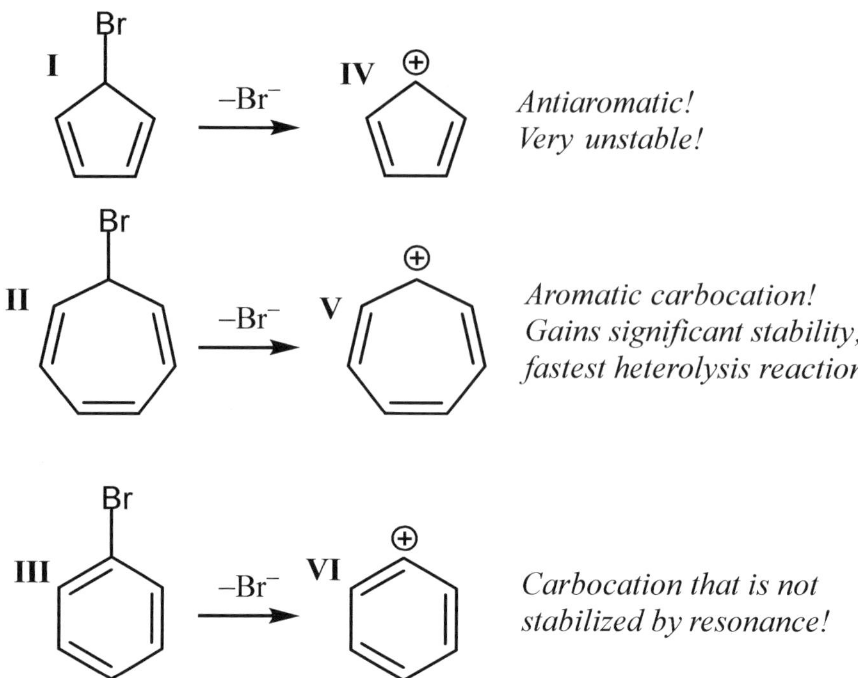


Only heterolysis of **II** leads to the formation of a highly-stabilized aromatic carbocation (**V**), so this is the fastest reaction.

## Lesson IV.6. Nomenclature of Monosubstituted Benzene Compounds

*Lesson IV.6.1 General Rules for Monosubstituted Benzene*

For many common substituents, the rules for naming a monosubstituted benzene are the same as for a monosubstituted cyclohexane (See Lesson I.14 in OC1 Primer), but with benzene as the parent molecule instead of cyclohexane. Consider these examples:

F Cl $NO_2$

*fluorobenzene* *chlorobenzene* *ethylbenzene* *nitrobenzene* *isobutylbenzene or (2-methylpropyl)benzene*

Note the new functional group, $NO_2$, termed a nitro group. In cases where a **benzene ring is itself a substituent of a larger molecule, it is called a "phenyl" group** and is given the abbreviation "Ph".

*Lesson IV.6.2 Common Names for Some Monosubstituted Benzene Derivatives*

Some monosubstituted benzene compounds have specific common names that do not conform to systematic nomenclature rules. Many of these names have been used for centuries and are simpler than the systematic names, so they are still widely-used today. Because these non-systematic names are still used by chemists, these names must be memorized. The non-systematic names for monosubstituted benzene compounds that you will be expected to know are shown below:

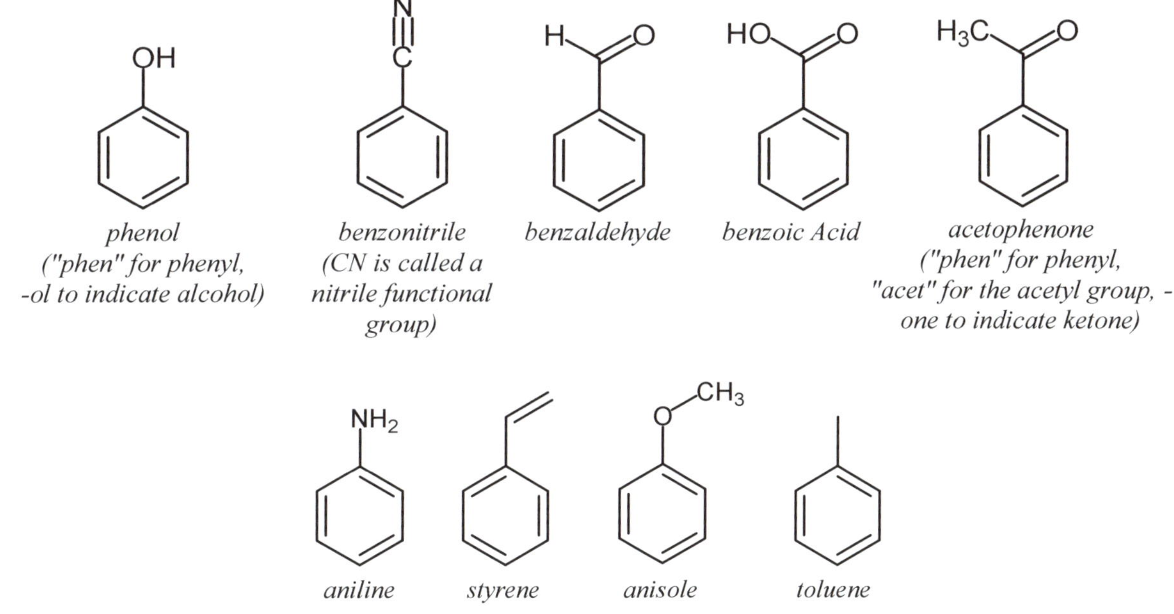


*phenol ("phen" for phenyl, -ol to indicate alcohol)* *benzonitrile (CN is called a nitrile functional group)* *benzaldehyde* *benzoic Acid* *acetophenone ("phen" for phenyl, "acet" for the acetyl group, -one to indicate ketone)*

*aniline* *styrene* *anisole* *toluene*

These are the most frequently encountered common names for monosubstituted benzenes that you are likely to encounter in introductory organic chemistry course. For some molecules, the non-systematic name is a logical reflection of the molecular structure (e.g., phenol is an alcohol on a phenyl group), whereas for others the name provides no such information.

Example IV.6.1

Provide IUPAC names for each of the molecules shown below:

I

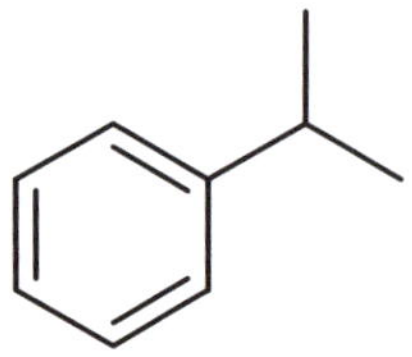

II

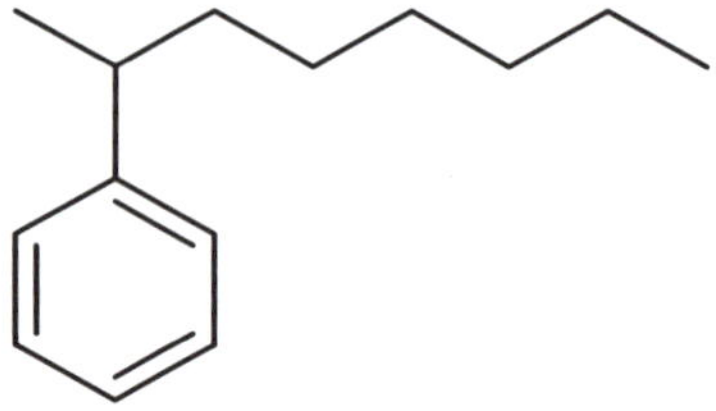

Solution IV.6.1

For molecule **I**, the parent is "benzene" and it has one substituent, which is an "isopropyl" group, so the molecule is properly named isopropylbenzene.

For molecule **II**, the octane chain has more carbon atoms than the benzene ring, so we must use "octane" as the parent. Recall that when benzene is a substituent on a larger molecule, it is called a "phenyl" group. So, molecule **II** is properly named 2-phenyloctane.

## Lesson IV.7. Electrophilic Aromatic Substitution I: Friedel–Crafts Alkylation and Acylation

### *IV.7.1 Electrophilic Aromatic Substitution is Electrophilic Addition then Electrophilic Elimination*

The C=C bonds in benzene are much less reactive than are isolated alkenes due to the extra stability that aromaticity affords (recall that stability and reactivity are inversely related). Because the π-bonds in benzene are less reactive (i.e., less nucleophilic) than those in non-conjugated alkenes, a more reactive electrophile is required to draw the π-bonding electrons out of a benzene ring to initiate a reaction. One type of reaction between an aromatic system and an electrophile is **electrophilic aromatic substitution (EAS).** As its name suggests, this reaction requires an electrophile and results in the substitution of an H on the benzene ring with the electrophile. EAS consists of two steps: 1) electrophilic addition of an electrophile to one of the π-bonds followed by 2) electrophilic elimination of $H^+$ from the C to which the electrophile added:

$E^+$ *electrophilic addition* → *electrophilic elimination* → + $H^+$

We will cover several specific types of EAS in this lesson, but they all have the same underlying mechanism. These EAS reactions differ only in the identify of electrophile that adds to the aromatic system in the first step. In the next several lessons, we will thus look at seven specific sets of reagents and see how they generate electrophiles that are sufficiently reactive to add to the π-bonds in benzene.

### *IV.7.2 Carbocations as Electrophiles in EAS – Friedel-Crafts Alkylation.*

The first class of electrophile we will consider that undergoes electrophilic addition to benzene is a carbocation. When a carbocation is the electrophile for EAS, the reaction is called **Friedel–Crafts alkylation**:

$R^+$ *electrophilic addition* → *electrophilic elimination* → + $H^+$

We have seen several ways to generate carbocations, and two of these methods are commonly used in the Friedel–Crafts alkylation: 1) action of a strong acid to an alcohol (for $S_N1$ and E1 reactions, see Lesson II.11), and 2) electrophilic addition of $H^+$ to the π-bond in an alkene. A third method we will now consider is the abstraction of $Cl^-$ from an alkyl chloride by $AlCl_3$. Here is a recap of these three ways that one can generate carbocations for use in Friedel–Crafts alkylation:

1) Alcohol in the presence of acid

$H_2SO_4$ ; $-H_2O$

2) Alkene in the presence of acid

$H_2SO_4$

most common*

3) Alkyl chloride in the presence of $AlCl_3$

Cl ; Al—Cl ; Cl ; + $[AlCl_4]^-$

*Al has 6 valence electrons*

*Al has an octet*

Chloride abstraction by $AlCl_3$ is the **most commonly-employed method to generate a carbocation** for Friedel–Crafts alkylation, and is so effective that **even primary carbocations can be generated in this way**. The various ways that the carbocation-generating reactions shown above may be harnessed for Friedel–Crafts alkylation reactions are shown below:

1) Carbocation is from an Alcohol in the presence of Sulfuric Acid

$H_2SO_4$ / $CH_3CH_2OH$ ; + $H_2O$

2) Carbocation is from an Alkene in the presence of Acid

$H_2SO_4$ / $H_2C{=}CH_2$

3) Carbocation is from Alkyl chloride in the presence of $AlCl_3$

$AlCl_3$ / $CH_3CH_2Cl$

Reaction 3 is the most-used EAS reaction in the world today, because the ethylbenzene product is a key intermediate in the synthesis of polystyrene (the main component of Styrofoam and other common plastics).

Regardless of which of these routes is used to generate the carbocation, remember that 1) a spontaneous carbocation rearrangement may occur if rearrangement affords a more stable carbocation and 2) any carbocation rearrangement will occur before electrophilic addition to an aromatic system, as illustrated in the example shown below:

Example IV.7.1

Provide the major organic product for each of these reactions:

A)

$(CH_3)_3COH$ + [benzene] $\xrightarrow{H_2SO_4}$

B)

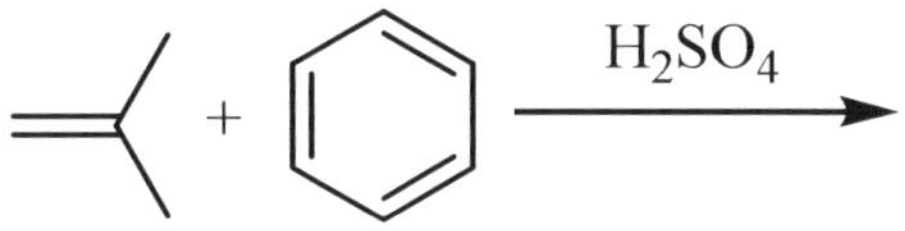

[isobutylene] + [benzene] $\xrightarrow{H_2SO_4}$

C)

Cl [1-chloro-2-methylpropane] + [benzene] $\xrightarrow{AlCl_3}$

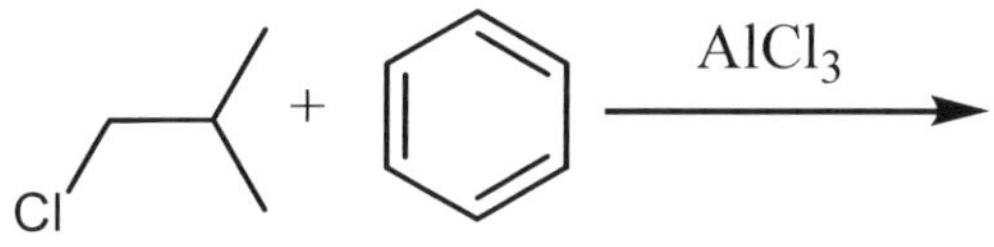

Solution IV.7.1

A) We must recognize that a carbocation will be generated by action of the acid with the alcohol, so first we need to check whether the carbocation rearranges:

$$(CH_3)_3COH + H_2SO_4 \longrightarrow (CH_3)_3C^{\oplus} + H_2O + HSO_4^-$$

In this case, the carbocation is tertiary and will not rearrange. So, we use it as the electrophile for electrophilic aromatic substitution with the benzene:

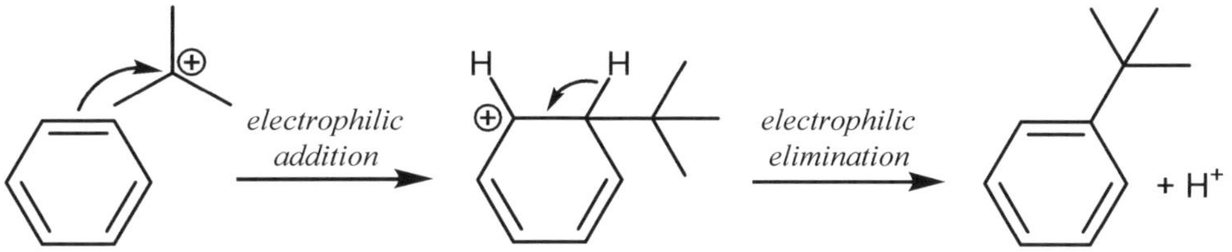

*electrophilic addition*

H H ⊕

*electrophilic elimination*

+ $H^+$

B) We see that the alkene will produce a cation upon electrophilic addition of a proton from the acid. The more substituted carbocation is formed:

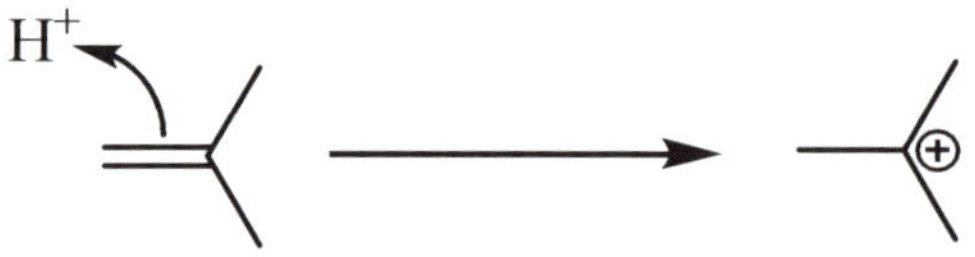

This is actually the same cation as was formed for part A, so the same product will form, and the EAS mechanism will be identical to that shown in the solution to part A.

C) The $AlCl_3$ will abstract a chloride from the alkyl chloride to generate a carbocation:

:Cl: Cl–Al–Cl (Cl, Cl) ⟶ ⊕ + $[AlCl_4]^-$

The carbocation formed initially is primary, but a 1,2-hydride shift from the adjacent carbon will afford a more stable tertiary carbocation. Therefore, the initial carbocation rapidly rearranges before it encounters a benzene molecule for further EAS reactivity:

This carbocation – the same cation we used for EAS in parts A and B – will then participate in the EAS mechanism as shown in the solution to part A.

### *IV.7.3 Acylium Cations as Electrophiles in EAS – Friedel-Crafts Acylation.*

Chloride abstraction by $AlCl_3$ also works with acid chlorides to generate acylium ions:

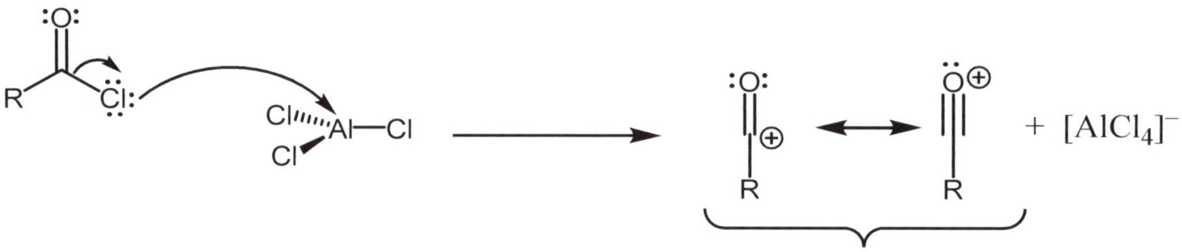

:O: R Cl: Cl–Al–Cl Cl ⟶ :O: ⊕ R ⟷ O⊕ R + $[AlCl_4]^-$

*Acylium ion*
*(two resonance contributors)*

Acylium ions are reactive enough to react with C=C bonds in benzene. When an acylium ion is used as the electrophile in an EAS reaction, the reaction is called **Friedel-Crafts acylation**:

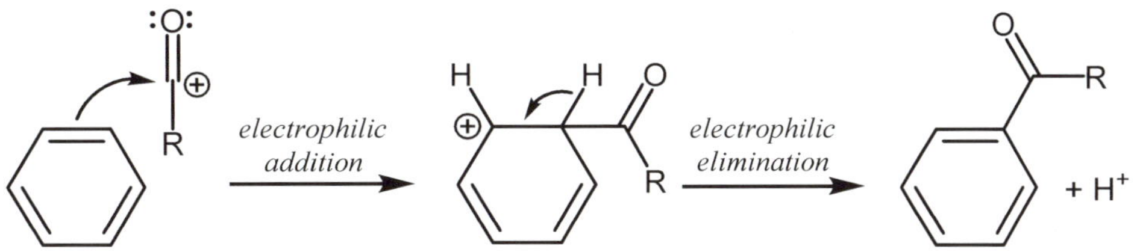

electrophilic addition

electrophilic elimination

+ $H^+$

Unlike alkyl carbocations, acylium cations do not undergo carbocation rearrangement because an octet can be formed at the carbon via donation from the oxygen, so after an acylium cation has been formed, it will participate in EAS without rearrangement.

Example IV.7.2

Draw the major organic product for each of the following reactions:

A) + $\xrightarrow{AlCl_3}$

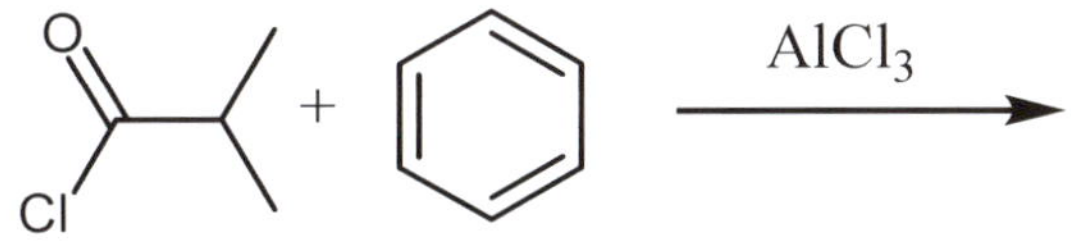

B) + $\xrightarrow{AlCl_3}$

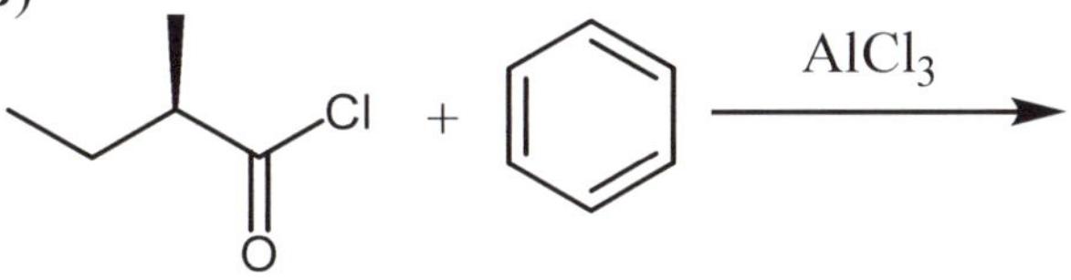

Solution IV.7.2

A) The $AlCl_3$ will abstract a $Cl^-$ from the acid chloride to form an acylium ion:

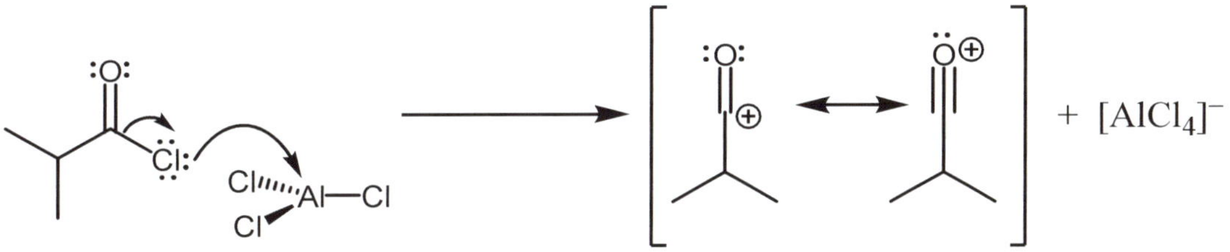

+ $[AlCl_4]^-$

The acylium ion then acts as the electrophile for an EAS reaction:

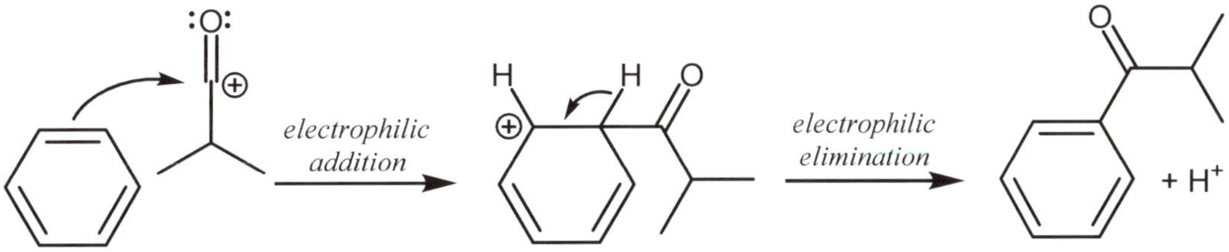

electrophilic addition

electrophilic elimination

+ $H^+$

B) The first step is again generation of the acylium cation:

:O: Cl: Cl Al Cl Cl Et :O: Et :O: + $[AlCl_4]^-$

Note that the chiral center in the acid chloride is retained in the acylium cation, because that carbon remains $sp^3$-hybridized throughout the reaction. It will maintain its configuration even after undergoing EAS:

:O: Et *electrophilic addition* H H O Et *electrophilic elimination* O Et + $H^+$

## Lesson IV.8. Electrophilic Aromatic Substitution II: Nitration and Sulfonation of Benzene

*IV.8.1 Nitration: EAS using a Nitronium Electrophile*

If benzene is reacted with nitric acid ($HNO_3$) and sulfuric acid ($H_2SO_4$), nitrobenzene is produced. This reaction involves a nitronium ion ($[NO_2]^+$) as the electrophile and the reaction is called **nitration**. The nitronium electrophile is generated by the protonation of nitric acid by sulfuric acid as follows:

$$O_2N{-}OH + HO{-}SO_2{-}OH \longrightarrow O_2N{-}OH_2^{\oplus} \longrightarrow O{=}N^{\oplus}{=}O + H_2O + HSO_4^-$$

*Nitric Acid* *Sulfuric Acid* *Nitronium Ion*

After the nitronium has been generated, it acts as the electrophile in the usual EAS mechanism:

*electrophilic addition* *electrophilic elimination*

NO2 + H+

*IV.8.2 Sulfonation: EAS using a Sulfonium Electrophile*

If benzene is reacted with either sulfuric acid ($H_2SO_4$) alone or a mixture of sulfuric acid and $SO_3$, benzenesulfonic acid ($C_6H_6SO_3H$) is produced. The $-SO_3H$ group is called a **sulfonic acid** functional group. The mixture of sulfuric acid/$SO_3$ is called "**fuming sulfuric acid**" because of the $SO_3$ fumes it releases. This reaction can involve either a sulfonium ion ($[HSO_3]^+$) or sulfur trioxide as the electrophile, and the reaction is called **sulfonation**. The sulfonium electrophile is generated by reaction of sulfuric acid as follows:

$$HO_3S{-}OH + HO{-}SO_2{-}OH \longrightarrow HO_3S{-}OH_2^{\oplus} \longrightarrow HO{-}S^{\oplus}O_2 + H_2O + HSO_4^-$$

*Sulfuric Acid* *Sulfuric Acid* *Sulfonium Ion*

After the sulfonium ion has been generated, it acts as the electrophile in the usual EAS mechanism:

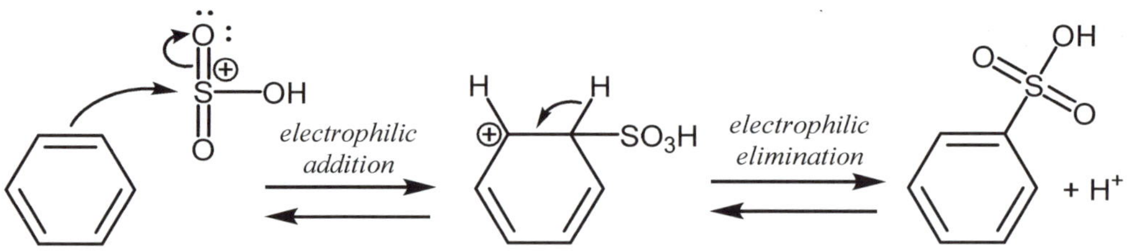
OH
electrophilic
addition
H
H
SO3H
electrophilic
elimination
OH
+ H+

When $SO_3$ is used as the electrophile in EAS, the mechanism can be shown as:

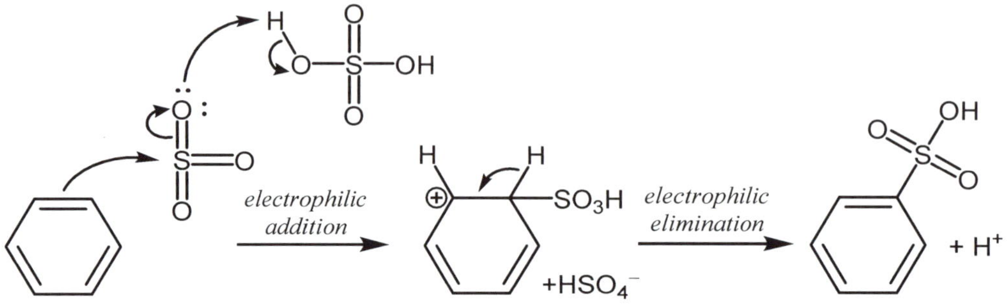
H
OH
OH
electrophilic
addition
H
H
SO3H
+HSO4−
electrophilic
elimination
+ H+

Because $SO_3$ itself does not have a proton, it must pick one up from the sulfuric acid to be able to form the $HSO_3$ group. Unlike the other EAS reactions we have seen, sulfonation is a highly reversible reaction, so if an aromatic compound with a sulfonic acid group is heated in the presence of water with trace acid, the sulfonic acid group is removed and replaced with a proton.:

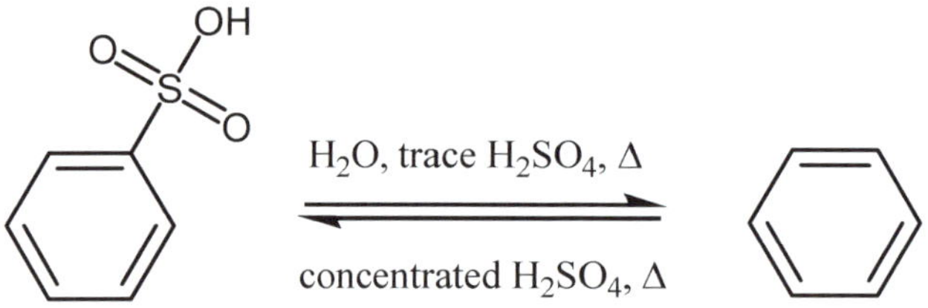
OH
H2O, trace H2SO4, Δ
concentrated H2SO4, Δ

Example IV.8.1

Draw the major organic product for each of these reactions:

A) 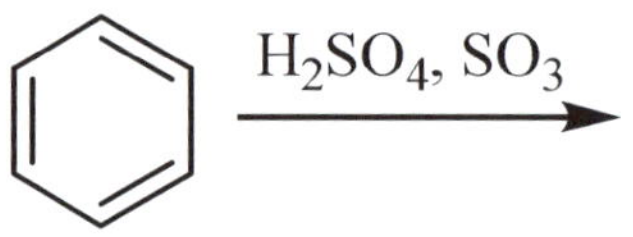
H2SO4, SO3

B) 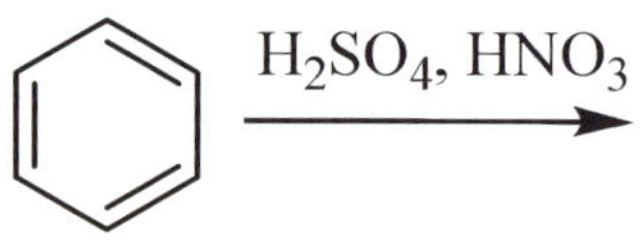
H2SO4, HNO3

C) 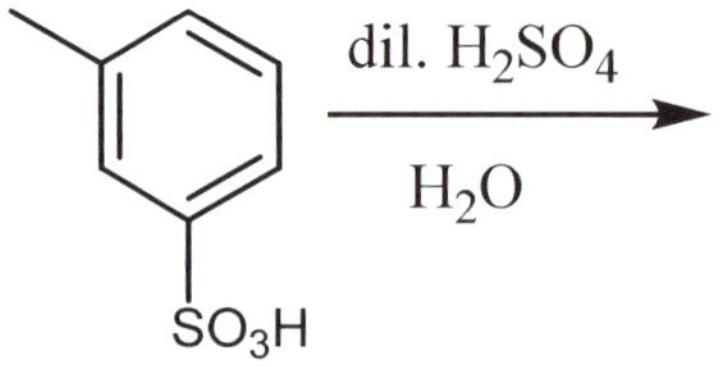
dil. H2SO4
H2O
SO3H

Solution IV.8.1

A) This is a sulfonation reaction. The product is:

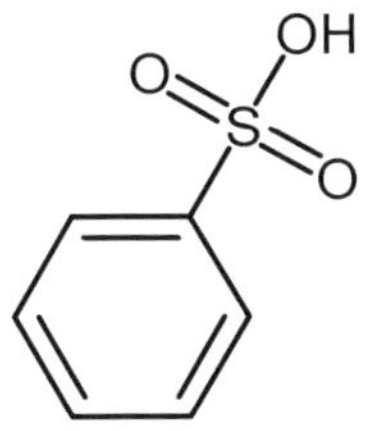

B) This is a nitration. The product is:

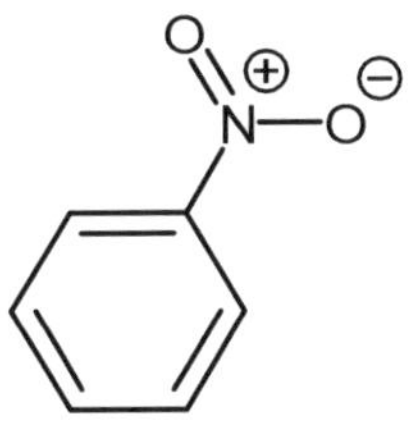

C) This is the reverse of sulfonation. The product will be:

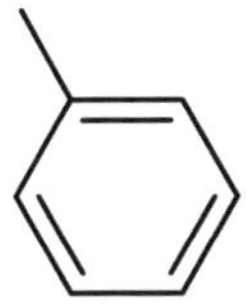

## Lesson IV.9. Electrophilic Aromatic Substitution III: Halogenation of Benzene

### *IV.9.1 Iodination of Benzene*

In Lesson IV.8, we saw that sulfuric acid can facilitate the formation of electrophiles such as nitronium and sulfonium ions. Sulfuric acid can also oxidize iodine to form iodenium ions ($I^+$):

$$2H^+ + H_2SO_4 + I_2 \rightarrow 2I^+ + H_2SO_3 + H_2O$$

An iodenium ion can then act as an electrophile for EAS with benzene to form iodobenzene:



### *IV.9.2 Chlorination and Bromination of Benzene*

Unlike iodine, chlorine and bromine cannot be oxidized as easily to form free chlorenium or bromenium ions. When an electrophilic chlorine or bromine is needed for a reaction, a good alternative is to polarize the halogen by reaction with $FeX_3$ (here, X = Cl or Br):

$$X{-}X \xrightarrow{FeX_3} \overset{\delta+}{X}{-}\overset{\delta-}{X}{\cdots}FeX_3$$

Here, the dashed line indicates an attractive force that has not become a full bond and the X can be Cl or Br. The mechanism using this polarized halogen as the electrophile can be represented like this:



In practice, chlorination and bromination reactions may employ Fe powder instead of $FeCl_3$ or $FeBr_3$, respectively, because $Cl_2$ and $Br_2$ will react with Fe to make $FeCl_3$ or $FeBr_3$. So, there are a couple of ways that you might see this reaction represented:

$$\text{C}_6\text{H}_6 \xrightarrow{Br_2,\ Fe} \text{C}_6\text{H}_5\text{Br} + H^+$$

$$\text{C}_6\text{H}_6 \xrightarrow{Cl_2,\ FeCl_3} \text{C}_6\text{H}_5\text{Cl} + H^+$$

In the above examples, $Cl_2$ and $Br_2$ can be interchanged depending on whether one wants to add a Cl or Br to the benzene ring, respectively.

Example IV.9.1

Draw the major organic product for each of these reactions:

A) benzene $\xrightarrow{H_2SO_4,\ SO_3}$

B) benzene $\xrightarrow[H_2SO_4]{CH_3CH_2OH}$

C) benzene $\xrightarrow{FeBr_3,\ Br_2}$

D) benzene $\xrightarrow[AlCl_3]{CH_3C(O)Cl}$

E) benzene $\xrightarrow{Fe,\ Cl_2}$

F) benzene $\xrightarrow{I_2,\ H_2SO_4}$

G) benzene $\xrightarrow[AlCl_3]{(CH_3)_2CHCl}$

H) benzene $\xrightarrow{H_2SO_4,\ HNO_3}$

Solution IV.9.1

A) This is a sulfonation reaction (Lesson IV.8). The product is phenylsulfonic acid ($PhSO_3H$).

B) This is a Friedel–Crafts alkylation (Lesson IV.7) and is the subset in which an ethyl carbocation is generated from ethanol by acid. The product is ethylbenzene.

C) This is a bromination reaction (Lesson IV.9). The product is bromobenzene.

D) This is a Friedel–Crafts acylation (Lesson IV.7). The product is acetophenone, ($PhC(=O)CH_3$).

E) This is a chlorination reaction (Lesson IV.9). The product is chlorobenzene.

F) This is an iodination reaction (Lesson IV.9). The product is iodobenzene.

G) This is a Friedel–Crafts alkylation (Lesson IV.7) and is the subset in which an isopropyl carbocation is generated from isopropyl chloride by $AlCl_3$. The product is isopropylbenzene.

H) This is a nitration reaction (Lesson IV.8). The product is nitrobenzene ($PhNO_2$).

# Lesson IV.10. Nomenclature of Polysubstituted Benzene Compounds

## *IV.10.1 Systematic Nomenclature*

Polysubstituted benzene derivatives can be named using the same approach as for polysubstituted cyclohexane derivatives, but without the need for any *cis-* or *trans-* labels, because benzene is planar. Here are some examples:

*1,2-difluorobenzene* *1-bromo-2-chlorobenzene* *1-chloro-3-ethyl-2-isopropylbenzene* *2-ethyl-4-iodo-1-propylbenzene*

## *IV.10.2 Special Parent Structures*

We saw in Lesson IV.6 that some monosubstituted benzene derivatives have historical common names that are now deeply rooted in the field of chemistry. When these structures are present in a polysubstituted benzene derivative, one must use the common name as the parent rather than using benzene as the basis for the name. In the common name parent structure, the substituent from which its common name arises is always given the number 1. In phenol, for example, the C bearing the OH group is assigned the number 1, in toluene, the C with the methyl group is assigned the number 1, etc. Here are a few examples to illustrate:

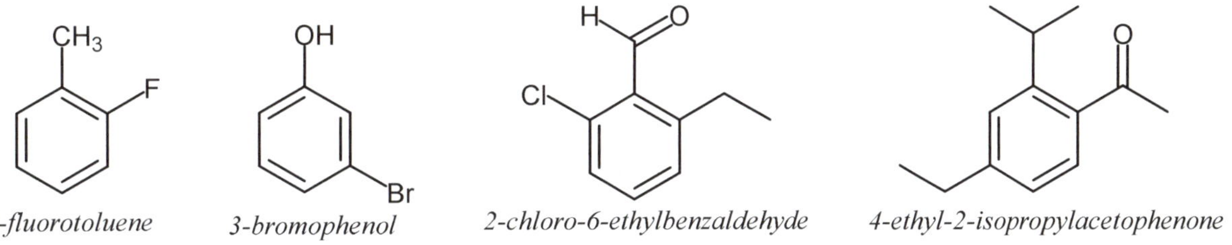


*2-fluorotoluene* *3-bromophenol* *2-chloro-6-ethylbenzaldehyde* *4-ethyl-2-isopropylacetophenone*

## *IV.10.3 Alternative Nomenclature for Disubstituted Benzene Derivatives*

For disubstituted benzene derivatives, one can use the systematic nomenclature as illustrated above. An alternative nomenclature system can also be used *for disubstituted benzene derivatives only*. Instead of using 1,2-, one can simply use *ortho-*. In the case of an *ortho*-disubstituted benzene derivative, the *o-* label is placed in front of the rest of the name, and the numbers are omitted. A similar convention allows us to use *m-* (for *meta-*) in place of numbers for 1,3- derivative or *p-* (for *para-*) in place of numbers for 1,4- derivatives.

*o-dichlorobenzene* *m-dibrombenzene* *o-fluorotoluene* *m-bromophenol* *p-ethylanisole*

As for monosubstituted benzene derivatives, there are some disubstituted benzene derivatives that have long-standing common names that must be committed to memory. The most frequently-encountered examples in organic chemistry are the "xylenes", which are "dimethylbenzene" derivatives:

*o-xylene* *m-xylene* *p-xylene*

Example IV.10.1

Provide the proper names for each of these molecules:

Solution IV.10.1

One challenge is to identify whether the parent has a non-systematic common name, such as xylene (leftmost molecule) or toluene (center molecule). The proper names should be:

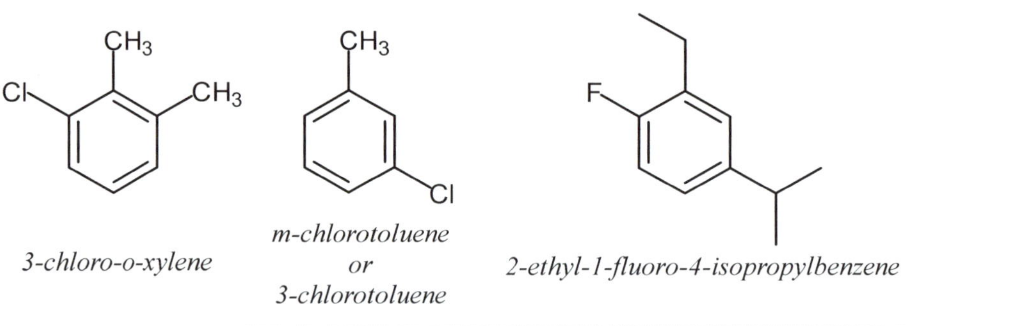


*3-chloro-o-xylene*

*m-chlorotoluene*
*or*
*3-chlorotoluene*

*2-ethyl-1-fluoro-4-isopropylbenzene*

# Lesson IV.11. Substituent Effects on the Rate of Electrophilic Aromatic Substitution

## *IV.11.1 More Stable Cations Form More Quickly in EAS*

The key intermediate in EAS is the carbocation formed by electrophilic addition to the aromatic π-system. The formation of this carbocation is the rate-limiting step of the two-step sequence. In general, the more stable a carbocation is, the more quickly it will form, so anything that stabilizes the carbocation intermediate will accelerate the reaction, as illustrated by this reaction coordinate diagram:

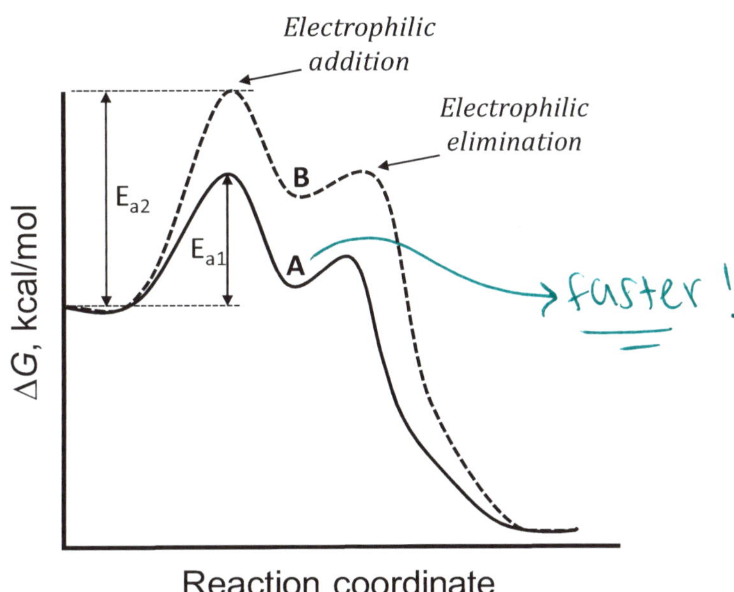


The energy of activation, $E_{a1}$, to form the more stable carbocation (intermediate **A**) is much lower than the energy of activation, $E_{a2}$, to form the less stable carbocation (intermediate **B**), so the EAS reaction proceeding through intermediate **A** will be faster.

## *IV.11.2 Resonance Donor Substituents Activate the Ring to Faster Reaction*

We know that resonance can stabilize carbocations. It thus follows that if a benzene ring has substituents that can donate electrons to stabilize the carbocation intermediate of an EAS reaction, the presence of these substituents will accelerate the reaction. Substituents with a lone pair-bearing atom directly adjacent to the benzene π-system will be the best resonance donors. For an atom to readily donate its lone pair into the π-system, however, that lone pair orbital must be comparable in size to the π-bond orbitals on C. To put it more succinctly, the lone pair-bearing atom should be from the same row as C (i.e., similar in size) to achieve optimum resonance donation. Electronegativity also plays a role in electron donating ability. Fluorine is so much more electronegative than carbon that it will not donate a lone pair via resonance. Taken together, this leaves us with O and N as elements that are in the same row as C and that commonly have lone pairs. The common **resonance electron-donating groups (EDGs)** that will be **Activating for EAS** are -OR and $-NR_2$ (where R can be H or alkyl).

*IV.11.3 Alkyl Groups are Inductive Donors and thus Slightly Activate a Ring to EAS*

We know from Lesson I.11 (OC1 Primer) that alkyl substituents can stabilize carbocations by hyperconjugation. It makes sense, then, that if a benzene ring has alkyl substituents which can stabilize the carbocation intermediate of an EAS reaction, the presence of these substituents will accelerate the reaction. Hyperconjugation is an example of an inductive effect, so **alkyl groups are classified as weakly inductive electron-donating groups (EDGs)**, which will be **Slightly Activating for EAS**.

*IV.11.4 Halogens are Weak Inductive Withdrawing and thus Slightly Deactivating for EAS*

Halogens substituents have lone pairs, but generally do not donate electrons to aromatic systems for two different reasons. Although fluorine is from the same row as C and will have an appropriate size match, it is significantly more electronegative than C, so it does not donate its lone pair electrons into a C-based π-system. Conversely, the heavier halogens are closer in electronegativity to C, but because they are so much larger, they are unable to overlap with the C-based π-system to donate electrons (i.e., a bad size match). Because each halogen is unable to participate in resonance donation (for different reasons), the inherent inductive electron-withdrawing effect of a halogens will dictate the rate of EAS for halogenated rings. Effects caused by nearby electronegative atoms are called inductive effect (Lesson I.10 in OC1 Primer), so halide groups are classified as **weakly inductive withdrawing groups (EWGs)**, which will be **Slightly Deactivating for EAS**.

*IV.11.5 Strong Inductive Withdrawing Groups are Deactivating for EAS*

When a substituent features a partial positive or formal positive charge on the atom immediately adjacent to the benzene ring, this charge will strongly repel any additional positive charge that would form on the aromatic system. As a result, this Coulombic repulsion means that a substituent with a partial positive or formal positive charge will significantly increase the energy of the carbocation intermediate formed during EAS. Substituents such as these are classified as **strongly inductive electron withdrawing groups (EWGs)**, which will be **Deactivating for EAS**. Common strong inductive withdrawing groups include $-NO_2$, –C(=O)R, –C(=O)OR, –C(=O)NR2, –C(=O)H, $-CF_3$, and $-SO_3H$.

Example IV.11.1

Rank these substrates from fastest (1) to slowest (5) rate of electrophilic aromatic substitution reaction:

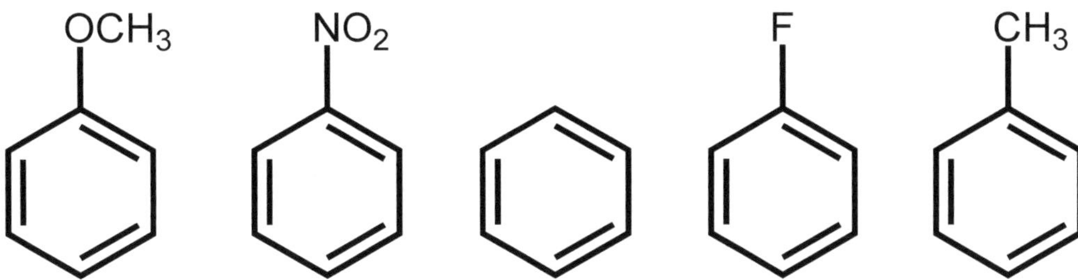


Solution IV.11.1

We start by determining what type of substituent (resonance donor, inductive donor, etc.) is present on each of the molecules.

From left to right:

The –$OCH_3$ is a resonance donor and is strongly activating.
The –$NO_2$ group has a formal 1+ charge on the N directly attached to the aromatic system. It is therefore strongly electron withdrawing and strongly deactivating
This is benzene without any substituents. This is the reference molecule or "baseline" to which each of the substrates is be compared.
The –F is a halogen and is thus a weakly inductive withdrawing group that slightly deactivates the ring with respect to EAS.
The –$CH_3$ is a weak inductive donor and is slightly activating.

With these assignments in hand, we can now rank the molecules from fastest to slowest reaction rate in EAS:

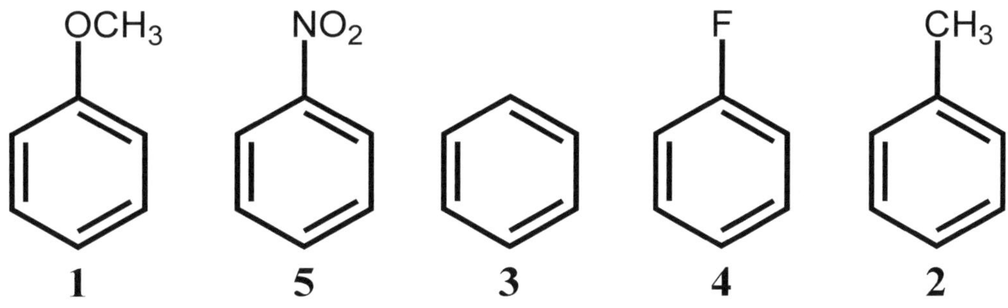

## Lesson IV.12. Substituent Effects on the Regiochemistry of Electrophilic Aromatic Substitution

### *IV.12.1 More Stable Intermediates Lead to Higher Product Yield*

The placement of substituents on a benzene ring not only influences the rate of the EAS reaction (Lesson IV.11), but also on the regiochemistry of the products. What does this mean? When a monosubstituted benzene derivative undergoes EAS, for example, there are three possible sites where the electrophile can go relative to the substituent "Z", i.e., to form *o*-, *m*- and *p*- isomers:

Z
$E^+$ (electrophile)
E
*o*-product
*m*-product
*p*-product

Each of these products derives from a different carbocation intermediate:

*electrophilic addition*
$E^+$ (electrophile)
*electrophilic elimination*
*o*-product
*m*-product
*p*-product

In our previous studies of alkene hydration and hydrohalogenation (Lesson III.5 in OC1 Primer), we saw that the major product was always derived from the most stable carbocation that could form after electrophilic addition of $H^+$ to the C=C bond. The EAS reaction is the same: **the major product(s) derive from the most stable carbocation(s)** that form in the electrophilic addition step. By analyzing in more detail each carbocation intermediate that gives rise to each possible product isomer, we will be able to assess what the major products will be. The rest of this lesson will examine how each type of substituent (resonance donor, etc.) influences the stability of the various carbocation intermediates and, as a result, which product isomers will be formed in the greatest yield.

### *IV.12.2 Resonance Donor Substituents are ortho-/para- Directors*

Resonance donor substituents have lone pairs that they can donate into an aromatic π-system via resonance. These substituents can provide additional resonance stabilization to the carbocation intermediate formed if the electrophile adds *ortho-* or *para-* relative to the substituent –Z, but not if the electrophile adds *meta-*:

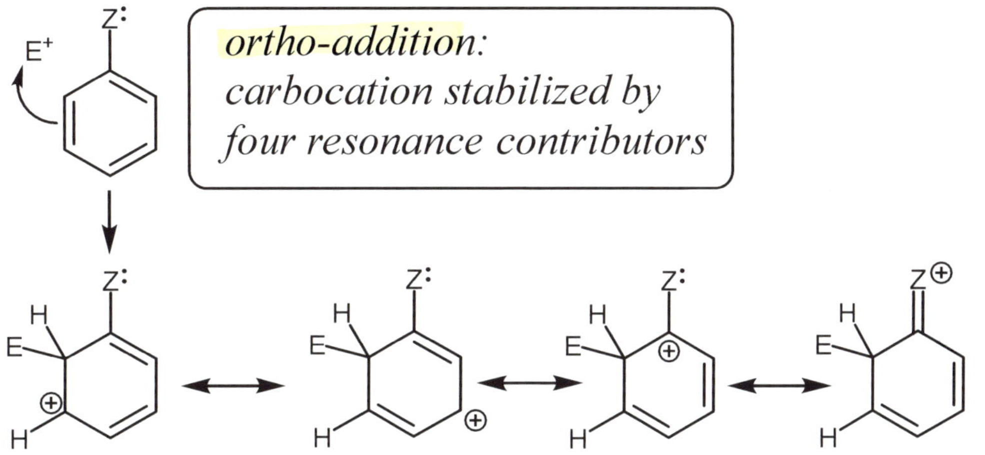

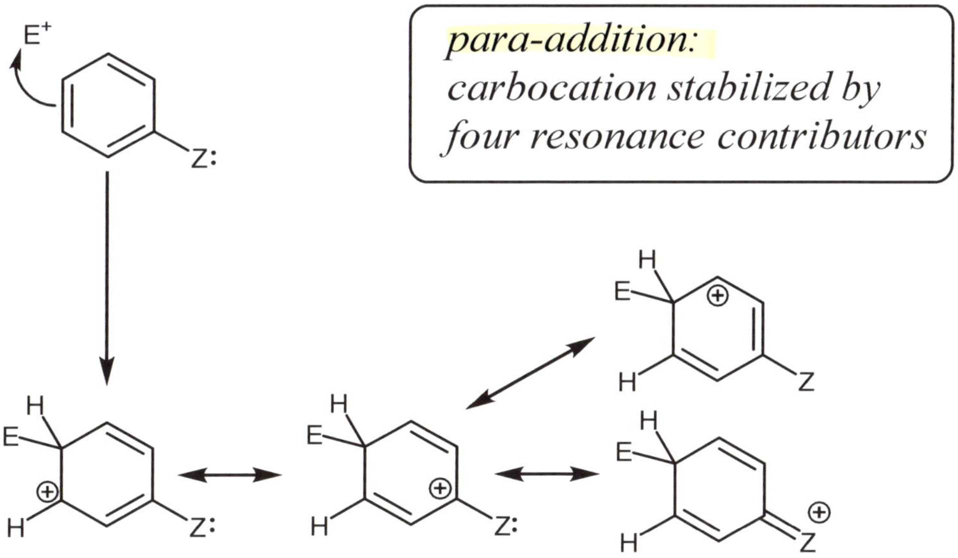

*ortho-addition:*
*carbocation stabilized by*
*four resonance contributors*

---

*para-addition:*
*carbocation stabilized by*
*four resonance contributors*

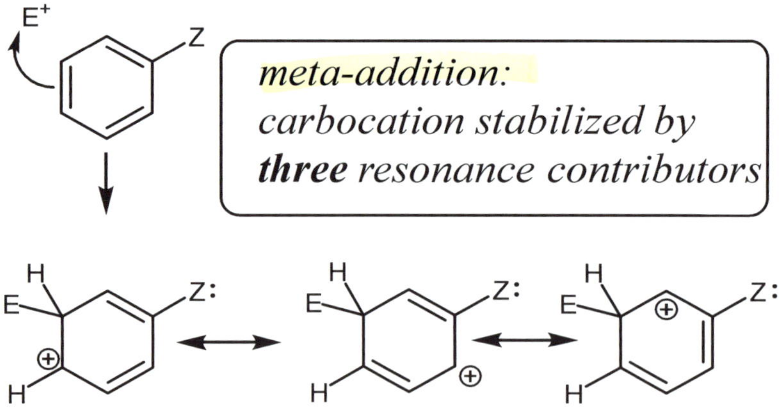

*meta-addition:*
*carbocation stabilized by*
***three** resonance contributors*

Because the intermediates on the path to the *ortho-* and *para-* products are more stable (due to more resonance contributors/more delocalization) than the intermediate on the path to the *meta-* product, **resonance donors are ortho-/para-directing groups** for electrophilic aromatic substitution reactions. The *ortho-* and *para-* products will be the major products of electrophilic aromatic substitution on a benzene ring having a resonance donor substituent.

*IV.12.3 Inductive Donor Substituents are ortho-/para- Directors*

The inductive donors that we discussed – alkyl groups – will provide added stabilization to the carbocation intermediates of EAS when the electrophile adds *ortho-* or *para-* to the substituent. This is because the carbocations will be tertiary (3°) instead of secondary (2°) in one of the resonance contributors to the carbocation intermediate in the case of *ortho-* or *para-* addition, thus providing stabilization due to hyperconjugation in those cases:

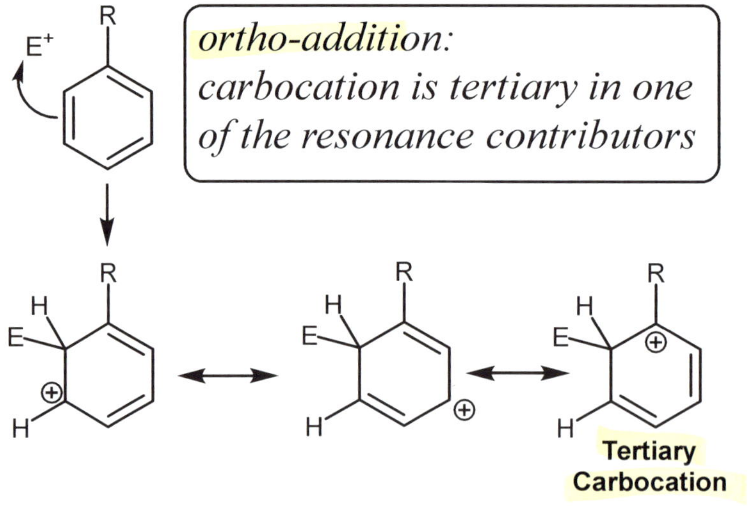

*ortho-addition:*
*carbocation is tertiary in one of the resonance contributors*

**Tertiary Carbocation**

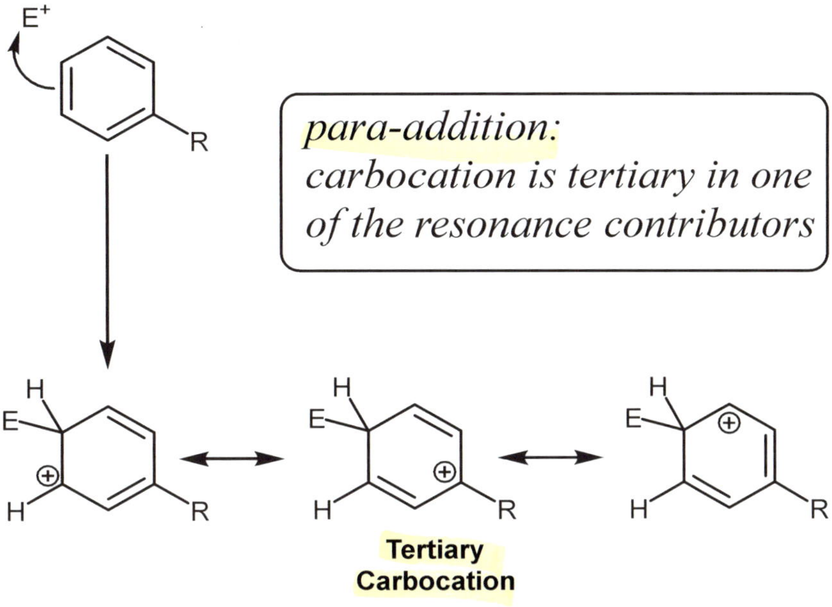

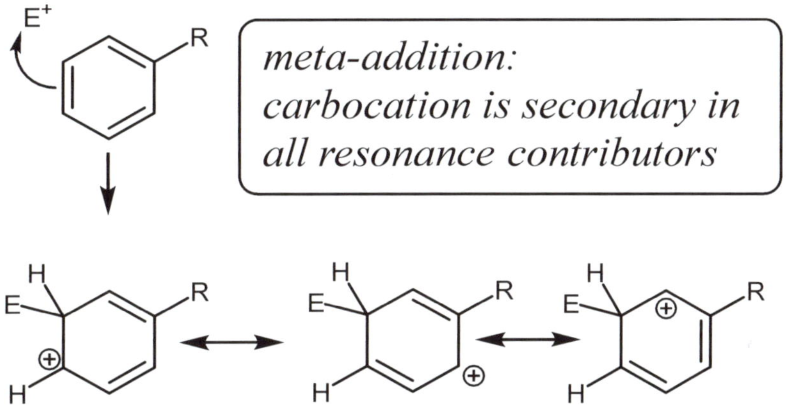

*para-addition:*
*carbocation is tertiary in one of the resonance contributors*

**Tertiary Carbocation**

---

*meta-addition:*
*carbocation is secondary in all resonance contributors*

Because of the extra stability provided by hyperconjugation in the intermediates leading to the *ortho-* and *para-* products, **inductive donors are *ortho-/para-*directing groups** for electrophilic aromatic substitution reactions.

### *IV.12.4 Halogens are ortho-/para- Directors*

Unlike resonance donors, halogen lone pairs (other than fluorine) are too poorly matched in size to effectively donate electrons to the benzene π system via resonance. Even fluorine, a good size match to C, does not effectively donate its lone pair by resonance. A lone pair adjacent to a cationic site, however, can stabilize a carbocation in a hyperconjugation-like fashion by through-space attraction between the lone pair and the empty *p*-orbital of a carbocation:

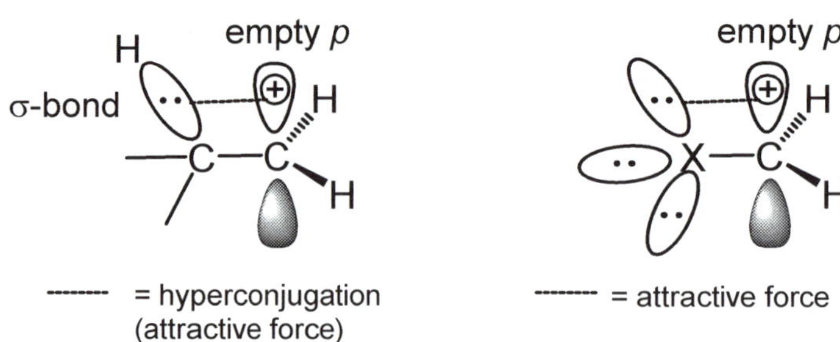


We also saw this hyperconjugation-like stabilization in dichlorocarbene (Lesson III.8 in OC1 Primer). In the current context, the presence of this hyperconjugation-like stabilizing effect in the carbocation intermediates that lead to the *o*- and *p*- products makes these intermediates more stable:

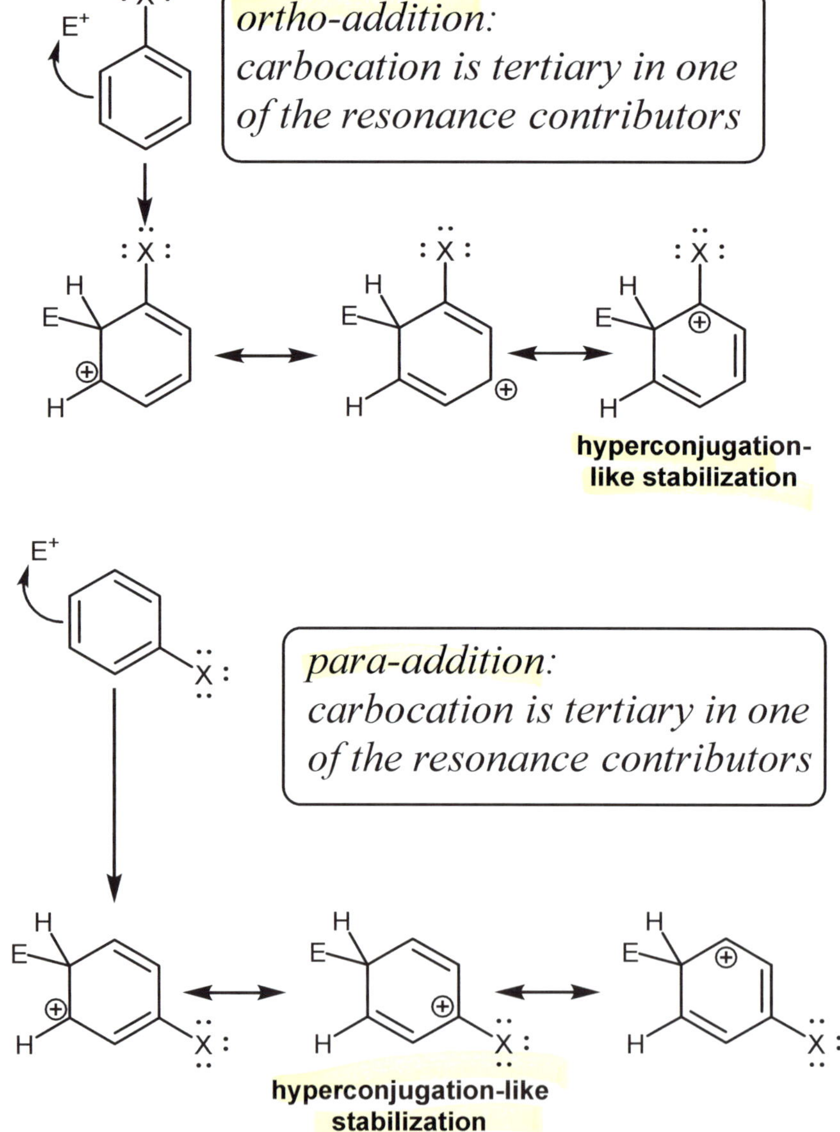

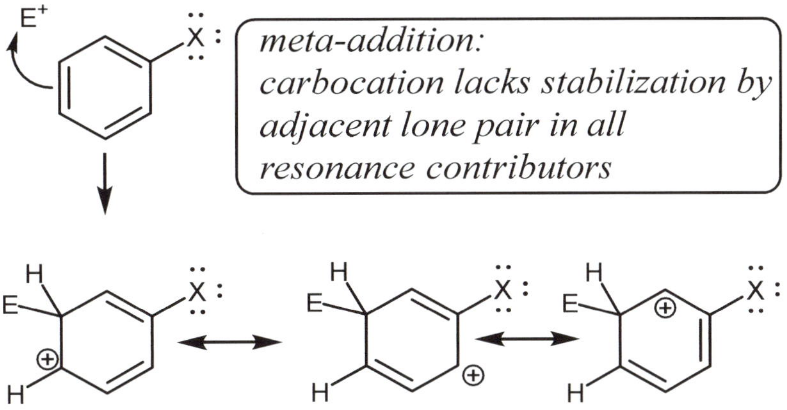
E+
X
meta-addition:
carbocation lacks stabilization by
adjacent lone pair in all
resonance contributors
H
E
H
X
H
E
H
X
H
E
H
X

Because of the extra stability provided by the adjacent lone pairs in the intermediates leading to the *o*- and *p*- products, **halogens are *ortho-/para*-directing groups** for electrophilic aromatic substitution reactions.

*IV.12.5 Strong Inductive Withdrawing Groups are the only meta- Directors*

The final class of substituents that we have to examine are those that feature a partial positive or formal positive charge on the atom adjacent to the benzene ring, the **strongly inductive electron withdrawing groups (EWGs**, i.e., –$NO_2$, –C(=O)R, –C(=O)H, –$CF_3$, and –$SO_3H$). Let us take a look at how this type of substituent can influence the stability of the various possible carbocation intermediates:

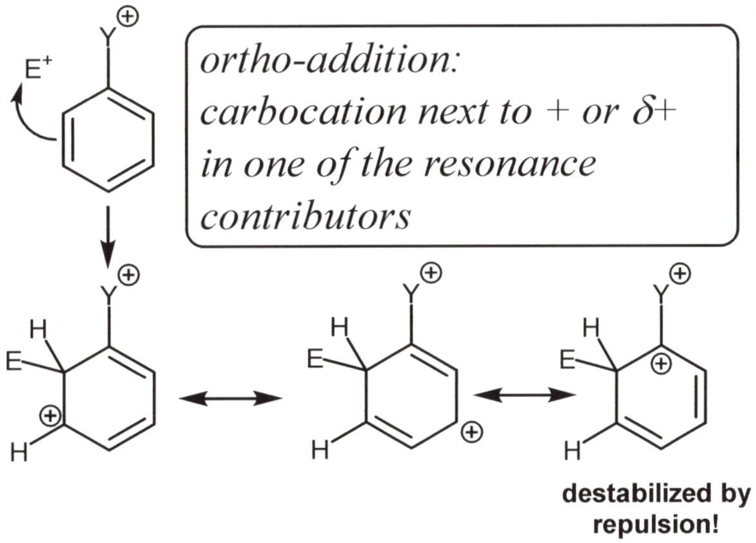
Y
E+
ortho-addition:
carbocation next to + or δ+
in one of the resonance
contributors
Y
H
E
H
Y
H
E
H
Y
H
E
H
destabilized by
repulsion!

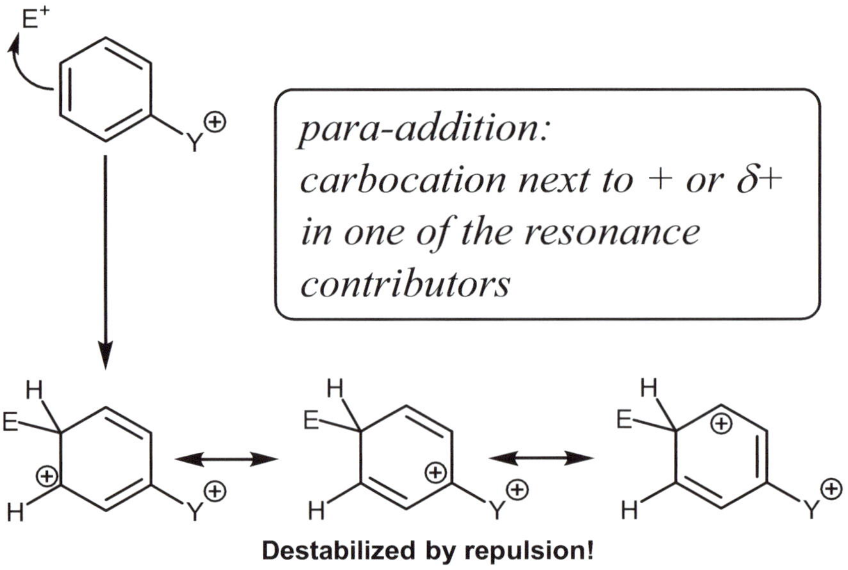

E+

Y⊕

*para-addition:*
*carbocation next to + or δ+ in one of the resonance contributors*

**Destabilized by repulsion!**

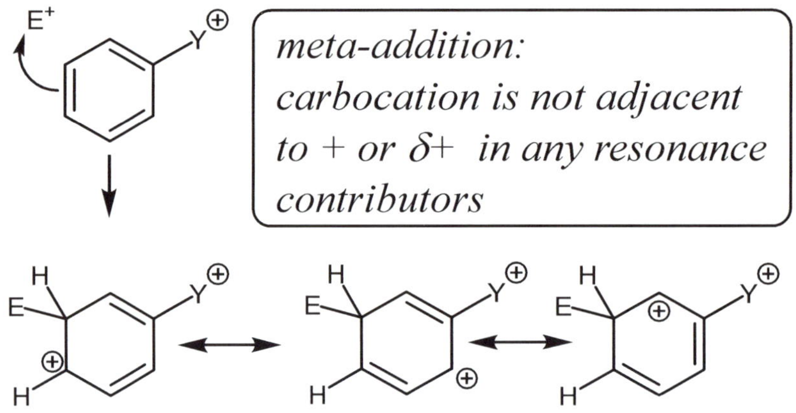

---

*meta-addition:*
*carbocation is not adjacent to + or δ+ in any resonance contributors*

In all of the other cases – resonance donors, inductive donors, and halogens – we observed phenomena that *stabilized* the intermediates leading to the *o*- and *p*- products relative to the intermediate leading to the *m*-product. In the case of electron-withdrawing groups, however, we see that the intermediates leading to the *o*- and *p*- products are *destabilized* relative to the intermediate leading to the *m*- product. This leads to the fact that **electron-withdrawing groups are *m*-directing**. The *m*-substitution product will be the major product of EAS for a monosubstituted benzene ring having an EWG on it.

Example IV.12.1

Provide the major product(s) of chlorination of each of these starting compounds:

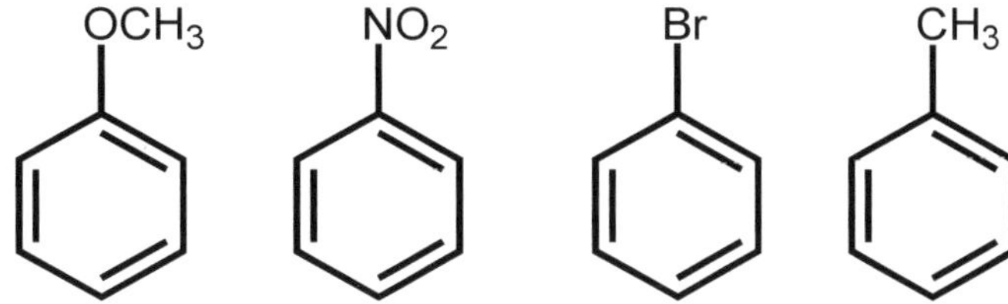


Solution IV.12.1

We start by determining what type of substituent (resonance donor, inductive donor, etc.) is present on each of the molecules.

From left to right:

The $-OCH_3$ is a resonance donor and is an *ortho-/para-* directing group.
The $-NO_2$ group is strongly electron withdrawing and *meta-* directing.
The -Br is a halogen, and these are weak inductive withdrawing groups that are *ortho-/para-* directing group.
The $-CH_3$ is a weak inductive donor and is an *ortho-/para-* directing group.

With these assignments in hand, we can now provide the major products of chlorination:

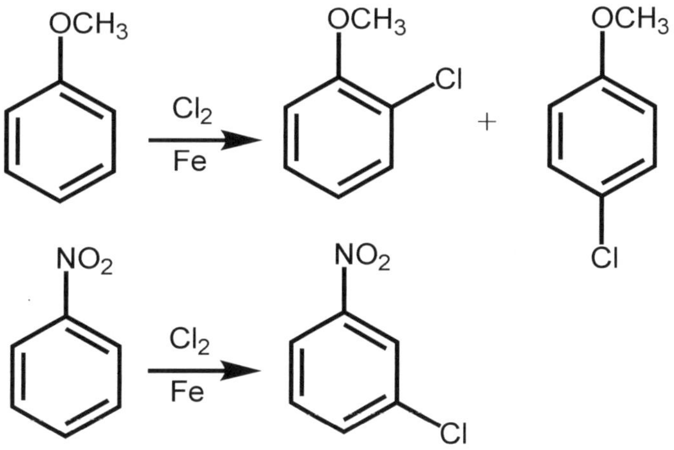

(cont'd)

Br $\xrightarrow[\text{Fe}]{Cl_2}$ Br, Cl + Br, Cl

$CH_3$ $\xrightarrow[\text{Fe}]{Cl_2}$ $CH_3$, Cl + $CH_3$, Cl

## Lesson IV.13. More on the Directing Effects of Substituents in EAS

### *IV.13.1 EAS on Polysubstituted Benzene*

In the previous Lesson, we saw that a substituent on a benzene ring influences the regiochemistry of the electrophilic addition step of a subsequent EAS reaction, directing the electrophile to add *o-*/*p-* or *m-* to the first substituent. What happens if *more than one* substituent is present on the benzene ring when we perform an EAS reaction? Consider an EAS on 2-ethylanisole. The -$OCH_3$ group would direct an electrophile to add to the sites indicated by the solid-line arrows in the diagram below:

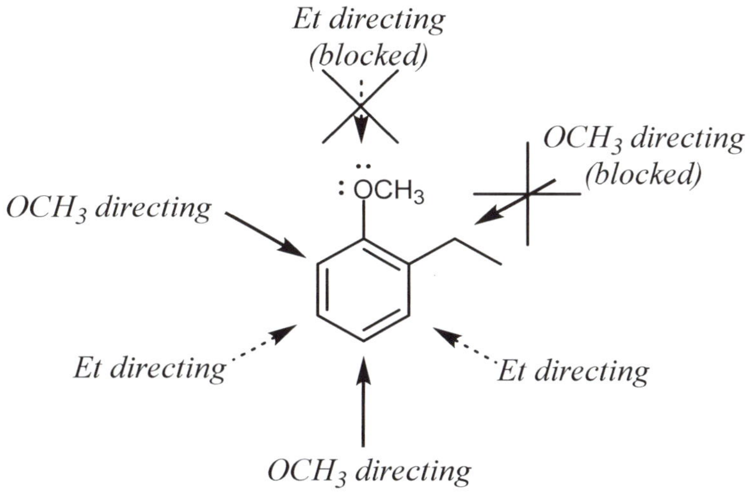

*Et directing (blocked)*

*$OCH_3$ directing (blocked)*

*$OCH_3$ directing*

$:\ddot{O}CH_3$

*Et directing*

*Et directing*

*$OCH_3$ directing*

One of these sites is blocked because it already has a substituent on it (remember that an electrophile can only replace an H atom, not any other substituent, in EAS). On the other hand, the ethyl group would direct an electrophile to add to the sites *ortho-* or *para-* relative to the ethyl (indicated by the dashed-line arrows). As before, one of the sites is blocked (i.e., has a non-hydrogen substituent). So, how do we predict which site will actually be substituted in the major product when a disubstituted benzene undergoes EAS? The answer is surprisingly simple. The major product of a reaction is always derived from the most stable intermediate, therefore **the substituent that best stabilizes a carbocation intermediate will dictate the site of EAS regardless of how many substituents are on the ring**. From most-stabilizing to least-stabilizing categories of substituents, we have:

1) Resonance donors (i.e., -$OCH_3$, -OH, -$NR_2$, -OC(O)R)
2) Alkyl groups
3) Halogens (-F, -Cl, -Br, -I)
4) Electron-withdrawing groups (i.e., -C(O)R, -$CF_3$, -$NO_2$)

Example IV.13.1

Provide the major product(s) of Friedel-Crafts Alkylation (using $CH_3Cl$ and $AlCl_3$) of each of these starting compounds:

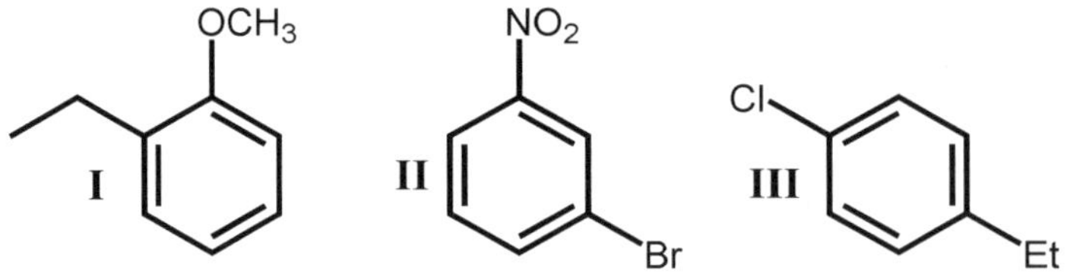

Solution IV.13.1

We start by determining what type of substituent (resonance donor, inductive donor, etc.) is present on each molecule and which type would best stabilize the carbocation intermediate (and thus dictates the regiochemistry of the EAS product).

For starting compound **I**, $-OCH_3$ is a resonance donor and the ethyl group is a weak inductive donor. The $-OCH_3$ dictates the regiochemistry of the next substitution to be *ortho-/para-* to the $-OCH_3$

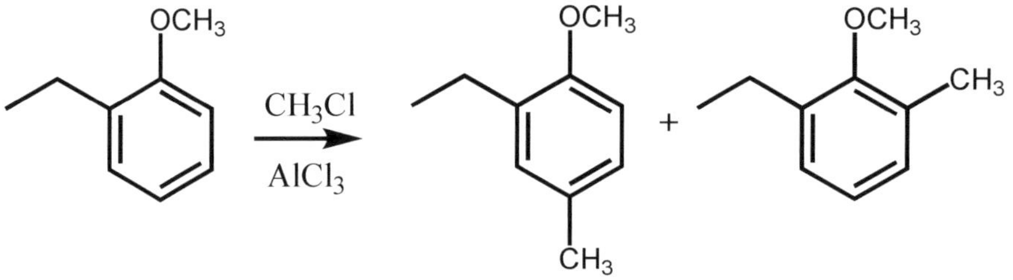

For starting compound **II**, the $-NO_2$ group is strongly electron withdrawing, whereas the –Br is a weak inductive withdrawing group. The –Br dictates the regiochemistry of the next substitution to be *ortho-/para-* to the –Br:

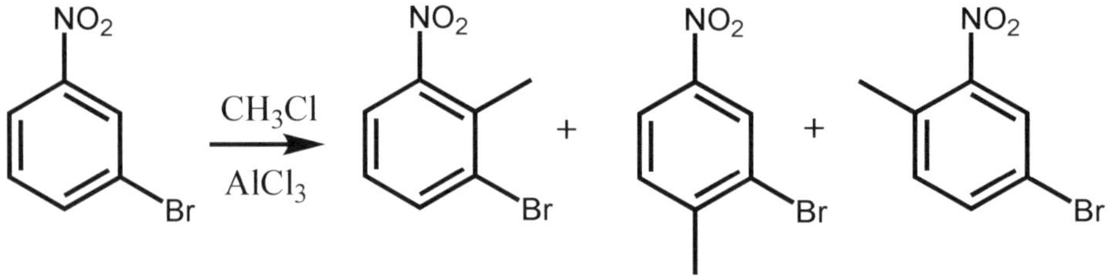

For starting compound **III**, the –Cl is a weak inductive withdrawing group, whereas the Et group is a weak inductive donor. The –Et dictates the regiochemistry of the next substitution to be *ortho-/para-* to the –Et. However, the *para* position is blocked by the Cl, and both *ortho* positions are chemically identical, so there is only one major product here:

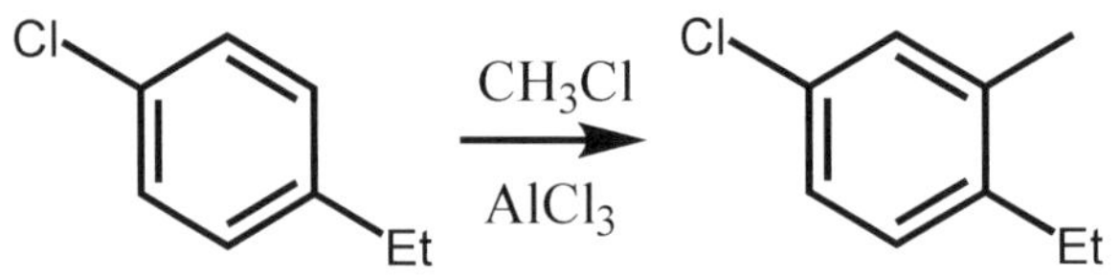

*IV.13.2 Other Factors Influencing Substitution Site in EAS on Polysubstituted Benzene*

So far, we have only considered the electronic effects that contribute to carbocation stability, and we did not account for steric effects. Steric repulsion means that it is more difficult to place a new substituent *ortho-* to an existing substituent than it is to place a new substituent adjacent only to H atoms. The bigger a substituent is, the more steric strain it will exert on neighboring substituents. So, if at all possible (and it may not be possible), the major product will be the one that has the minimum amount of steric strain. However, steric effects do **not** generally supersede the electronic effects we learned for the directing ability of substituents. When predicting reaction products, you should always **first** identify the sites to which the electrophile will be directed and only then evaluate which of these "directed-to" sites has the least amount of steric strain.

Recall that, for EAS on a monosubstituted benzene, a number of substituents will direct the electrophile to yield a mixture of *o-* and *p-* products. This may seem to contradict the idea of minimizing steric strain, which should favor making exclusively *para*-products. After all, the *o-* product places the second substituent immediately adjacent to the first, whereas the *p-* product places the second group as far from the first as possible. The observation that a mixture of *o-* and *p*-products are formed can be explained by the fact that there are *two* possible *ortho-* substitution sites but only *one* possible *para-* substitution site. So, statistics would favor formation of the *ortho-* product, but sterics we would favor formation of the *para-* product. These two factors balance each other out (roughly speaking), which affords a mixture of the *ortho-* and *para-* products. An exception would be if one of the two substituents in the product is bulky (has three non-H branches off of the point of attachment to the benzene ring, for example). In such cases we will get predominantly the *p-* product.

Finally, it is generally quite a bit more difficult to place a substituent between two substituents than next to one substituent with an H on the other side. We should consider such sites only when we have no other choice based on the directing ability of the existing substituents.

Example IV.13.2

Provide the major product(s) of nitration of each of these starting compounds:

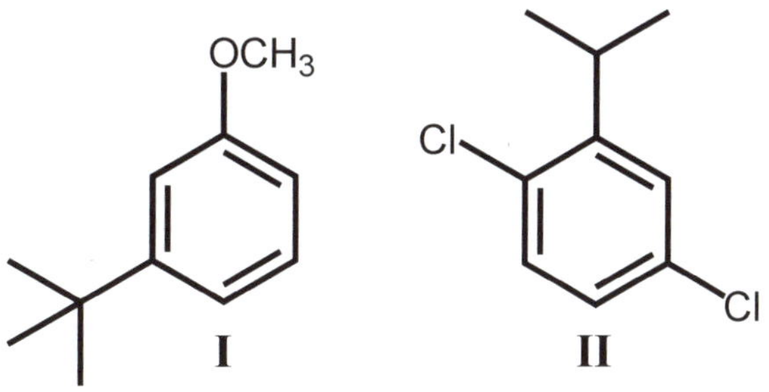

$OCH_3$

Cl

Cl

**I** **II**

Solution IV.13.2

For molecule **I**, $-OCH_3$ is a resonance donor and the *t*-butyl group is a weak inductive donor. The $-OCH_3$ dictates the regiochemistry of the next substitution to be *ortho-/para-* to the $-OCH_3$. Two of these sites are directly adjacent to the bulky *t*-butyl group (has three non-H branches), so steric effects will strongly disfavor the electrophile reacting at these sites. The major product, then, is the only one in which substitution is *both ortho-* to the $-OCH_3$ and not directly adjacent to the bulky *t*-butyl group:

$OCH_3$

$NO_2$

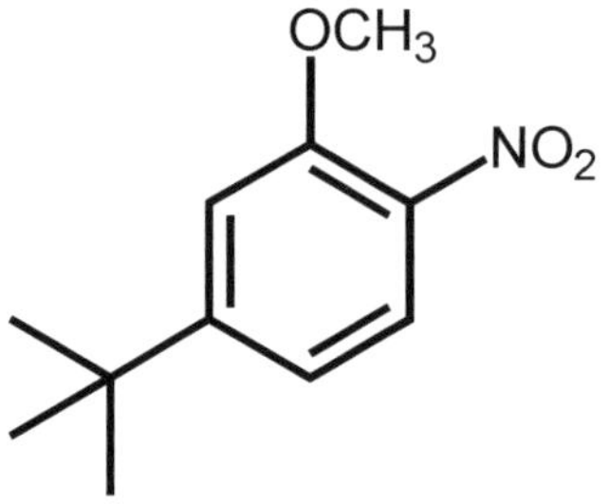

For molecule **II**, the $-NO_2$ group is strongly electron withdrawing, the –Cl is a weak inductive withdrawing group, and the isopropyl is a weak inductive donating group. The isopropyl dictates the regiochemistry of electrophilic addition to be *ortho-/para-* to itself. One *ortho-* position is already occupied by a Cl. The other *ortho-* site is between the isopropyl and the Cl, so steric effects disfavor electrophilic addition at that. As a result, substitution *para-* to the isopropyl is the major product:

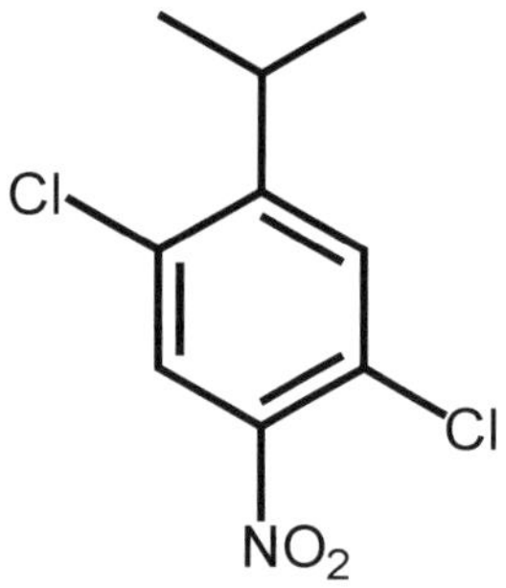

Cl

Cl

$NO_2$

## Lesson IV.14. Oxidation and Reduction of Substituents on Benzene Rings

### *IV.14.1 Oxidation by Chromium and Manganese Reagents*

In Lesson II.13, we saw that chromium reagents can be used to oxidize alcohols. However, chromium reagents can be either (1) weak oxidizing agents (like PCC and PDC) that facilitate only one unit of oxidation (remove only one H from the C and one H from the OH) or (2) strong oxidizing agents ($H^+/CrO_4^{2-}$, $H^+/Cr_2O_7^{2-}$, and $CrO_3/H_2SO_4$) that facilitate full oxidation of a carbinol to a carboxylic acid. These reactions were covered in Lesson II.13. Some manganese reagents can also oxidize alcohols. Two commonly-used manganese-containing reagents for oxidizing benzylic sites (a carbon adjacent to a benzene ring) are manganese dioxide ($MnO_2$) and potassium permanganate ($KMnO_4$). Potassium permanganate facilitates full oxidation. Like PCC, however, $MnO_2$ is capable of only one unit of oxidation:

OH H H → PCC, PDC or $MnO_2$ → O H aldehyde

OH R' H → PCC, PDC or $MnO_2$ → O R' ketone

A benzylic position is more susceptible to chemical reactions than positions in an alkyl or aryl unit. The extra reactivity of the benzylic site means that it can undergo oxidation with $KMnO_4$/dilute base (followed by acid workup) or one of the stronger chromium oxidizing agents, even if there is no OH group on the benzylic site. Upon reaction with a strong oxidizing agent, **a non-quaternary benzylic carbon will be oxidized to a carboxylic acid,** even if it involves breakage of multiple C–C bonds:

*Any of these starting materials*

OH H H; $CH_3$; O H; OH R' H; H H R; H R R'

$H^+/CrO_4^{2-}$, $H^+/Cr_2O_7$, $CrO_3/H_2SO_4$ → or 1. dil. $KMnO_4$/base, 2. $H_3O^+$ → O OH

The only requirement is that the benzylic carbon must have at least one H for the reaction to proceed:

R is not H

$H^+/CrO_4^{2-}$, $H^+/Cr_2O_7$, $CrO_3/H_2SO_4$ or 1. dil. $KMnO_4$/base, 2. $H_3O^+$ → *No reaction*

### *IV.14.2 Reduction of Benzylic Carbonyls to form Alcohols*

Functional groups at benzylic sites are also more susceptible to reduction. Benzylic aldehydes or ketones can be reduced to alcohols using the same conditions we saw for reducing alkenes (i.e., $H_2$ gas in the presence of Ni/Pd/Pt metal):

$H_2$, Pd or Pt catalyst

The R group with a bond drawn to the center of a benzene ring indicates that there could be substituents at *any* position around the ring and the reaction will still work.

### *IV.14.3 Reduction of Nitro to form Amines*

Like benzylic aldehydes and ketones, nitro groups that are directly attached to a benzene ring (making the N effectively benzylic) can also be reduced by $H_2$ in the presence of catalytic Pd or Pt metal:

$H_2$, Pd or Pt catalyst

An alternative is to use HCl as the hydrogen source in conjunction with stoichiometric Sn metal:

HCl, Sn

Note that both the oxygen atoms are removed from the N (in the form of $H_2O$), which converts the nitro into an amine functional group.

### *IV.14.4 Reduction of Nitrile to form Amines*

Benzylic nitriles are yet another functional group class that is susceptible to reduction by $H_2$ in the presence of catalytic Pd or Pt metal. Nitriles comprise a carbon triply bonded to a nitrogen, so an analogy can be made with the reactivity of alkynes, which comprise a carbon triply bonded to another carbon. Recall that $H_2$/Pd (or Pt) will reduce both π bonds in an alkyne by the addition of two H atoms to each carbon. Similar reactivity will occur with a nitrile:

N

$H_2$, Pd or Pt catalyst

$NH_2$

R

R

If we subject an alkyne to $H_2$/Lindlar's catalyst (Lesson III.13), only one π-bond is reduced upon *syn* addition of two H atoms, which affords a *cis*-alkene. One π bond of a nitrile can likewise be reduced to yield an imine (a functional group of the form RC(=NR')R") under the same conditions:

N

$H_2$, Lindlar's catalyst

NH

R

R

The oxidation and reduction reactions presented in this Lesson provide a way to convert functional groups on benzene to other functional groups, some of which cannot be installed directly via EAS. In the following lessons, we will learn reactions other than EAS by which substituents can be installed or modified to access a diverse range of substituted aromatics.

## Lesson IV.15. Radical Halogenation of Allylic and Benzylic Compounds

### *IV.15.1 Radicals Abstract Hydrogen Atoms*

In Lesson III.18 (OC1 Primer), we saw that we can generate radicals by heating a radical initiator (often a peroxide, RO–OR). Once generated, radicals can abstract H atoms even from typically unreactive molecules like alkanes:

H H H H •X ⟶ H H H + H–X

We also saw that the H atom abstracted most rapidly is generally the one that will leave behind the most stable radical possible. **Hydrogen atoms at allylic and benzylic sites are very readily abstracted to form radicals** because these sites provide significant resonance stabilization for the radical. An allylic site is the term for a C adjacent to a C=C bond. A benzylic site is the term for a C adjacent to a benzene ring:

*benzylic site*

R — Radical initiator / Halogen (X) source ⟶ X, R

*allylic site*

R — Radical initiator / Halogen (X) source ⟶ X, R

### *IV.15.2 Radicals React with Halogens*

In Lesson III.18, we saw that alkyl radicals can react with halogens to make alkyl halides:

H H H H H X—X ⟶ H H H H H X + X•

Allylic and benzylic radicals can also react with halogens to yield allyl and benzyl halides. When carrying out allylic or benzylic halogenation, there are some side reactions with halogens that we need to avoid. For example, if a high concentration of halogen is around when we attempt to do a radical halogenation of the allylic site in 1-hexene, we would expect to also see some halogenation of the alkene:

*allylic site can be halogenated*

*alkene can be halogenated*

Radical initiator

$X_2$

*Multiple products!*

For this reason, chemists do not use the halogen directly. Instead, they have discovered some compounds that can slowly produce a very low concentration of the halogen during the course of the reaction. This allows the halogen to be consumed by the radicals as they are generated, so that a high yield of the desired reaction is accomplished. The compound used for slow generation of halogen for these reactions is *N*-halosuccinimide. We will focus on the use of *N*-bromosuccinimide (NBS) in this book because it is the most-commonly used.

*Reagent for slow production of $Br_2$*

*N*-bromosuccinimide
(NBS)

### *IV.15.3 Radical Halogenation using NBS*

Now we have all the information in place to understand how allylic or benzylic halogenation works. The mechanism is provided here:

*Initiation*

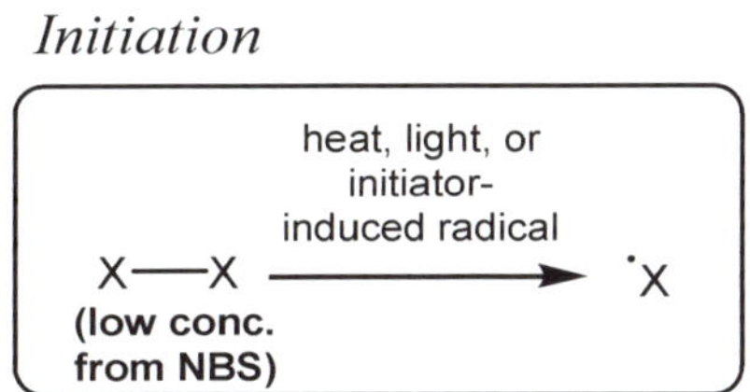


*Propagation*

*Net Reaction*

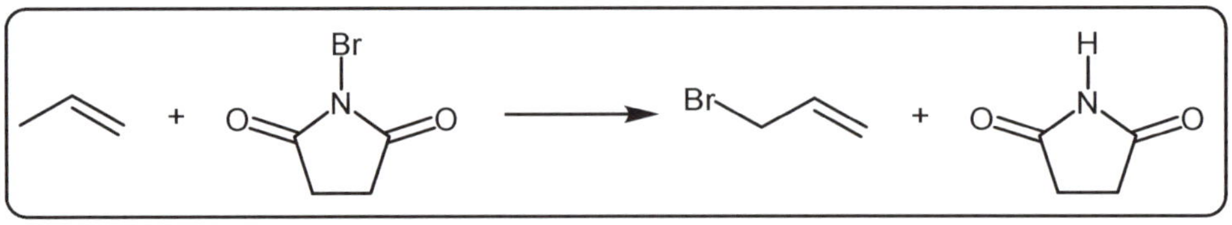

In this case, the initiator is often (ROOR), where R = –C(O)Ph, and the molecule is called **benzoyl peroxide (BPO)**. Benzoyl peroxide is a good initiator because it is relatively stable at room temperature and forms radicals readily when heated to the boiling temperature of common organic reaction solvents. It will dissociate into two RO• radicals, which can abstract an H from allylic or benzylic sites to form the first radical.

The mechanism for benzylic substitution is identical to the one shown above for allylic bromination, but with the benzylic site acting as the allylic site does in the mechanism above. There is one notable difference between allylic and benzylic substitution, however. For allylic halogenation, one must consider all possible resonance contributors to the structure of the radical formed upon H atom abstraction by the Br radical, so there is the potential for multiple substitution sites:

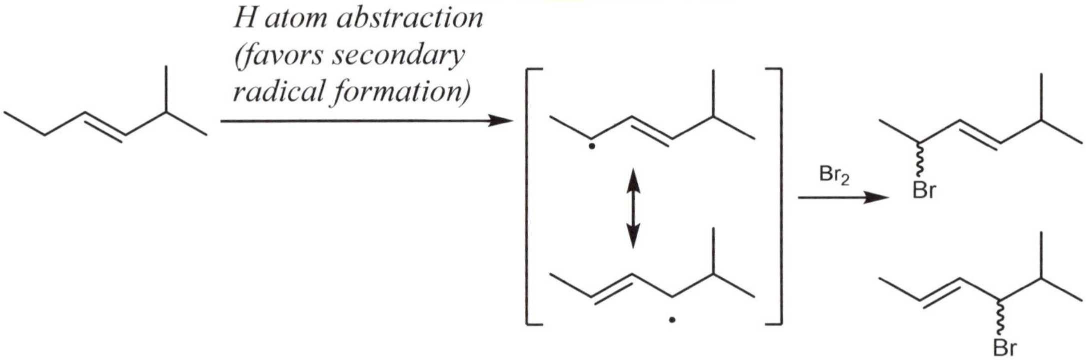

*H atom abstraction (favors secondary radical formation)*

$Br_2$

**To identify the major product of allylic halogenation**, we first make the most stable radical and then add the Br to the resonance contributor having the more substituted radial (there is more radical character at this site). For benzylic halogenation, however, the benzene ring will remain intact because of the very stabilizing aromaticity it possesses, so that **in the case of benzylic halogenation there is one site of substitution**.

Allylic/benzylic halogenation is an incredibly effective way to replace an H with a Br. The Br is a good leaving group and can subsequently be converted to other functional groups by the various substitution and elimination reactions we have already learned. This makes the radical halogenation a very powerful tool to open up access to a multitude of other species.

Example IV.15.1

Provide the major organic product for each of these reactions:

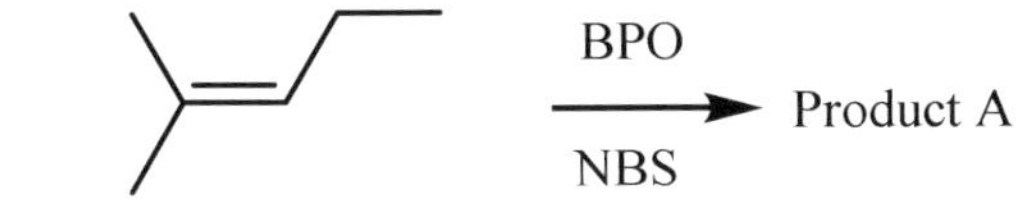

BPO

NBS

Product A

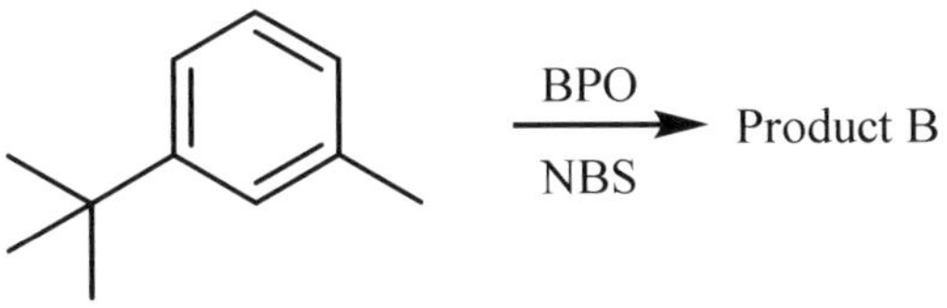

BPO

NBS

Product B

Solution IV.15.1

For the first reaction, there are three allylic sites: the two methyl groups to the left of the C=C and the $CH_2$ to the right of the C=C. The more stable radical results from H atom abstraction from the $CH_2$. The resultant radical has two resonance contributors:

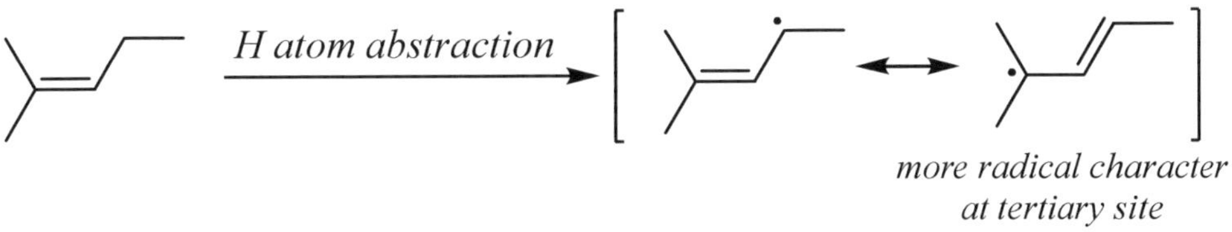

*H atom abstraction*

*more radical character at tertiary site*

There is more radical character on the tertiary carbon (illustrated by the resonance contributor on the right), so the major product features bromination at this site:

Product A

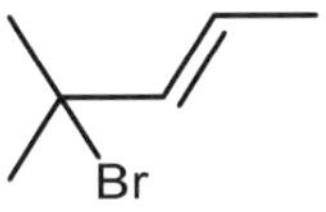

Br

For the second reaction, the analysis is simpler. H atom abstraction of the benzylic H on the methyl group (the *t*-butyl group does not have a benzylic H and is thus unreactive) leads to a radical that is brominated upon reaction with the $Br_2$ generated from NBS:

Product B

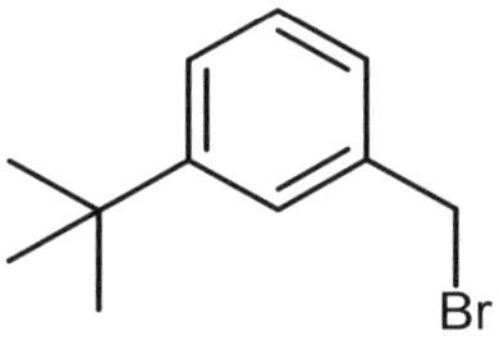

Br

Example IV.15.2

Provide *all* of the possible products of allylic bromination of 3-methylcyclopentane.

Solution IV.15.2

There are two sites with allylic hydrogens that can be abstracted:

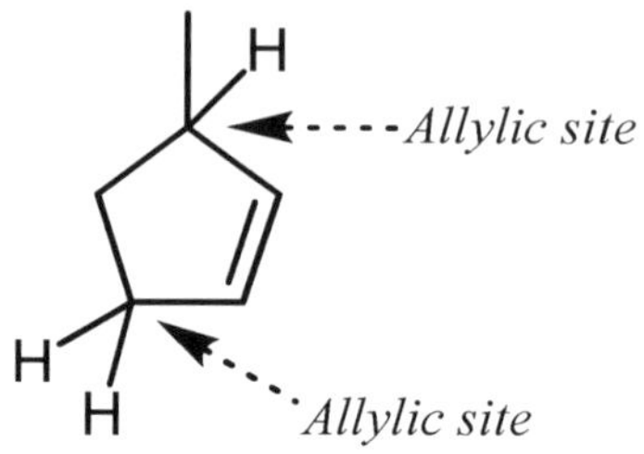

H
Allylic site
H
H
Allylic site

This leads to two different radical intermediates, each of which has two resonance contributors.

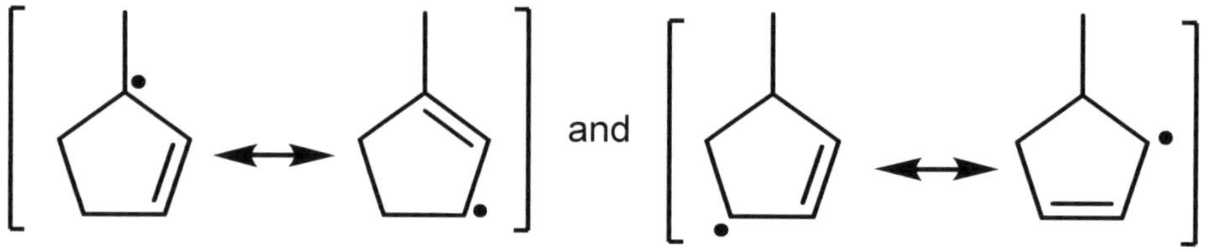

and

Each of the four sites having radical character could be brominated, so that the net reaction showing all of the possible products would be:

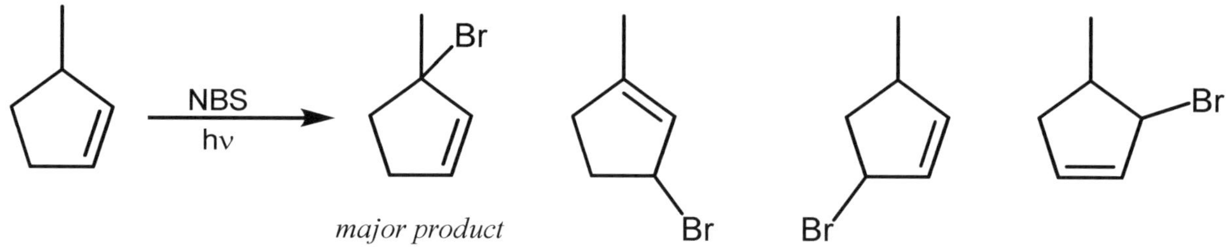

Br
NBS
hν
Br
*major product*
Br
Br
Br

## Lesson IV.16. Nucleophilic Aromatic Substitution

A benzene ring is not reactive to typical nucleophiles. Indeed, we have seen that benzene reacts with *electrophiles* because there are π-bonding electrons that are held loosely enough to be pulled away. For the reactions of nucleophiles that we have seen before, there is usually a positive charge or a partial positive charge to which the nucleophile is attracted. We can create a partial positive charge on a benzene ring by substituting it with an electron-withdrawing group (EWG). We learned about electron-withdrawing groups in Lesson IV.11. If a benzene ring has one of these substituents, it is possible that the benzene ring will react with a nucleophile. For a substitution reaction like $S_N2$, a nucleophile is able to substitute because there is a good leaving group present such as a halogen substituent. In order to do a substitution on a benzene ring, we will likewise need a good leaving group. Consider the example of 2-bromonitrobenzene reacting with hydroxide:

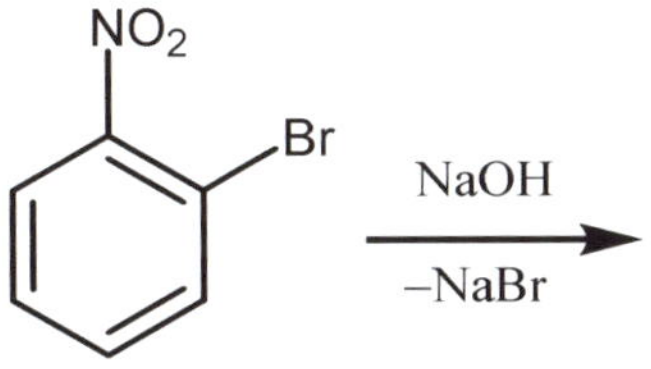

$NO_2$ Br —NaOH / –NaBr→ $NO_2$ OH

You can see that the net result is that the leaving group is replaced by the nucleophile. We know, however, that the nucleophilic substitution reactions that we have seen up to now ($S_N1$ and $S_N2$) do not work on $sp^2$ hybridized C atoms. So, what is the mechanism here? The first step is nucleophilic addition to a C–C bond:

*ortho- nucleophilic addition*

$NO_2$ Br ⊖OH → [ $NO_2$ OH Br ⊖ ↔ $NO_2$ OH Br ⊖ ↔ ⊖O ⊕ N O ⊖ OH Br ↔ ⊖O ⊕ N O OH Br ]

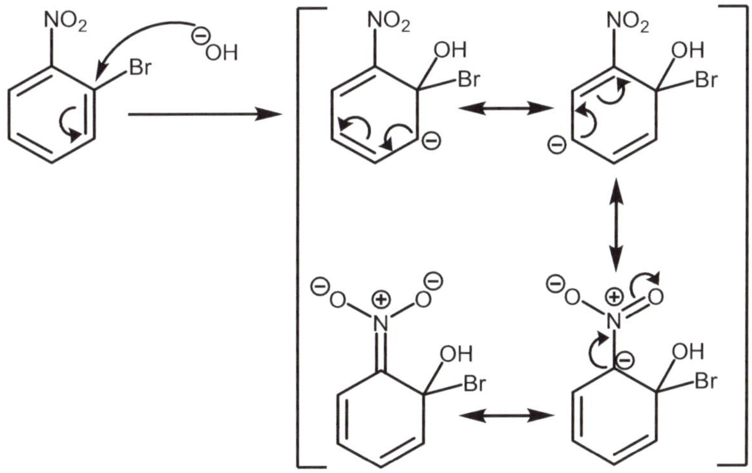

*anion intermediate is stabilized by EWG*

**This step is *only* possible if the anion can be stabilized by the presence of an EWG adjacent to the anionic carbon in at least one of the resonance contributors**. When we substitute a nucleophile for a leaving group that is *ortho-* to the EWG, this criterion is met. What about the cases in which we attempt to substitute for a leaving group that is *meta-* to the EWG:

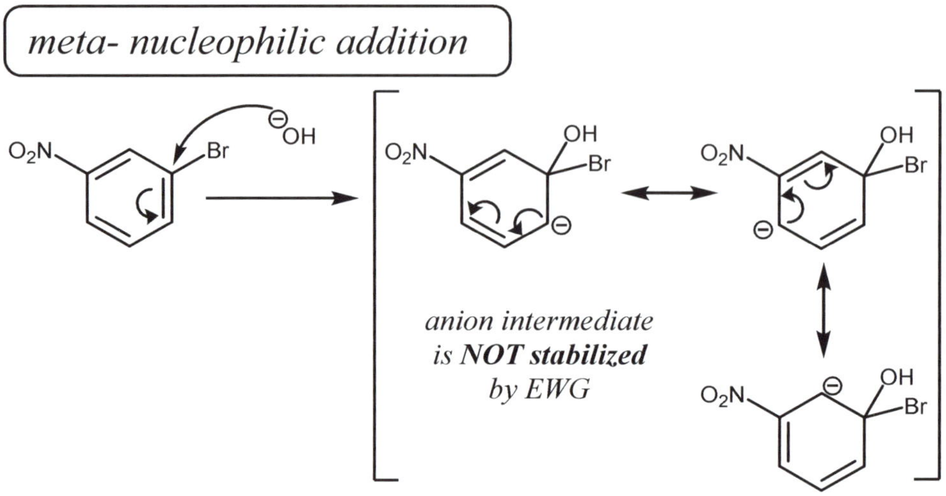

*meta- nucleophilic addition*

$O_2N$ Br ⊖OH

$O_2N$ OH Br ⊖

$O_2N$ OH Br ⊖

*anion intermediate is **NOT stabilized** by EWG*

$O_2N$ ⊖ OH Br

Or *para-* to the EWG:

*para- nucleophilic addition*

Br ⊖OH $O_2N$

OH Br $O_2N$ ⊖

OH Br $O_2N$ ⊖

*anion intermediate is stabilized by EWG*

OH Br ⊖O N ⊕ O ⊖

⊖ OH Br $O_2N$

*anion intermediate is stabilized by EWG*

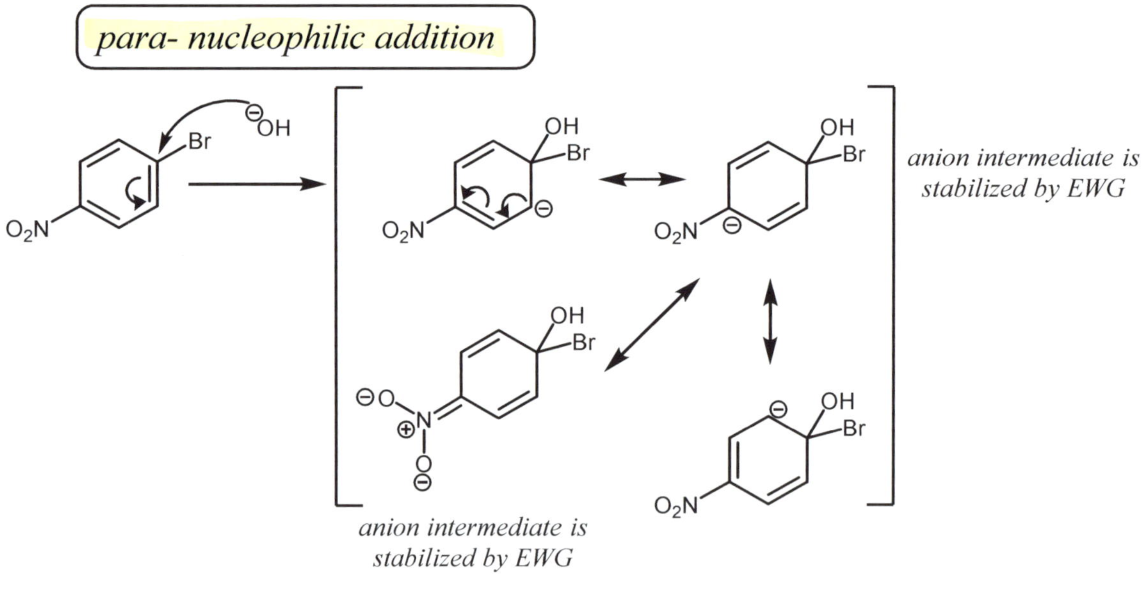

As we see from the resonance contributors, the EWG can only stabilize the anionic intermediate when it is *ortho-* or *para-* to the leaving group. For this reason, **nucleophilic addition is only possible when a good leaving group is *ortho-* or *para-* to the EWG on a benzene ring**. Following nucleophilic addition, nucleophilic elimination completes the mechanism of the **nucleophilic aromatic substitution reaction** (often abbreviated **SNAr**):

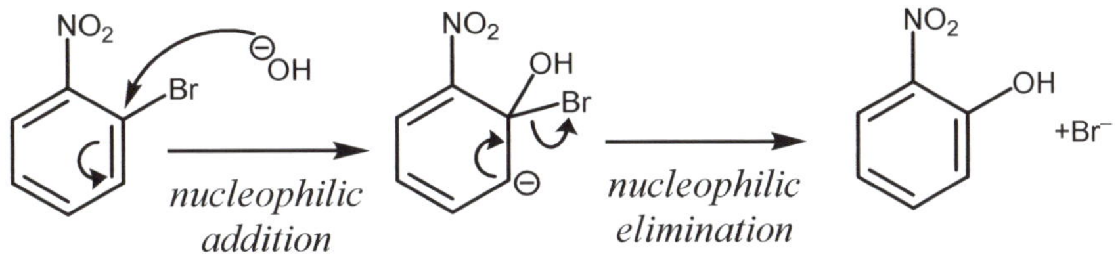

NO2 ... Br ... ⊖OH → *nucleophilic addition* → NO2, OH, Br, ⊖ → *nucleophilic elimination* → NO2, OH +Br⁻

Example IV.16.1

Which starting aryl bromide is required to prepare the target upon reaction with $HN(Et)_2$?

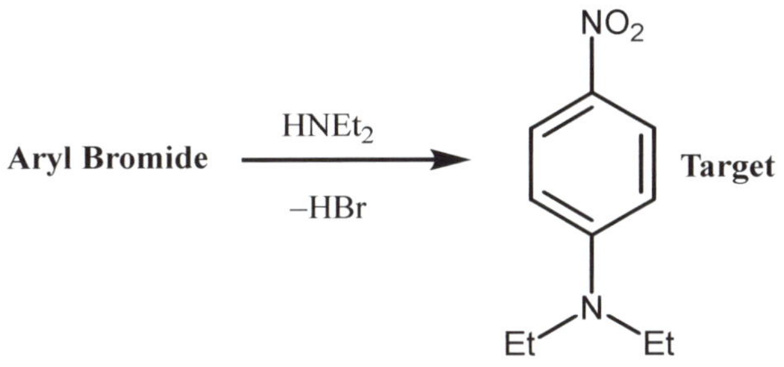

**Aryl Bromide** —($HNEt_2$, –HBr)→ **Target** ($NO_2$, N, Et, Et)

Solution IV.16.1

The leaving group Br must be in a position *para-* to the electron-withdrawing nitro group, so the appropriate starting material is 4-bromonitrobenzene.

## Lesson IV.17. Formation and Reaction of Diazonium Salts

### *IV.17.1 Formation of Diazonium Salts*

A diazonium salt, $[RN_2]^+Y^-$, can be formed from an aniline derivative upon reaction with $NaNO_2/H^+$ at low temperatures ($Y^-$ is the conjugate base of the acid source). The $N_2$ unit of the diazonium salt is a very good leaving group (as $N_2$ gas), so it can be replaced by any number of nucleophiles, as we will see soon. The mechanism of diazonium salt formation is as follows:

$$NaNO_2 + 2H^+ \longrightarrow N^+{=}O + H_2O + Na^+$$

–H⁺

–H⁺

–H₂O

*Diazonium Salt*

The R group with a bond drawn to the center of a benzene ring indicates that there could be substituents at *any* position(s) around the ring and the reaction will still work.

### *IV.17.2 Replacing the Diazonium Group*

Once a diazonium salt is formed, it can be replaced by various "X" units:

*Diazonium Salt* —MX→ + $N_2$ (gas)

The common "MX" units used to supply the X nucleophiles are CuCl (X = Cl), CuBr (X = Br), KI (X = I), $HBF_4$ (X = F), $Cu_2O$ (X = O, where the H comes from the acidic reaction mixture), $H_3PO_2$ (X = H). The diazotization reaction is exceedingly useful not only for installing groups that we previously did not know how to install by EAS (for example -OH and -F units), but it also allows us to transform the existing $–NH_2$ to other groups that have different reactivity and endow the molecule with different properties.

Example IV.17.1

Provide a synthesis of fluorobenzene from nitrobenzene using reactions in section IV of this book.

Solution IV.17.1

The only reaction that we know for placing a fluoro substituent onto an aromatic ring is by diazotization, and we need an $-NH_2$ to do a diazotization reaction, so our final step must be:

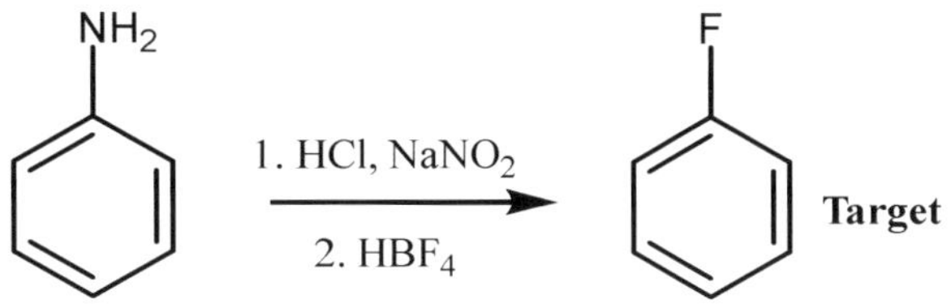

$NH_2$ — 1. HCl, $NaNO_2$ / 2. $HBF_4$ → F **Target**

*Required intermediate to get to target*

Fortunately, we know how to make aniline from nitrobenzene (see Lesson IV.14), our given starting material:

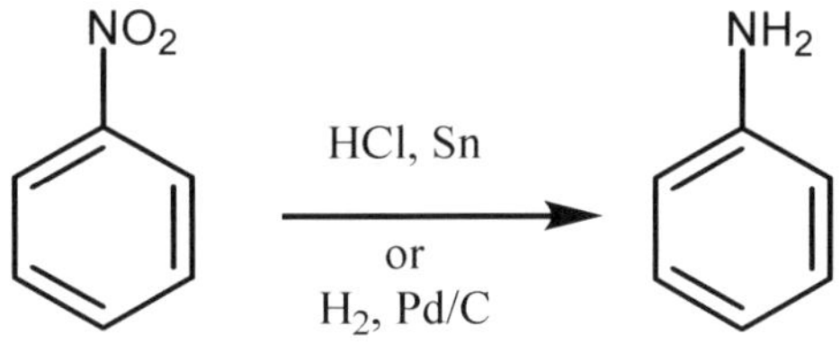

$NO_2$ — HCl, Sn or $H_2$, Pd/C → $NH_2$

So, the complete synthesis would be:

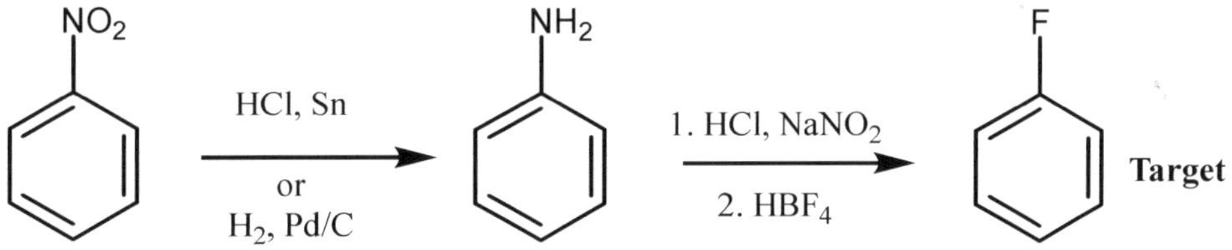

$NO_2$ — HCl, Sn or $H_2$, Pd/C → $NH_2$ — 1. HCl, $NaNO_2$ / 2. $HBF_4$ → F **Target**

**PART V: Organometallic Compounds and Metal Hydrides**

V.1 Introduction to Organometallics and Metal Hydrides

V.2 Preparation of Organolithium, Grignard and Gilman Reagents

V.3 Reaction of Organometallics/Metal Hydrides with RX and Epoxides

V.4 Palladium-Catalyzed C–C Bond-Forming Reactions

V.5 Introduction to Alkene Metathesis

V.6 Applications of Alkene Metathesis

# Lesson V.1. Introduction to Organometallics, Metal Hydrides and Carbenes

## *Lesson V.1.1 The Need for Nucleophilic Carbon*

Most of the reactions we have seen up to this point in the course involve break heteroatom bond and forming another carbon–heteroatom bond or a π bond. Very few reactions that we have seen involve forming a new carbon–carbon σ-bond. Notable exceptions are the Friedel-Crafts alkylation/acylation reactions (Lesson IV.7) or $S_N2$ reactions using a cyanide ($N{\equiv}C^-$) or acetylide ($H{-}C{\equiv}C^-$) anion as the nucleophile (Lesson III.16, OC1 Primer).

Remember that an *sp*-hybridized C – like the negatively-charged C in cyanide or acetylide – is much more electronegative than an $sp^3$-hybridized C (Lesson I.10, OC1 Primer), so an *sp*-hybridized C is better able to stabilize a negative charge. Having access to an anion with a negative charge on an $sp^3$-hybridized C would allow us to make new bonds to C without having triple bonds.

The lower stability of an $sp^3$-hybridized-C anion makes them harder to prepare and handle than the *sp*-hybridized-C anions. Ideally, we could access the desired nucleophiles as salts having an ionic bond between an $sp^3$-C and a metal like Li, Na, or Mg. Molecules having a metal–carbon bond are called **organometallic compounds**. Organometallic compounds are versatile sources for a wide range of carbon-centered nucleophiles. We will briefly assess how several important organometallic species are made and used in the following lessons.

## *Lesson V.1.2 Oxidative Addition, Reductive Elimination and Transmetallation*

One way that organometallic salts are formed is by **oxidative addition**. In the oxidative addition reaction, the metal oxidation state increases by "X" when "X" new groups (groups attached to metals are often called **ligands**) attach to the metal (X is 1 or 2 for reactions in this book). Many of the oxidative addition reactions we will encounter in this course will be the addition of a carbon–halogen bond across a metal (breaking a carbon–halogen bond and forming a metal–carbon and a metal–halogen bond):

$$R{-}X \;+\; M^{n} \longrightarrow R{-}\underset{\substack{|\\X}}{M^{n+2}}$$

The reverse of oxidative addition is called **reductive elimination**. In reductive elimination, two anionic ligands come off of the metal and the metal oxidation state goes down by two. Many of the reductive elimination reactions we will encounter will involve the formation of a new C–C bond:

$$R{-}\underset{\substack{|\\R'}}{M^{n+2}} \longrightarrow R{-}R' \;+\; M^{n}$$

Similar to how carbon can undergo nucleophilic substitution reactions, a metal can undergo **ligand exchange reactions**, in which one or more ligands on the metal are substituted by one or more new ns. If one of the anionic ligands undergoing exchange has a C bound to the metal (i.e., R = a drocarbon ligand), then the reaction is classified as a **transmetallation** reaction. A transmetallation reaction has the general form:

$$M–X + M'–R \rightarrow M–R + M'–X$$

Ligand exchange reactions, particularly transmetallation, occur in most of the stoichiometric and catalytic chemical reactions of organometallic complexes with organic compounds, so it is absolutely essential for the student to understand these reactions.

*Lesson V.1.3 The Need for Nucleophilic Hydrogen*

We have now seen how organometallic compounds could be useful for preparing new C–C bonds. In a similar fashion, sources of nucleophilic hydrogen could be useful to make new C–H bonds. Anionic hydrogen, $H^-$, called a **hydride** anion, is extremely unstable and reactive. This is because H is the smallest atom, and we know that larger atoms are better at stabilizing negative charge (Lesson I.10, OC1 Primer).

One especially useful application of hydride nucleophiles is nucleophilic addition of hydride to a carbonyl. If nucleophilic addition is followed by protonation of the O, an alcohol will result:

R''C(=O)R/H + $H^-$ —*nucleophilic addition*→ R''C(O$^-$)(Y)R/H + $H^+$ —*protonation*→ R''CH(OH)R/H

For arrow-pushing mechanisms using metal hydrides, **the reactive species is often abbreviated as "H$^-$"** (as in the above example), but we will see that there are several possible sources of nucleophilic H, and they do react differently than this simplification might make it appear.

*Lesson V.1.4. Sources of Hydride: $NaBH_4$ and $LiAlH_4$*

Metal–hydride complexes – often simply called **metal hydrides** – can be easy to make for some metals simply by combining the elemental metal with hydrogen gas (e.g., Na + ½ $H_2$ → NaH). In the metal hydrides, the hydrogen has a substantial $\delta^-$ charge, so many metal hydrides will spontaneously decompose upon exposure to the moisture in air, to yield hydrogen gas and a lot of heat, often catching fire. This is why ***handling many metal hydrides is dangerous and requires special training***.

The more polar the metal–hydrogen bond, the more reactive the metal hydride generally is. One common hydride source used by chemists is **sodium borohydride ($NaBH_4$)**. Boron has an electronegativity value (E.N. = 2.04) that is lower than, but still very close to, the value for hydrogen (E.N. = 2.2). As a result, a boron–hydrogen bond will still be polarized ($B^{\delta+}$–$H^{\delta-}$), but it will not be as reactive as, say Na–H or Mg–H bonds. Because it is so easy to handle, $NaBH_4$ is one of the most-widely used metal hydrides in organic synthesis. The borohydride anion $BH_4^-$ features a tetrahedral B-atom at the center connected to 4 H-atoms, all of which carry a $\delta^-$ charge that render them nucleophilic.

Some groups are less reactive to nucleophilic addition than others, so some are not reactive enough to react with $NaBH_4$. For this reason, chemists developed **lithium aluminum hydride** ($LiAlH_4$, sometimes abbreviated as **LAH**). This compound is similar to sodium borohydride, but has more polar Al–H bonds instead of B–H bonds, making it more reactive. LAH is reactive enough to facilitate nucleophilic addition to all of the carbonyl functional groups discussed in this text, as we will see in Part VI of this Primer.

*Lesson V.1.5: Cyclopropanation – Simultaneous Nucleophilic and Electrophilic Addition of "$CH_2$"*

In our study of alkenes, we learned that sometimes a single atom in a reagent molecule will react as both an electrophile and nucleophile toward a C=C bond, affording an intermediate that contains a 3-membered ring (i.e., halonium and mercurinium, respectively). By carefully selecting an appropriate reagent, we can convert an alkene to a 3-membered ring-containing molecule (a cyclopropane derivative) that is a stable, neutral product.

The simplest atom to consider that can react as both electrophile and nucleophile is an atom that has one empty orbital (electrophilic) and one lone pair (nucleophilic). Although this sounds like a very exotic species, one can easily be generated by treating $CHCl_3$ with $KO^tBu$. As a strong base but a poor nucleophile, $^tBuO^-$ can only deprotonate $CHCl_3$. As the H is being removed as $H^+$, one of the Cl atoms begins to leave the carbon as $Cl^-$, which produces "$:CCl_2$". This carbon has only 6 valence electrons but is neutral, a species known as a **carbene.** The carbene carbon in $:CCl_2$ is $sp^2$-hybidized, with an empty *p*-orbital and a lone pair in an $sp^2$-orbital. Normally, a carbon without an octet would be very unstable, but the lone pairs on the Cl atoms donate into the empty *p*-orbital and stabilize the carbene via hyperconjugation (Lesson I.11).

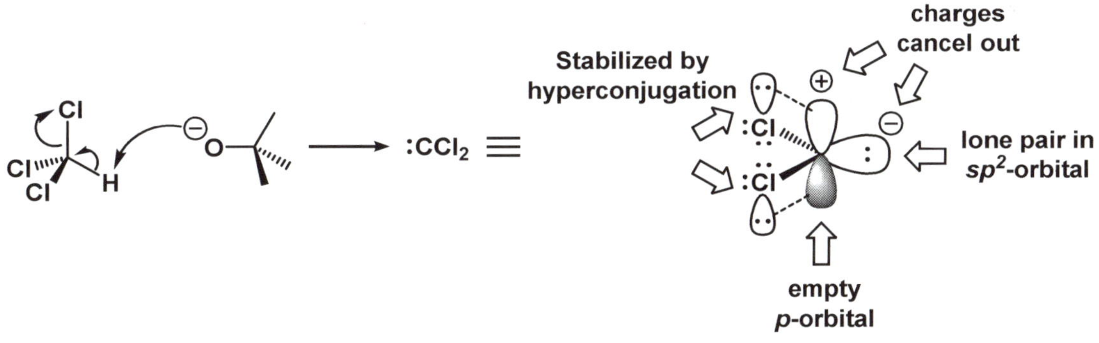

Once $:CCl_2$ has been generated, the $\pi$-electrons in a C=C bond will attack the empty *p*-orbital to form a bond between the less substituted alkene carbon and the carbene carbon. At the same time as this, the lone pair on the carbene carbon will attack the more substituted alkene carbon to form another C–C bond. The product is a cyclopropane ring, and this reaction is referred to as **cyclopropanation**.

Because everything happens in a single, concerted step, the relative spatial orientation of the alkene substituents is retained. If the C=C bond in the reactant has *trans*-stereochemistry, then its substituents will end up *trans*- to each other on the cyclopropane ring. If the C=C bond has *cis*-stereochemistry, the substituents on cyclopropane ring will be *cis*. Remember that $:CCl_2$ can approach the C=C bond from either above or below, and if the two approaches are identical in energy, then a 1:1 mixture of stereoisomers will be formed.

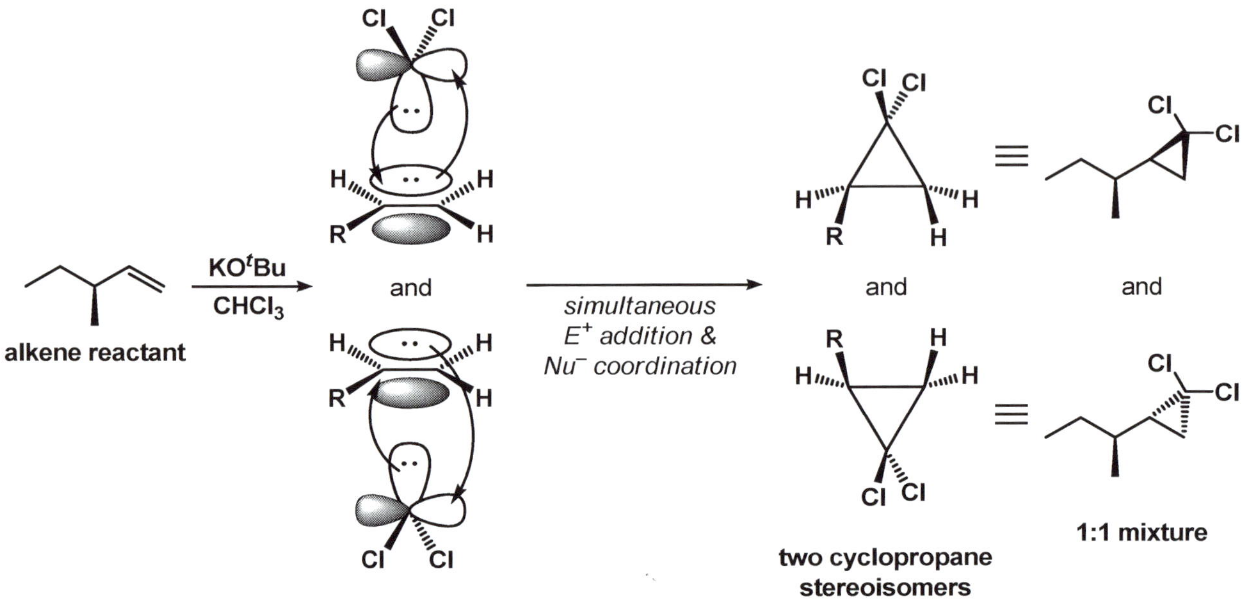

What do we do if we want a cyclopropane ring without the two Cl atoms? This would require a carbene with the structure ":$CH_2$", but this carbene is too unstable to form, so we must use a species of the form X–$CH_2$–Y that is not a carbene but does react like one. This type of species is called a "carbenoid".

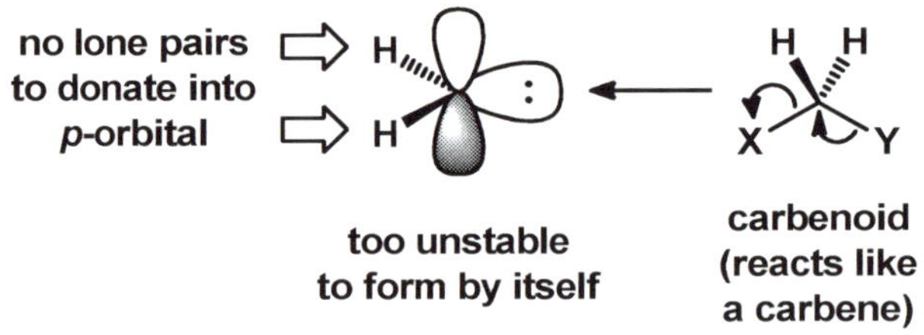

The reaction of $CH_2I_2$ with zinc in the presence of copper (written as "Zn(Cu)") can generate a carbenoid. When the I–$CH_2$–ZnI carbenoid is generated, it will react like ":$CH_2$" with the C=C bond. The reaction of this carbenoid to form a cyclopropane derivative is called the **Simmons-Smith** reaction. Again, everything happens in a single concerted step, so if we begin the reaction with a *trans*-alkene, those substituents in the cyclopropane ring will also be *trans* to each other. Keep in mind that the carbenoid can react either above or below the plane of the alkene, and if there is no difference in energy between the two approaches, a 1:1 mixture of stereoisomers will be produced.

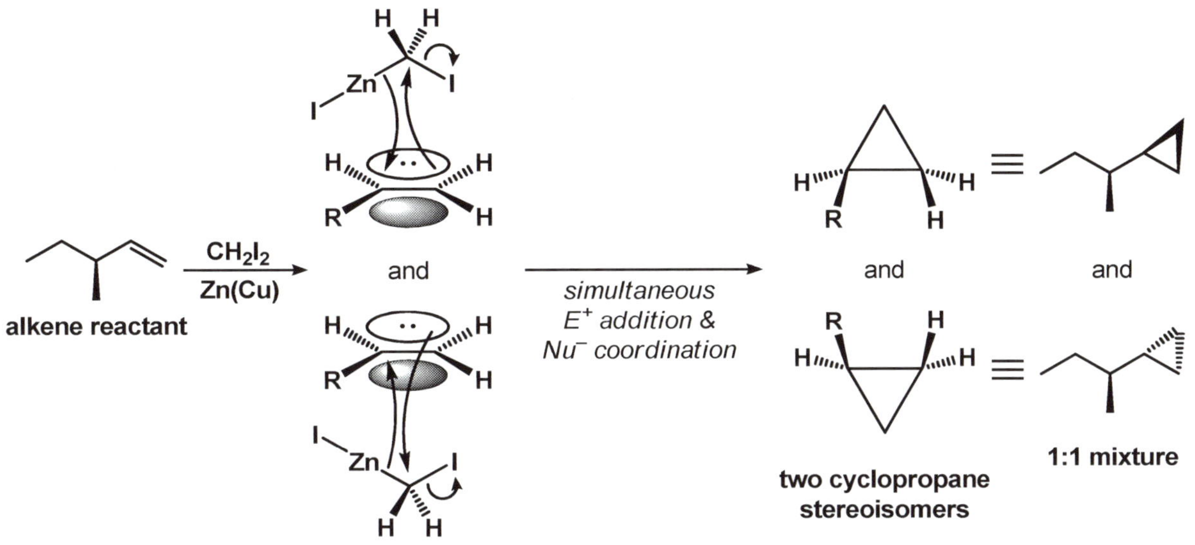

alkene reactant

$CH_2I_2$ / Zn(Cu)

and

*simultaneous $E^+$ addition & $Nu^-$ coordination*

two cyclopropane stereoisomers

1:1 mixture

Example V.1.1

What is the major product of the reaction shown below?

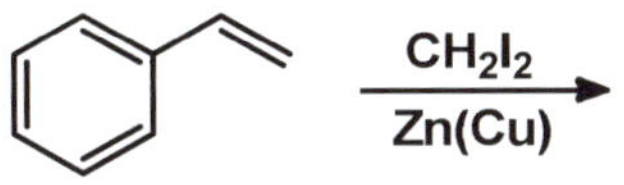

$CH_2I_2$ / Zn(Cu)

Solution V.1.1

The reactant $CH_2I_2$, in the presence of Zn(Cu), is converted to the carbenoid I–$CH_2$–ZnI, which acts as a source of ":$CH_2$" (but does not actually generate free ":$CH_2$"). In the presence of styrene, this reacts to add ":$CH_2$" to the π-bond, which can occur from either above or below the C=C plane. Because this alkene is not symmetric, the addition of ":$CH_2$" does not afford a meso compound, and instead a 1:1 mixture of cyclopropane stereoisomers will be formed (in this case a racemate).

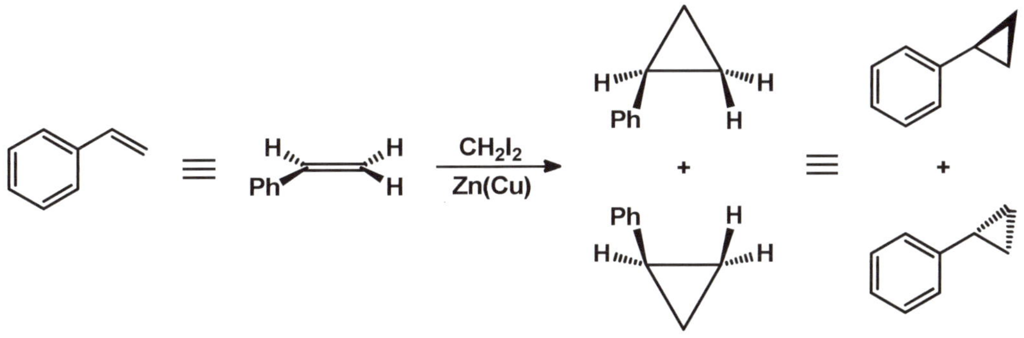

$CH_2I_2$ / Zn(Cu)

## Lesson V.2. Preparation of Organolithium, Grignard and Gilman Reagents

### *Lesson V.2.1 Synthesis of Organolithium Reagents*

A compound having a C–Li bond is called an **organolithium reagent**. Organolithium reagents can be prepared by a variety of methods. One approach is to react an alkyl halide (R–X) with 2 equivalents of Li metal:

$$R\text{—}X + 2\,Li \longrightarrow R\text{—}Li + LiX$$

This route involves oxidative addition because each Li oxidation state increases from 0 to +1.

One significant safety concern for organolithium reagents is that they are **pyrophoric**, meaning that they can spontaneously catch fire upon exposure to air. They are also very water sensitive and react exothermically with humidity in the air or in a reaction solvent. For these reasons, *organolithium reagents must be used under rigorously air- and water-free conditions by specially–trained individuals only.* **For arrow-pushing mechanisms using organolithium reagents, the reactive species is often abbreviated as "R⁻".**

Example IV.2.1

Write a balanced equation for the reaction shown below. How could you abbreviate the nucleophile produced when the organic product is added to solution?

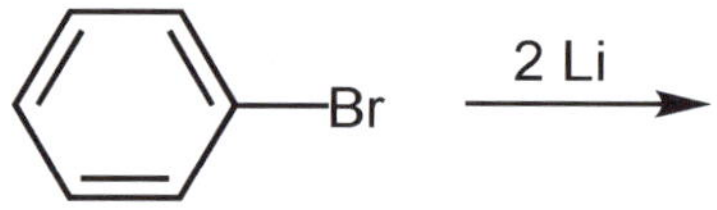

$$C_6H_5\text{—}Br \xrightarrow{2\,Li}$$

Solution IV.2.1

One lithium atom will oxidatively add to the benzene ring to form phenyl lithium. The other Li atom will form a salt with the Br:

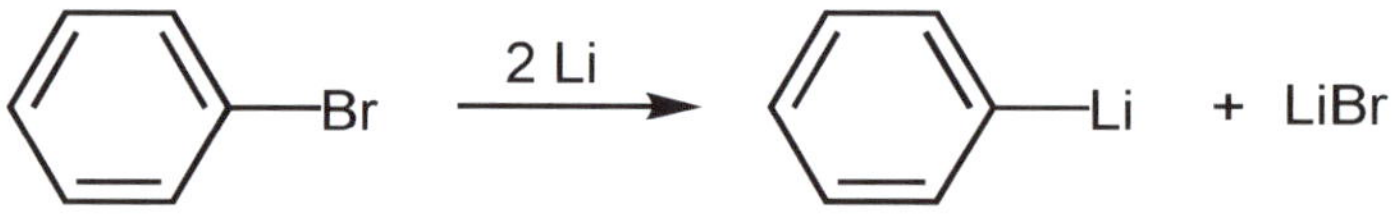

$$C_6H_5\text{—}Br \xrightarrow{2\,Li} C_6H_5\text{—}Li + LiBr$$

The nucleophilic part of phenyl lithium could be abbreviated:

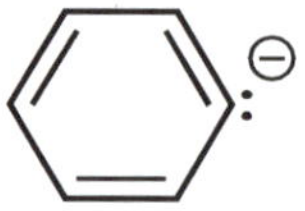

$$C_6H_5{:}^{\ominus}$$

*Lesson V.2.2 Synthesis of Grignard Reagents*

Organomagnesium reagents, more commonly known as **Grignard reagents (R–Mg–X)**, are safer to handle than are organolithium reagents, but are still highly reactive with air and water and must be handled with care. Grignard reagents are prepared by the oxidative addition of Mg to R–X:

$$\text{R—X} + \text{Mg} \longrightarrow \text{R—Mg—X}$$

An interesting property of Grignard reagents is that they can exist as multiple species in solution in equilibrium with each other: 2 R–Mg–X $\rightleftharpoons$ $MgX_2$ + $MgR_2$. For Grignard reagents, this process is termed the **Schlenk equilibrium** and is strongly dependent on the solvent used in the preparation of the Grignard. The Schlenk equilibrium could be a problem because when you do a chemical reaction you want one well-defined species, otherwise you might get more than one reaction happening at once, giving a mixture of products. Fortunately, if an ether solvent such as $Et_2O$ is used, the Grignard reagent actually ends up having solvent molecules bound to the Mg center via their oxygen atoms, stopping the Schlenk equilibrium:

$$2\ \text{R—Mg—X} \rightleftharpoons \text{X—Mg—X} + \text{R—Mg—R}$$

$$2\ \text{R—Mg(THF)}_2\text{—X} \not\rightleftharpoons MgX_2 + MgR_2$$

**For arrow-pushing mechanisms using Grignard reagents, the reactive species is often abbreviated as "R⁻".**

*Lesson V.2.3 The Limitations of Organolithium and Grignard Reagents*

Organolithium and Grignard reagents are strong bases as well as good nucleophiles. Carbon centers that have three or more alkyl branches are the exception; they are too bulky to be good nucleophiles, but are still strong bases. We know that reagents that are both strong bases and good nucleophiles can produce mixtures of $S_N2$ and E2 products with some substrates (Lesson II.10, OC1 Primer). For example, reaction of $^n$BuMgBr with 1-bromobutane produces some octane, but also substantial amounts of the E2 product 1-butene:

MgBr + Br → + 

**$S_N2$ product** **E2 product**

For this reason, scientists sought to develop other sources of nucleophilic carbon that are less basic. One of the most successful classes of less basic C nucleophile sources are the **Gilman reagents ($LiCuR_2$)**.

*Lesson V.2.4 Synthesis of Gilman Reagents*

Whereas organolithium and Grignard reagents can be prepared by direct reaction of alkyl halides with Li or Mg metal, this approach does not work with Cu metal. To circumvent this problem, the standard approach to making a Gilman reagent is to use an organolithium or Grignard reagent as the C ligand source and to transfer the ligands to copper by transmetallation. This is typically achieved by reacting the organolithium with CuBr or CuI:

*Preparing Gilman Reagents form Organolithium Reagents:*

$$2\,RLi \quad + \quad CuX \longrightarrow LiCuR_2 \quad + \quad Li{-}X$$

$$Li^{\oplus}\,[R{-}Cu{-}R]^{\ominus}$$

***Gilman Reagent***

In subsequent lessons, we will examine the reactivity of the Gilman reagents with various organic functional groups.

## Lesson V.3. Reaction C or H Nucleophiles with RX, Epoxides and Carbonyls

*Lesson V.3.1. Reaction of Organometallics with Carbon–Halogen Bonds*

As we discussed in Lesson V.2, the basicity of organolithium or Grignard reagents often causes mixtures of $S_N2$, E2, and other products with alkyl halides. In contrast, the less basic **Gilman reagents provide an effective and general method for substituting a C–X bond with a C–R bond**:

$$\underset{\textit{Gilman Reagent}}{LiCuR_2} + R'{-}X \longrightarrow R{-}R' + Li[X{-}Cu{-}R]$$

In contrast to $S_N2$ reactions (which can only substitute on $sp^3$-hybridized C), substitutions using **Gilman reagents work on $sp^3$, $sp^2$ *and* $sp$-hybridized carbon atoms**. This is because the mechanism of Gilman substitution is different from the $S_N2$ mechanism. The reaction involves oxidative addition of R'–X to Cu, then reductive elimination of the R–R' product:

*Oxidative Addition*

$$[R{-}Cu{-}R]^{\ominus} + R'{-}X \longrightarrow [R{-}Cu(R')(X){-}R]^{\ominus}$$

*Reductive Elimination*

$$[R{-}Cu(R')(X){-}R]^{\ominus} \longrightarrow R{-}R' + [X{-}Cu{-}R]^{\ominus}$$

Both oxidative addition and reductive elimination are concerted, so **stereochemistry at both R groups is retained**. Impressively, this reaction works with R'–X substrates spanning a remarkable variety of R' groups (alkyl, alkenyl, alkynyl, and aryl) and halides (X = Cl, Br, or I).

*Lesson V.3.2 Reactions with Epoxides*

We saw in Lesson II.15 that nucleophiles can cause ring-opening of epoxides. We also saw that when nucleophilic attack occurs in the absence of an acid, the nucleophile preferentially attacks the less-substituted side of the ring. Organometallics and metal hydrides are basic nucleophiles, so they can also be used to ring open epoxides by an $S_N2$-like attack on the less substituted side of the ring. In fact, we can treat *all* of the organometallic reagents we have discussed (organolithium, Grignard and Gilman reagents) as $R^-$, and the metal hydrides ($NaBH_4$ and $LiAlH_4$) as $H^-$ nucleophiles following the pattern we have seen for other basic nucleophiles:

*Ring-Opening by Organometallic Nucleophile:*

(from RLi, RMgX, or $LiCuR_2$)    (M = Li or Mg)    $H^+$, $H_2O$ or $H_3O^{\oplus}$

*Ring-Opening by Hydride Nucleophile:*

(from $NaBH_4$ or $LiAlH_4$)    (M = Na or Li)    $H^+$, $H_2O$ or $H_3O^{\oplus}$

The acidic workup is needed to protonate the O to give a neutral alcohol product.

*Lesson V.3.3 Reduction of Carbonyls*

Some organometallics and metal hydrides can readily undergo nucleophilic addition to certain carbonyl-based functional groups. The specific chemical identity of the carbonyl-based functional group dictates its reactivity with organometallics and metal hydrides. These reactions are covered in more detail in Part VI, but the first step of such reactions is the nucleophilic addition of the carbon-centered nucleophile ($R^-$):

*further reaction depends on identity of Y, M and R*

Note that because the number of C–O bonds is reduced in this step, this is a **reduction reaction**. An entirely analogous reaction of carbonyls with metal hydrides is possible, but using a hydride ($H^-$) nucleophile:

*further reaction depends on identity of Y, M and R*

The identity of "Y" dictates the name and the further reactivity of the carbonyl functional group. For example, if Y = H, the functional group is an aldehyde; if Y = R" (another hydrocarbon group like R'), the functional group is a ketone. The aldehyde and ketone functional groups are generally protonated following the nucleophilic addition step, giving alcohol final products:

M—Nu

Nu = H or R

Proton source

(often $H_2O$ or $H^+$ or $H_3O^+$)

Y = H (aldehyde)

or

Y = R" (ketone)

More details on the reactions of carbonyl functional groups with a wide range of nucleophiles is the subject of Part VI of this text.

# Lesson V.4. Palladium-Catalyzed C–C Bond-Forming Reactions

## *Lesson V.4.1. Organopalladium Reagents*

Although organolithium, Grignard and Gilman reagents provide useful carbon nucleophiles, they all suffer the limitation that a pyrophoric, dangerous and reactive species is involved in their preparation. Even the Gilman reagents, that are themselves much less reactive than RLi or RMgX, require that one use pyrophoric RLi in their preparation. Obviously, it would be much safer and more convenient to use carbon nucleophiles that can tolerate air and moisture. Fortunately, the group 10 elements (Ni, Pd, and Pt) undergo well-behaved oxidative addition reactions with aryl halides (Ar–X) and display excellent tolerance to air, moisture, and many organic functional groups. Among the Group 10 metals, organopalladium complexes can be formed the most readily and most cleanly, and they undergo the widest variety of subsequent reactions. As a result, there has been a tremendous amount of work devoted to the generation and reactivity of organopalladium complexes, and these complexes are among the most widely used reagents in synthetic chemistry today.

Because it is relatively cheap and easy to make, $[Pd(PPh_3)_4]$ is commonly used as the source of $Pd^0$ to undergo oxidative addition by an aryl halide:

*Oxidative Addition*

$[Pd(PPh_3)_4]$

Y can be alkyl, aryl, carbonyl, amine, alcohol

This oxidative addition reaction produces a metal-carbon bond, making the C nucleophilic. This first step can be harnessed in a variety of very useful catalytic C–C bond-forming reactions. These catalytic reactions follow quite similar patterns: 1) oxidative addition; 2) ligand exchange (or transmetallation if Pd exchanges a ligand with another metal); 3) reductive elimination. The reaction of the Pd catalyst with an aryl halide and a Grignard Reagent illustrates this general reaction sequence:

$PdL_4$ ⇅ $PdL_2$ — *oxidative addition* (Ar–X) → $L_2Pd(Ar)(X)$ — *transmetallation* (R–MgX → $MgX_2$) → $L_2Pd(Ar)(R)$ — *reductive elimination* → **target** Ar–R

Notice that the reductive elimination to release the target product with the new C–C bond also *regenerates the active catalyst [$PdL_2$]*. Once formed, R–Ar does not react with the catalyst, but more Ar–X can oxidatively add to the Pd. This is thus a **Pd-catalyzed C–C bond-forming reaction.** This particular combination of sources of carbon groups that get coupled together – an aryl halide and a Grignard reagent – is known as the **Kumada Reaction**.

*Lesson V.4.2. The Heck Reaction*

The **Heck Reaction** employs an alkene as the transmetallating reagent, allowing us to attach an aryl group in place of a H on an alkene carbon:

*The Heck Reaction*

$[Pd(PPh_3)_4]$, $NEt_3$ (weak base); products: + $[HNEt_3]X$

Y can be alkyl, aryl, carbonyl, amine, alcohol
X = Cl, Br, I, OTs, OMs or OTf

The series of steps by which this occurs is a bit more complicated than what we saw for the Kumada reaction:

Target; $Pd^0$; X–R; *oxidative addition*; X–$Pd^{2+}$–R; *ligand exchange*; HX; R–$Pd^{2+}$; *reductive elimination*

Note that in the major product the aryl group will attach to the less-substituted alkene carbon because there is less steric congestion at that position. The Pd catalyst in these catalytic cycles is sometimes shown as just "$Pd^0$", as above, or as Pd with some ligands "L", as in the other cycles in this lesson. The reaction steps are the same, regardless of the representation of the catalyst.

Example V.4.1

Draw the major product for the reaction shown below:

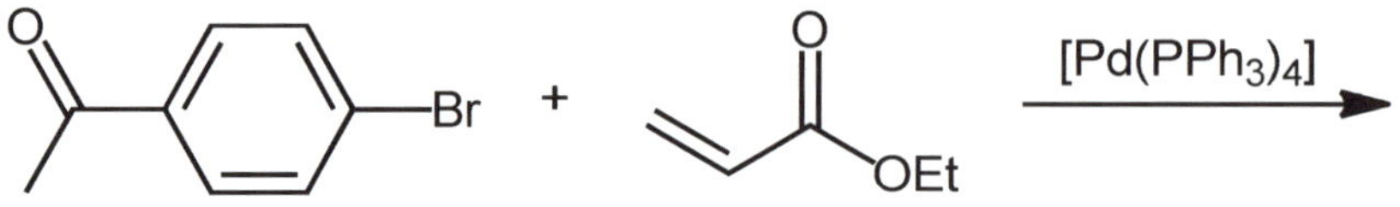

$[Pd(PPh_3)_4]$

Solution V.4.1

First, the C–Br bond will oxidatively add to $Pd^0$ to make the $[PdX(Ar)(PPh_3)_2]$ intermediate. Next, the alkene will bind to the $Pd^{2+}$ center, and the Ar group will do nucleophilic attack at the less substituted alkene carbon. Subsequent β-hydride elimination will afford a disubstituted internal alkene with *trans* stereochemistry (shown below):

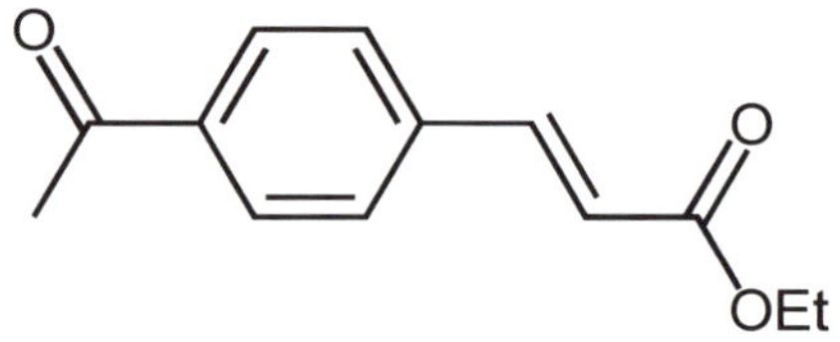

*Lesson V.4.3. The Stille Reaction*

In the **Stille reaction**, the transmetallating agent is typically a trialkylstannane ($R_3Sn–R'$, R is an alkyl group), where the R' ligand is attached to the Sn at an $sp^2$-hybridized C:

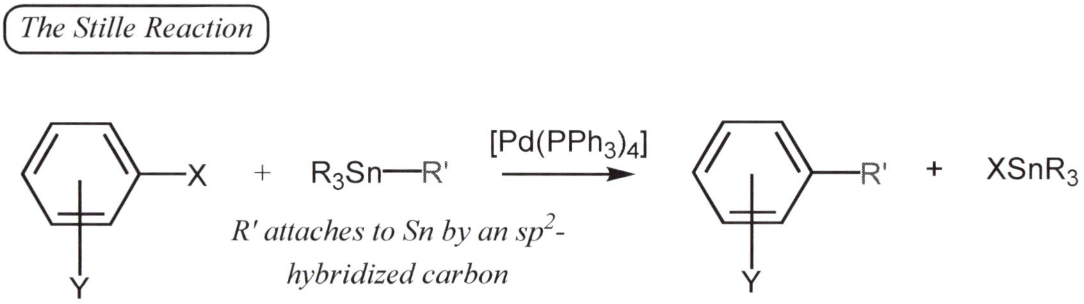

*The Stille Reaction*

$$\text{Ar(Y)–X} + R_3Sn\text{—}R' \xrightarrow{[Pd(PPh_3)_4]} \text{Ar(Y)–R'} + XSnR_3$$

*R' attaches to Sn by an $sp^2$-hybridized carbon*

Y can be alkyl, aryl, carbonyl, amine, alcohol
X = Cl, Br, I, OTs, OMs or OTf

Note that the alkyl groups on Sn stay on the Sn. Only the $sp^2$-C on Sn goes into the product. The series of steps by which this occurs is completely analogous to what we saw for the Kumada reaction, with the stannane in the Stille reaction serving the role that the Grignard reagent served in the Kumada coupling:

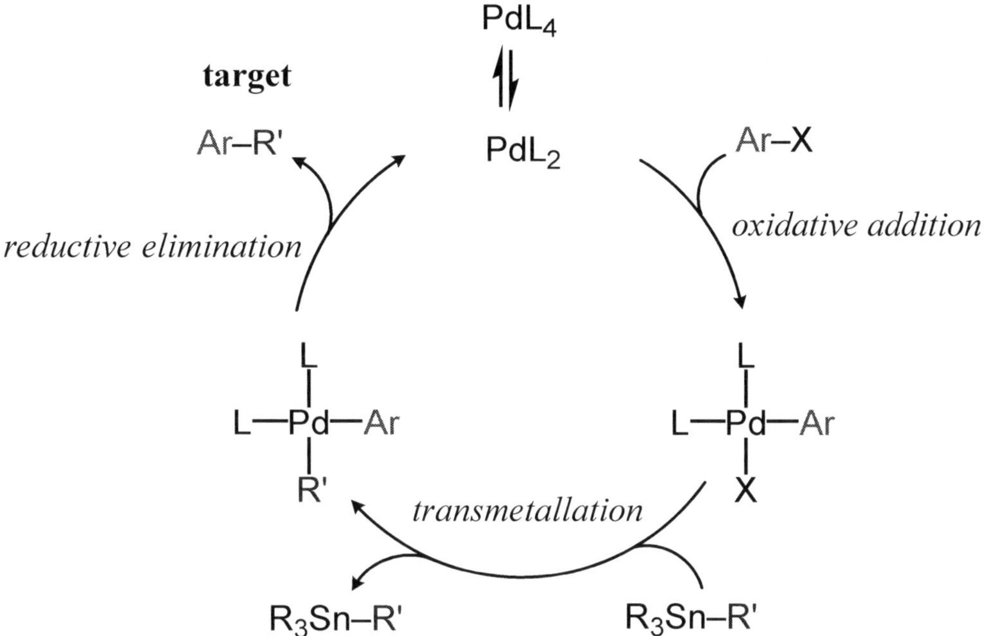


One of the primary disadvantages is that the stannane reagents can be difficult to prepare, and both the stannane reagents and stannyl halide byproducts are extremely toxic to handle in the lab and harmful for the environment.

Example V.4.2

Draw the major product for the reaction shown below:

+ 2 $Me_3Sn$– $\xrightarrow{[Pd(PPh_3)_4]}$

Solution V.4.2

Our substrate has 2 C–I bonds and we are adding 2 equiv. of $Me_3SnPh$, so both C–I bonds will react under these conditions. After oxidative addition of a C–I bond to $Pd^0$, transmetallation will transfer the Ph from $Me_3SnPh$ to $Pd^{2+}$ to afford the $[Pd(Ph)(Ar)(PPh_3)_2]$ intermediate and $Me_3Sn$–I. Subsequent reductive elimination will form a C–Ph bond. The presence of 2 equiv. of $Me_3SnPh$ ensures that both C–I bonds will be converted to C–Ph groups:

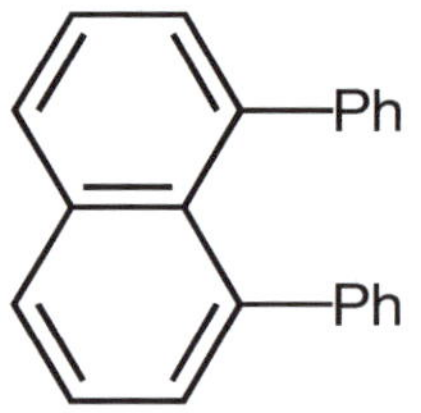

Ph
Ph

*Lesson V.4.4 The Sonogashira Reaction*

In the Sonogashira reaction, the transmetallating agent is a copper–alkynyl species of the general form Cu–C≡C–R. The copper–alkynyl is generated from a terminal alkyne, a base (usually an amine such as $R_3N$, $^iPr_2NH$, or $^iPr_2NEt$), and CuX (X = I or Br):

*The Sonogashira Reaction*

X
R
H
[Pd(PPh3)4]
CuX, R'3N
R
+
[NR'3]X
Y
Y

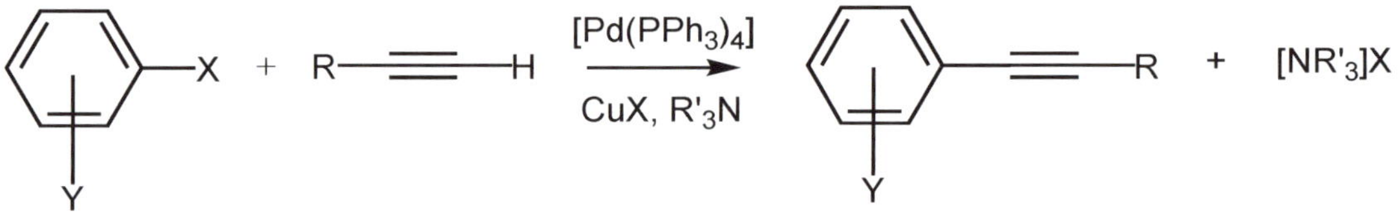

Y can be alkyl, aryl, carbonyl, amine, alcohol
X = Cl, Br, I, OTs, OMs or OTf

The series of steps by which this occurs is completely analogous to what we saw for the Stille and Kumada reactions, with the organocopper species (Sonogashira) serving the role that the stannane (Stille) or Grignard reagent (Kumada) served in those cases:

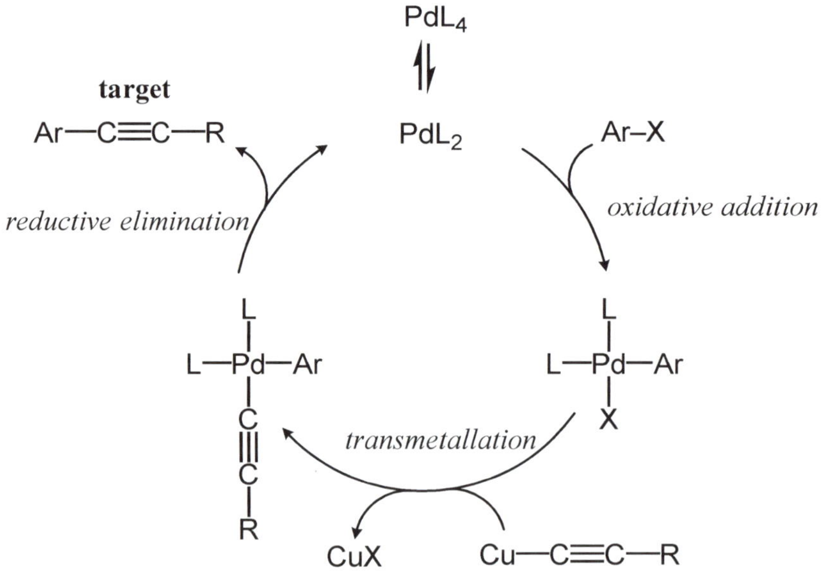

PdL4
PdL2
target
Ar—C≡C—R
reductive elimination
Ar–X
oxidative addition
L—Pd—Ar
X
L—Pd—Ar
C
C
R
transmetallation
CuX
Cu—C≡C—R

One of the drawbacks with Sonogashira coupling is the need to prepare a terminal alkyne (which can be labor-intensive/expensive in some cases) and terminal alkynes are relatively unstable, so they must be used shortly after preparation.

Example V.4.3

Draw the major product for the reaction shown below:

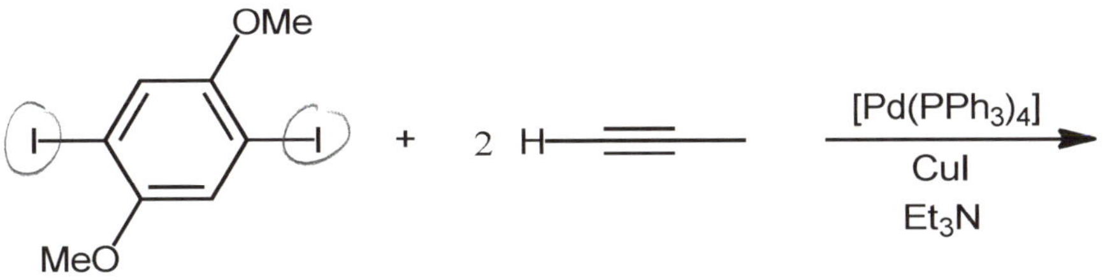

OMe

I I + 2 H——≡——

$[Pd(PPh_3)_4]$

CuI

$Et_3N$

MeO

Solution V.4.3

Whenever you see a Pd-catalyzed coupling reaction that has a terminal alkyne as a reactant and CuX and base as reagents, you should immediately think "Sonogashira". Remember that the CuX and base convert the alkyne into an organocopper transmetallating agent:

H——≡—— $\xrightarrow[Et_3N]{CuI}$ Cu——≡—— + $Et_3\overset{\oplus}{N}H$ $I^{\ominus}$

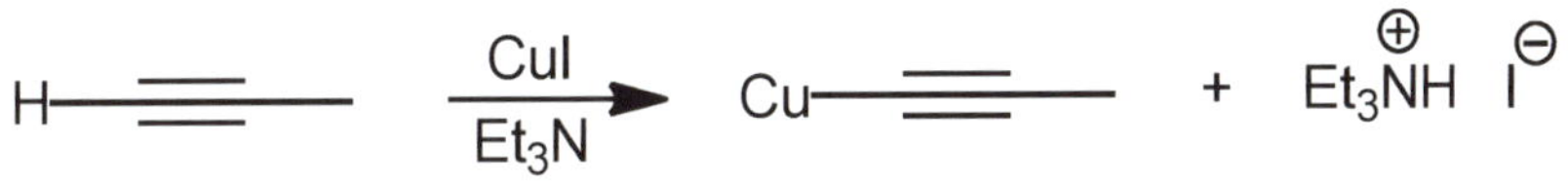

After oxidative addition of a C–I bond to $Pd^0$, transmetallation will transfer the alkynyl group from $Cu^{1+}$ to $Pd^{2+}$, and subsequent reductive elimination will form a C–alkynyl bond. The 2 equiv. of alkyne will ensure that both C–I bonds are converted into C–alkynyl groups:

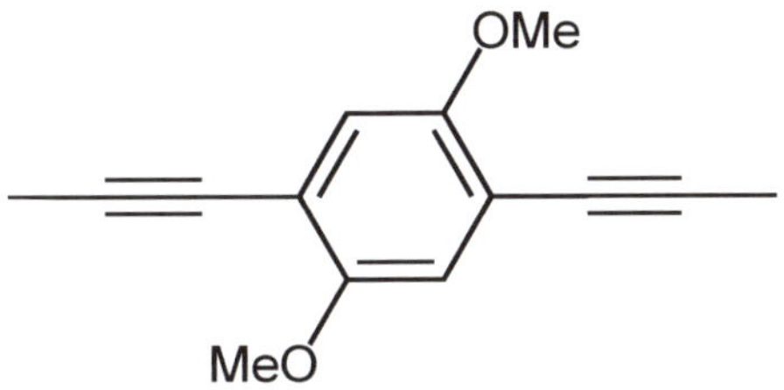

OMe

MeO

*Lesson V.4.5 The Suzuki Reaction*

In the Suzuki reaction, the transmetallating agent is typically either a boronic ester or a boronic acid, which have the general formulae $R–B(OR')_2$ or $R–B(OH)_2$, respectively:

The Suzuki Reaction

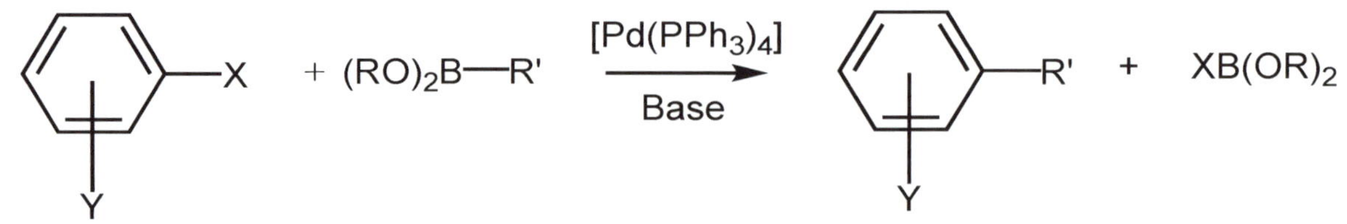

Y can be alkyl, aryl, carbonyl, amine,
X = Cl, Br, I, OTs, OMs or OTf

The series of steps by which this occurs is again analogous to what we saw for the Stille, Kumada, and Sonogashira reactions:

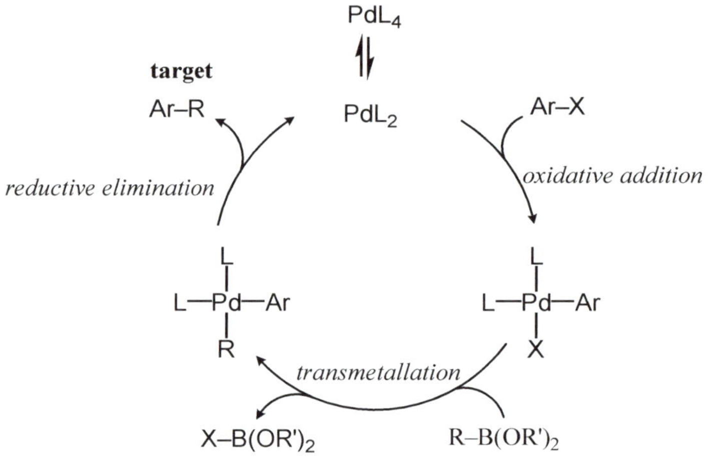

The exact mechanism of the Suzuki reaction is still not fully-understood, so all you need to be concerned with is that a base is required (e.g., $KO^tBu$, $Cs_2CO_3$) and the boronic ester/acid transfers the R group to Pd. Reductive elimination then releases the product, Ar–R, and re-generates the active catalyst [$Pd(PPh_3)_2$]. A major advantage of the Suzuki reaction is that it is tolerant to a wide variety of functional groups and to protic solvents as well. In fact, Suzuki reactions can even be performed in water! Challenges associated with the Suzuki reaction often involve poor solubility of R–$B(OH)_2$ in organic solvents and the cost of commercially-available R–$B(OR')_2$ species.

Example V.4.4

Draw the major product for the reaction shown below:

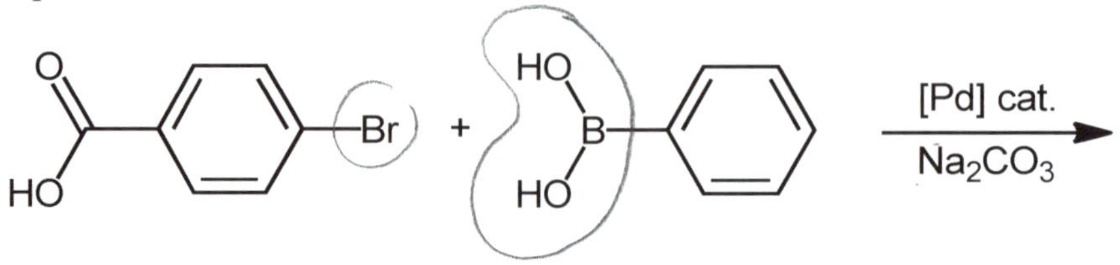

Solution V.4.4

In the first step, the [Pd] catalyst undergoes oxidative addition by the C–Br bond, and then transmetallation from $PhB(OH)_2$ affords the [Pd(Ph)(Ar)] intermediate. Subsequent reductive elimination releases the product shown below and regenerates the active [Pd] catalyst:

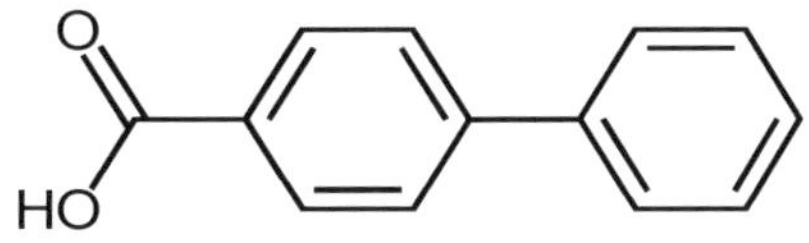

O

HO

## Lesson V.5. Introduction to Alkene Metathesis

### *Lesson V.5.1 What is Alkene Metathesis and Why Would it be Useful?*

If we are trying to make a new alkene that is not readily available from natural sources, from a simpler alkene, we are relatively limited in the synthetic routes available based on what we have learned so far.

*Problem:*

?

*Solution using only reactions to date in this text:*

$Br_2$ Br Br 2 KO$^t$Bu $NaNH_2$ ⊖ Br $NH_3$

This example emphasizes the significant difficulty that early chemists faced when they wanted to make even relatively simple alkenes. More recently, chemists discovered a process called **alkene metathesis** (or olefin metathesis) that made the transformation above exceedingly simple. Metathesis refers to a double replacement or double substitution reaction (two molecules swapping parts). In alkene metathesis, the two halves of an alkene are swapped:

2 H(H)C=C(R)R —*catalyst*→ H(H)C=C(H)H + R(R)C=C(R)R

This reaction often works in just one step and at room temperature. Some of the catalysts used are even quite tolerant of other functional groups being present (carbonyls, alcohols, amines, etc.). This reaction is so impressive and important that three of its discoverers (Schrock, Chauvin, and Grubbs) won the Nobel Prize in Chemistry in 2005. We will now briefly explore the contribution of each Nobel Laureate to developing alkene metathesis.

### *Lesson V.5.2. Schrock, Chauvin and Grubbs*

In 1974, while working for the DuPont Corporation, Richard R. Schrock discovered that some metal compounds that have a C=M double bond (called alkylidene complexes) can ***catalytically*** convert a short chain alkene into a longer alkene by alkene metathesis. Schrock eventually developed highly active catalysts having W=C or Mo=C bonds. These types of catalysts are called **Schrock catalysts**.

Schrock and others laid the foundation for alkene metathesis as a useful process, but, the exact processes by which these transformations occurred remained elusive for many years. Based on an extensive compendium of detailed mechanistic work, Yves Chauvin proposed what ended up being the correct mechanism for the various alkene metathesis reactions, which is shown below:

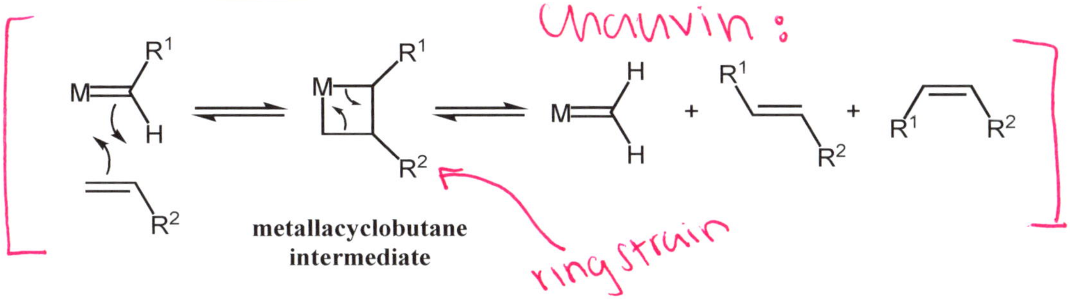

Chauvin:

**metallacyclobutane intermediate**

ring strain

First, the C=C bond aligns with the M=C bond. In a single concerted step, the existing M–C and C–C π-bonds are broken while new M–C and C–C σ-bonds are formed to make the four-membered ring. We know that a four-membered ring has a lot of ring strain (Lesson I.16, OC1 Primer), so a second concerted step relieves the ring strain and generates the new C=C bond and re-generates a M=C bond. The M=$CH_2$ unit can then react with more alkene, thus making the reaction catalytic.

One of the challenges of working with organometallic complexes is that they can be extremely sensitive towards $O_2$ and water. The Schrock catalysts suffer from this same problem. Robert Grubbs, however, developed a ruthenium complex that was (1) capable of performing alkene metathesis and (2) was stable in the presence of protic media like water. These stable catalysts are called **Grubbs' catalysts**, and are the most-used catalysts for alkene metathesis today. Some of Grubbs' catalysts are shown here:

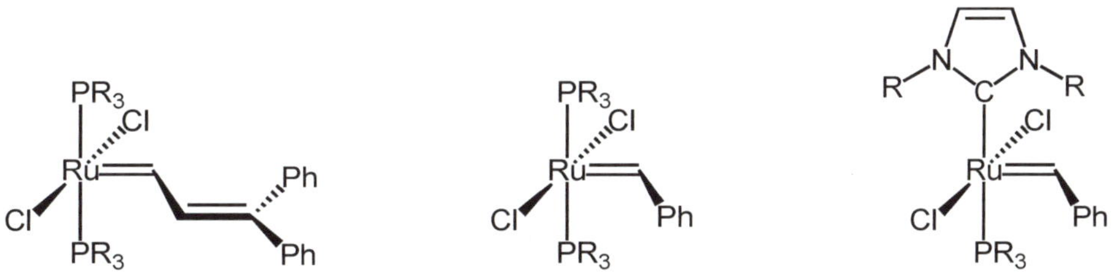

**first ruthenium alkylidene catalyst**

**first-generation Grubbs catalyst**

**second-generation Grubbs catalyst**

In the next lesson, we will see several useful applications of the alkene metathesis reaction.

## Lesson V.6. Applications of Alkene Metathesis

*Lesson V.6.1 Cross Metathesis and Ethenolysis*

Cross metathesis, in its simplest form, involves the conversion of a monosubstituted alkene into an internal, disubstituted alkene:

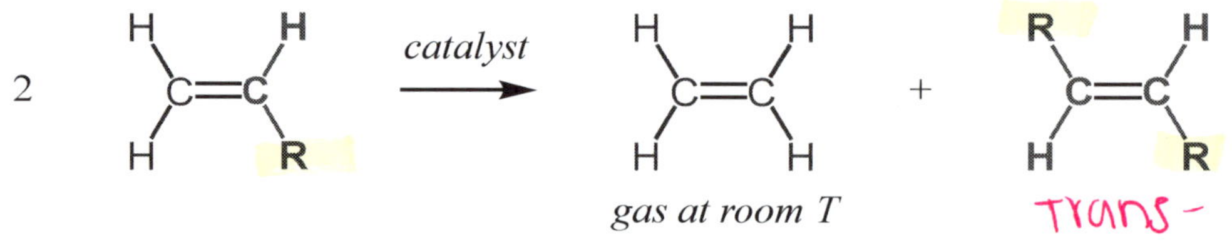

$$2\ H_2C{=}CHR \xrightarrow{catalyst} H_2C{=}CH_2 + RHC{=}CHR$$

gas at room *T* (ethylene); trans- (disubstituted alkene)

Note that ethylene is a gas and will rapidly escape from solution, which provides the driving force (by Le Chatelier's principle) for the reaction to push to the right and form the disubstituted alkene, often in close to 100% yield. We expect the *trans*-alkene to form as the major product because it is more stable than the *cis*-isomer.

more stable

Example IV.6.1

Draw all possible products for the reaction shown below:

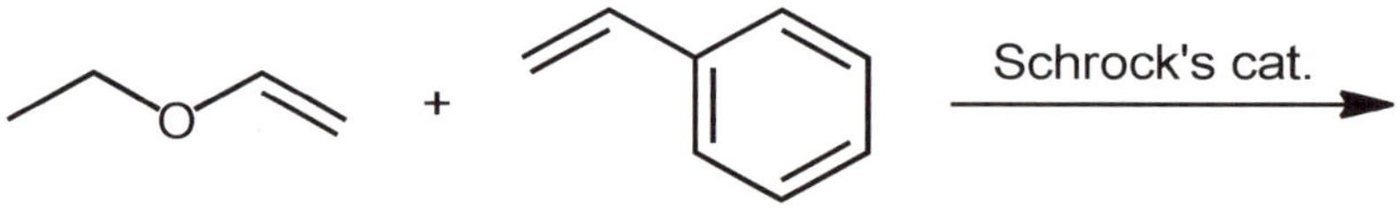

ethyl vinyl ether + styrene —Schrock's cat.→

Solution IV.6.1

Remember that, in a cross-metathesis reaction, the driving force is often the release of ethylene formed from a "$CH_2$" fragment from one alkene and a "$CH_2$" fragment from the other alkene.

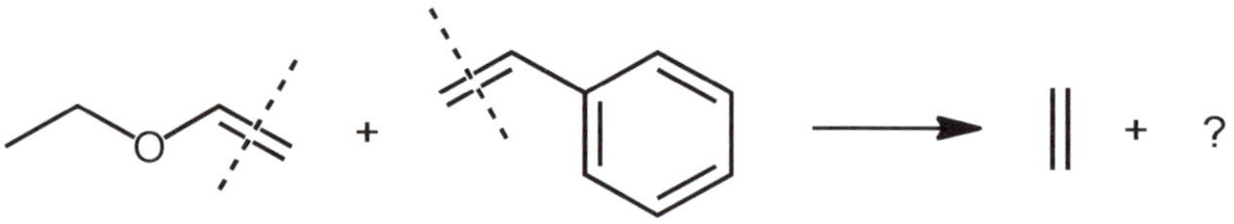

ethyl vinyl ether + styrene → ethylene + ?

In essence, we are chopping each alkene group in half and using all the "$CH_2$" fragments to make $C_2H_4$. Thus, the other products are all the possible combinations of the other fragments, i.e., there are 2 homocoupling products (left & right) and one heterocoupling product (center) that can be formed. Whether a cross-metathesis reaction gives one product more than another is an area of active research and is not something that can be easily predicted a priori. Keep in mind that cross-metathesis will afford a mixture of *E*- and *Z*-isomers.

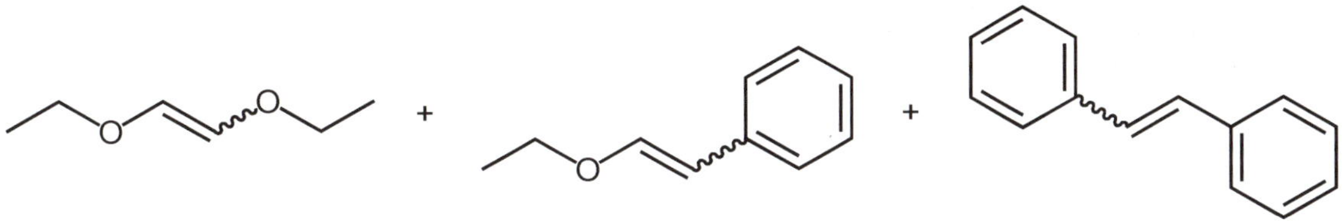

*Lesson V.6.2 Ring-Closing Metathesis*

We have seen that metal–alkylidene complexes can catalyze the cross metathesis of monosubstituted alkenes $H_2C{=}C(H)R^1$ to form disubstituted alkenes $R^1(H)C{=}C(H)R^1$ (mixture of *cis* and *trans*) and ethylene. If a molecule contains a monosubstituted alkene at each end of its chain (an **α,ω-diene**), the cross-metathesis reaction can lead to a cycloalkane and ethylene:

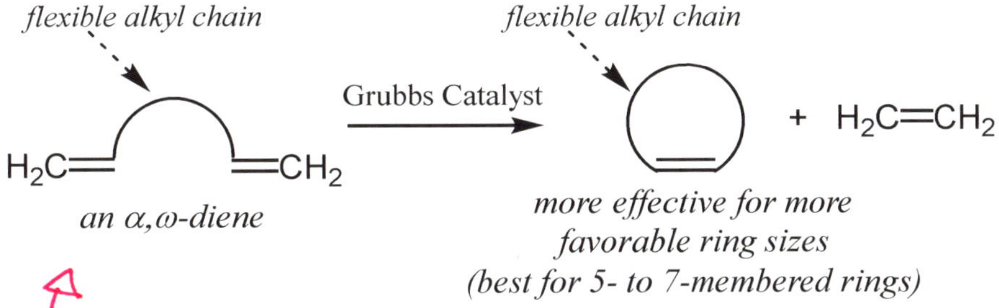

*flexible alkyl chain*

$H_2C{=}$ ⌒ $={CH_2}$ — Grubbs Catalyst → (ring) + $H_2C{=}CH_2$

*an α,ω-diene*

*more effective for more favorable ring sizes (best for 5- to 7-membered rings)*

This is called a **ring-closing metathesis** reaction. The catalytic cycle and reaction mechanism are the same as for other cross metathesis reactions, and the equilibrium is driven to the product by removing the ethylene gas as it forms. As with any other ring-forming reaction, it would be too difficult to form very strained rings (three- or four-membered rings), so we usually would use this reaction to make rings of five atoms or more.

Example V.6.2

Draw the organic product for the reaction shown below:

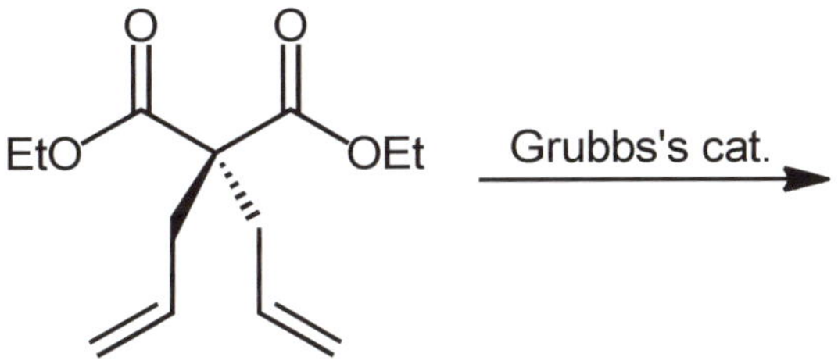

EtO, OEt, O, O — Grubbs's cat. →

Solution V.6.2

Sometimes a reactant will be drawn in such a way that it is harder to visualize the product after an alkene metathesis reaction. A good strategy when you are first exploring these reactions is to emphasize the "alkene" functional groups in the molecule and replace everything else with abbreviations (e.g., "R"), then draw your "chopping" lines across the C=C bonds:

R R

The "$CH_2$" fragments combine to make ethylene and, in ring-closing metathesis, the other fragments from an α,ω-diene combine with each other to make a ring. Thus, our 7-carbon α,ω-diene is converted into a 5-carbon cycloalkene:

$EtO_2C$ $CO_2Et$

*Lesson V.6.3 Ring-Opening Metathesis Polymerization*

Every alkene metathesis reaction is reversible, and the favored side of the equilibrium is the one with the more thermodynamically stable product (or we can drive it in one way by removing ethylene gas). We have seen that you can close a ring using alkene metathesis if a favorable (not very strained) ring results. Of course, the reverse of ring-closing metathesis is ring-opening metathesis. This type of reaction is favorable if it opens up a strained ring, thus relieving the angle and eclipsing strain. We can exploit the thermodynamically favorable ring-opening of strained or unfavorable ring sizes in a reaction called **ring-opening metathesis polymerization (ROMP)**.

2
1 3
ring strain!

**catalyst**
*(relieves ring strain)*

2 1 3 2
1 3 n 2 1 3

A polymer (long molecule consisting of the same unit repeating over and over again) results because when the ring opens, each end has to attach to the end of another of the opened rings (there are no other molecules around).

Example V.6.3

Draw the structure of the polymeric product for the reaction shown below:

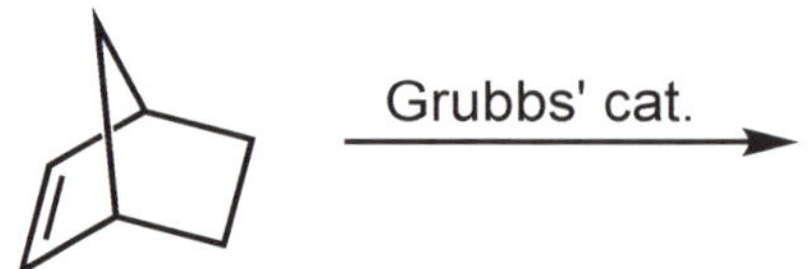

Solution V.6.3

As with any alkene metathesis reaction, the first step is to identify the reactive C=C bonds and the unreactive other parts of the substrate. In this example, there is one reactive C=C bond that is bridging across 2 positions of a cyclopentane ring that is unreactive:

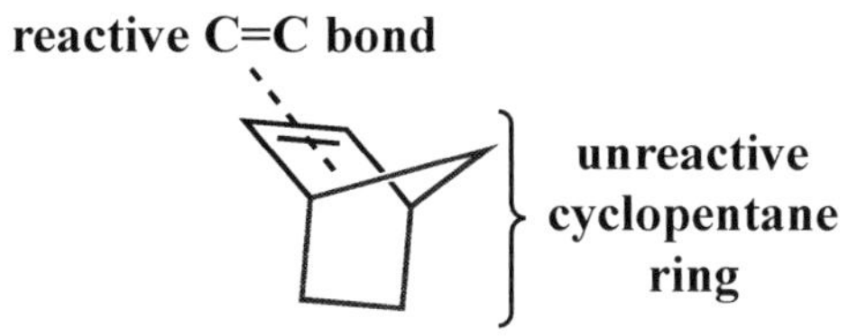

**reactive C=C bond**

**unreactive cyclopentane ring**

Keep in mind that there is no ethylene released in ROMP. The two fragments generated by metathesis will react with similar fragments in other molecules to make a polymer with the structure shown below. Note that the cyclopentane ring has not been affected by the reaction.

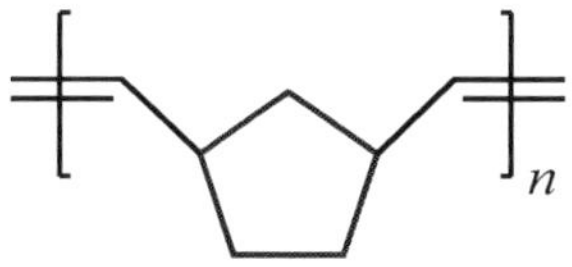

$n$

**PART VI. Property and Reactivity Trends of Carbonyl Compounds**

Lesson VI.1. Nomenclature of Carbonyls

Lesson VI.2. Review: Preparation of Carbonyls from Alcohols, Alkynes and Alkenes

Lesson VI.3. Classifying Reactions of Carbonyls

Lesson VI.4. Relative Rates of Nucleophilic Attack on Carbonyl Functional Groups

Lesson VI.5. Addition of Organometallics/Metal Hydrides to Aldehydes/Ketones

Lesson VI.6. Addition of Water or Alcohols to Aldehydes/Ketones

Lesson VI.7. Addition of Amine Derivatives to Aldehydes/Ketones

Lesson VI.8. The Wittig Reaction

Lesson VI.9. Nucleophilic Acyl Substitution of Acid Chlorides and Anhydrides

Lesson VI.10. $S_NAc$ of Carboxylic Acids to form Acid Chlorides

Lesson VI.11. $S_NAc$ Reaction of Oxygen Nucleophiles with Carboxylic Acids and Esters

Lesson VI.12. Amide Formation, Amide Hydrolysis and the Gabriel Synthesis

Lesson VI.13. Reaction of Carboxylic Acid Derivatives with H and C Nucleophiles

Lesson VI.14. Preparation and Reaction of Nitriles

Lesson VI.15. Preparation of Enolates and Alkylation

Lesson VI.16. Alpha-Halogenation and Haloform Reactions

Lesson VI.17. Aldol Addition and Condensation

Lesson VI.18. Claisen Condensation

Lesson VI.19. Decarboxylation and Synthetic Applications

Lesson VI.20. Addition of Nucleophiles to $\alpha,\beta$-Unsaturated Carbonyls

## Lesson VI.1. Nomenclature of Carbonyls

### *Lesson VI.1.1 Naming Aldehydes*

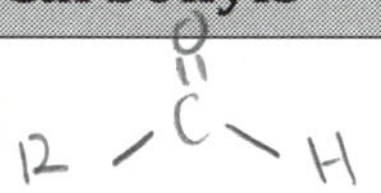

When naming an aldehyde, the aldehyde carbon is always assigned the number but the rest of the name is determined using the same rules we learned with alkan "-*e*" suffix to "-*al*" (similar to how the "-*e*" suffix is replaced by "-*ol*" for alco covered in Lesson I.14 (OC1 Primer) are used in a similar manner (i.e., substituent names, ordering, etc.), for example:

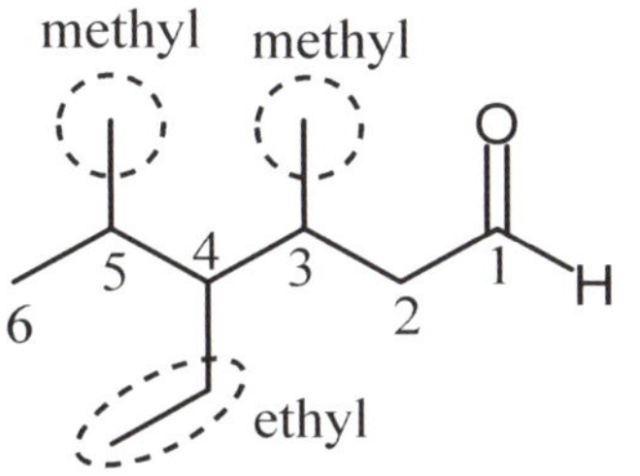

4-ethyl-3,5-dimethylhexanal

What happens if the molecule contains other functional groups in addition to the aldehyde? For naming, aldehydes have priority over both alcohol and ketone functional groups. When an OH group is a substituent of an aldehyde-containing alkyl chain, the OH group is termed a "hydroxy" substituent. Similarly, when a ketone is present in an aldehyde-containing alkyl chain, the ketone group is termed a "keto" substituent. Consider the following example:

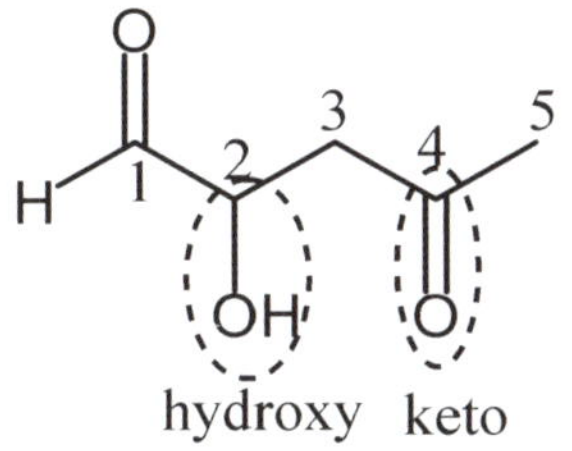

2-hydroxy-4-ketopentanal

The aldehyde carbon is numbered "1", which means the OH group is on carbon "2", and the ketone carbon is numbered "4". After determining the numbering, substituents are then listed alphabetically.

If the –C(=O)H functional group is a substituent of a cycloalkane, the suffix "carbaldehyde" is used. Consider this molecule:

cyclopentanecarbaldehyde

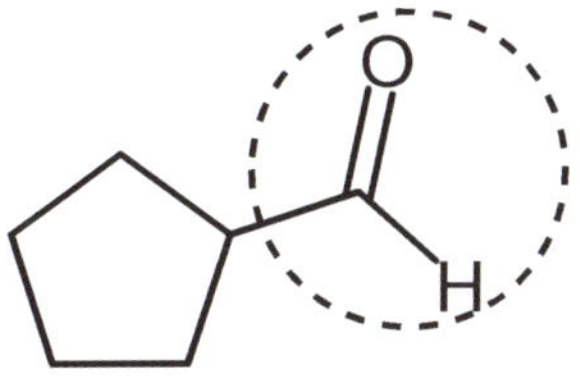

…he –C(=O)H group is a substituent of cyclopentane, the parent name "cyclopentane" is … with "carbaldehyde", and there should be no space between the two parts of the name.

*…n VI.1.2 Naming Ketones*

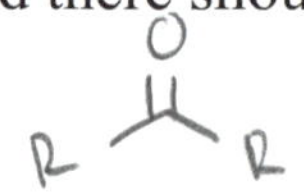

Nomenclature rules for ketones mirror the rules we learned for alkanes, but for ketones we replace the *-e* with the suffix *-one*. The main chain should be numbered such that the ketone carbon is assigned the smallest possible number. The remaining rules are applied as before. Consider this molecule:

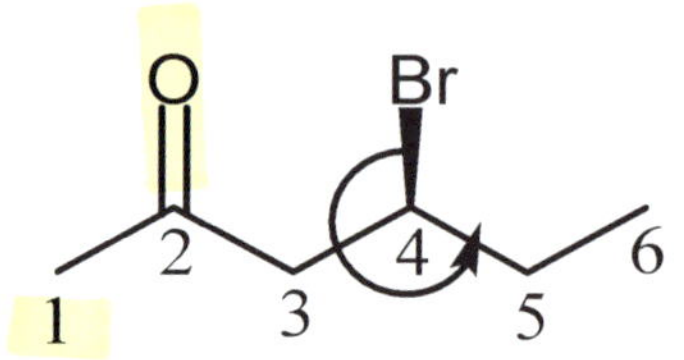

(*S*)-4-bromo-2-hexanone

In this molecule, we start numbering starting from the carbon closer to the carbonyl giving the carbon of the ketone number 2, after which we assign the configuration of the chiral center (carbon number 4) as we would for any other chiral molecule.

*Lesson VI.1.3 Naming Carboxylic Acids*

Nomenclature rules for carboxylic acids again mirror the rules for naming alkanes, but for carboxylic acids we use *-oic acid* in place of the *-e* at the end of the name. Because the carboxylic acid is necessarily at the end of a chain, we always give the carbon of the –C(O)OH group the number 1. All other *naming* rules apply as usual. Consider this molecule:

methyl

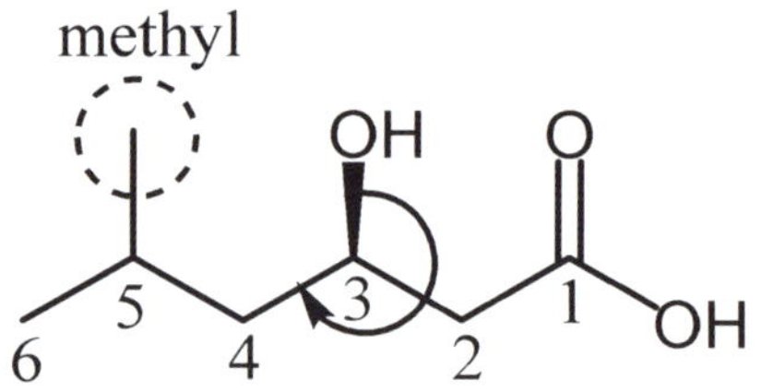

(*R*)-3-hydroxy-5-methylhexanoic acid

The carbon in the –C(=O)OH group must be assigned "1", and the –OH group on carbon 3 will be listed as a "hydroxy" substituent (a carboxylic acid takes priority over an alcohol). The remaining substituents and stereochemistry are determined using the standard conventions.

*Lesson VI.1.4 Naming Esters*

Naming an ester is slightly more involved than naming a carboxylic acid. When naming esters, we first list the name of the alkyl group "R–" that is attached to the oxygen. We then name the main chain (R"–C(O) portion), assigning the C of the –C(O)O unit as "1" and substituting the "-e" in the suffix by "-oate". The figure below shows the general formula for naming the ester:

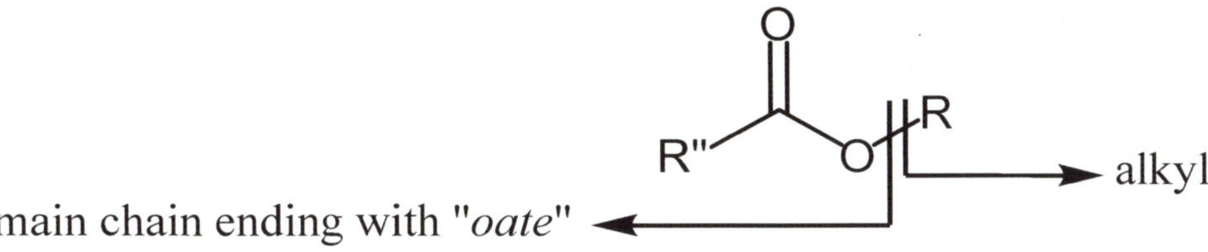


Consider a specific example:

methyl

3-methylbutanoate

isopropyl

In this molecule, the alkyl group on the oxygen is an isopropyl group, the main chain (starting with the carbonyl carbon) has four carbons, and carbon "3" has a methyl substituent. Therefore, the main chain is 3-methylbutanoate. The name of this ester will be given as **isopropyl 3-methylbutanoate** (there must be a space between the oxygen substituent and the ester chain names).

### *Lesson VI.1.5 Naming Acid Chlorides*

When naming acid chlorides, we use the suffix "*-oyl chloride*" and start numbering from the terminal –C(O)Cl group. All other IUPAC rules apply. Consider this example:

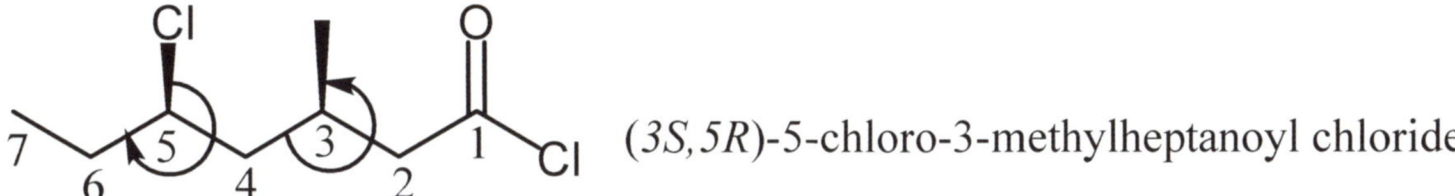


The carbon in the –C(=O)Cl group is "1", and we can then assign the configurations at the two chiral centers then list the substituents alphabetically.

## Lesson VI.2. Review: Preparation of Carbonyls from Alcohols, Alkynes and Alkenes

### *Lesson VI.2.1 Oxidation of Alcohols Using Chromium and Manganese Reagents*

Primary and secondary alcohols can react with inorganic reagents containing Cr or Mn to give aldehydes and ketones, respectively. Manganese dioxide ($MnO_2$) is the most common Mn reagent for this purpose, while two of the most common Cr(VI) salts used are **P**yridinium **C**hloro**c**hromate (PCC) and **P**yridinium **D**i**c**hromate (PDC). These reactions are called *oxidation reactions*, which are sometimes abbreviated as [oxid], [OX], or [O] (see Lesson II.12 for an overview of redox reactions in organic chemistry). In these cases, the carbinol (the C with the –OH on it) C–H and the O–H σ-bonds break and a C–O π-bond forms, producing a carbonyl-containing functional group (see example below):

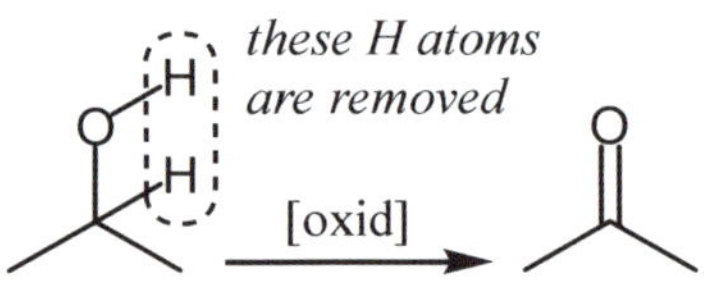

*these H atoms are removed*

[oxid]

[oxid] = $MnO_2$, PDC or PCC

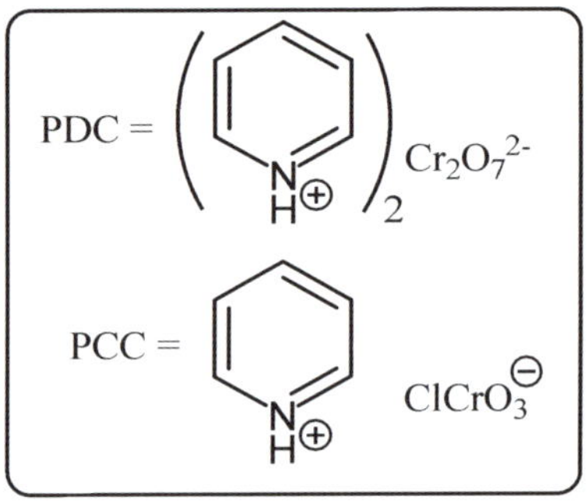

PDC = (pyridinium)$_2$ $Cr_2O_7^{2-}$

PCC = pyridinium $ClCrO_3^{\ominus}$

Tertiary alcohols cannot undergo oxidation under these conditions because there is no way to form a C–O π-bond (it would violate the octet rule).

Example VI.2.1

Draw the major product for each of the following reactions:

A. (cyclohexanol) $\xrightarrow[CH_2Cl_2]{PCC}$

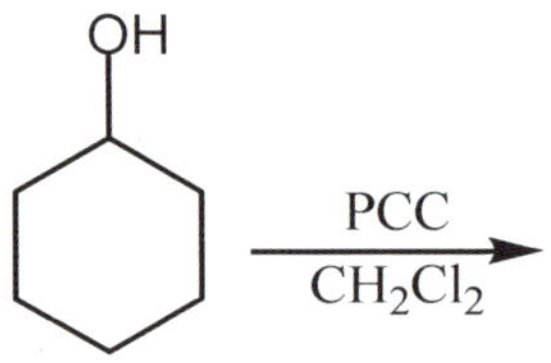

B. (3-methylbutan-1-ol) $\xrightarrow[CH_2Cl_2]{PDC}$

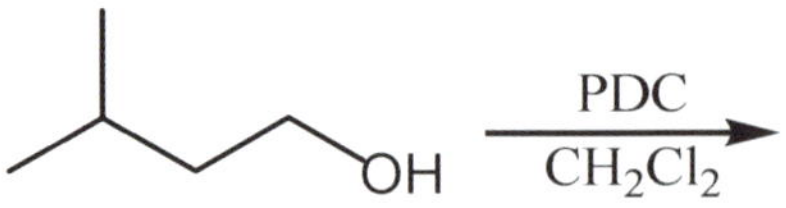

Solution VI.2.1

A) PCC will oxidize the 2° alcohol to a ketone, so the product is cyclohexanone.
B) PDC will oxidize the 1° alcohol to an aldehyde, so the product is 3-methylbutanal.

A. PCC / $CH_2Cl_2$ — ketone product

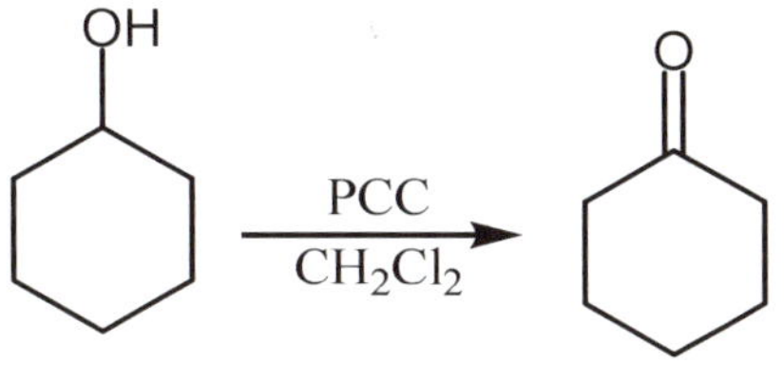

B. PDC / $CH_2Cl_2$ — aldehyde product

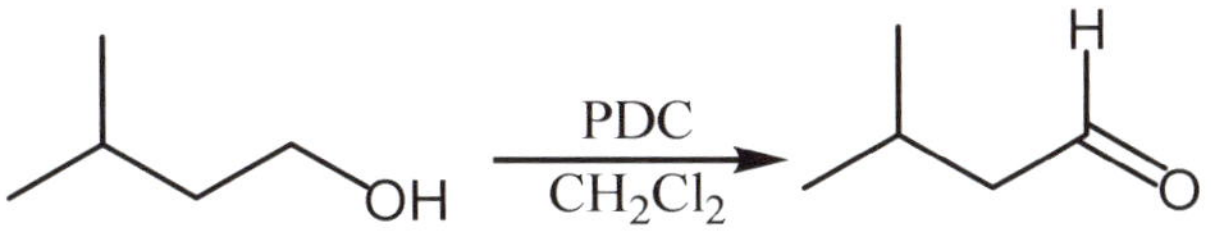

*Lesson VI.2.2 Oxidation of Alcohols Using Other Manganese or Chromium Reagents*

In Lesson II.13, we saw that $H^+/CrO_4^{2-}$, $H^+/Cr_2O_7^{2-}$, and $CrO_3/H_2SO_4$ (reagents for the **Jones Oxidation**) are stronger oxidants than are PCC or PDC. In Lesson IV.14, we saw that $KMnO_4$ can perform a similar function. Since these reagents are powerful oxidants, they will transform primary alcohols into carboxylic acids by further oxidizing the initially-formed aldehydes. These oxidizing reagents can be used to produce ketones from secondary alcohols, but aldehydes cannot be produced using these reagents.

Example VI.2.2

Draw the major product for each of the following reactions:

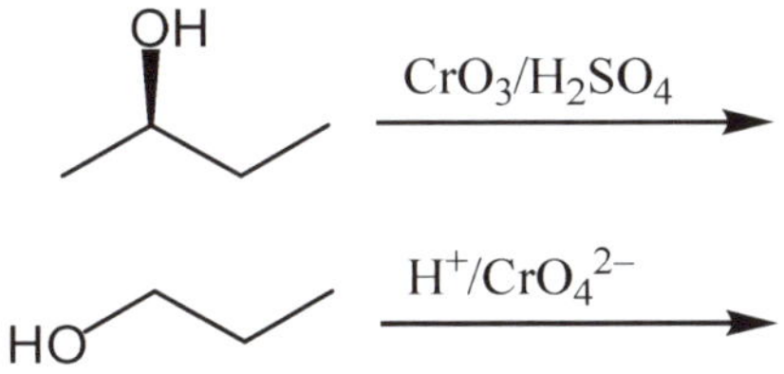

$CrO_3/H_2SO_4$

$H^+/CrO_4^{2-}$

Solution VI.2.2

These oxidizing agents are strong enough to oxidize an aldehyde to a carboxylic acid, so the 1° alcohol in the bottom reaction is converted to a carboxylic acid. However, these reagents do not oxidize ketones, so the 2° alcohol in the top reaction is converted to a ketone:

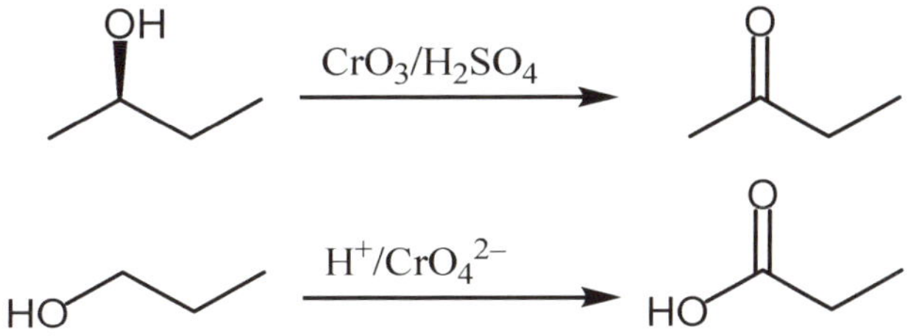

$CrO_3/H_2SO_4$

$H^+/CrO_4^{2-}$

1° alcohol → carboxylic acid
2° alcohol → ketone

*Lesson VI.2.3 Oxidation of Alcohols Using Swern Oxidation*

Treating a 1° or 2° alcohol with trifluoroacetic anhydride (abbreviated TFAA, with formula $(CF_3C(O))_2O$) in DMSO at temperatures below –50 °C will afford an aldehyde or ketone, respectively. These reaction conditions are known as the **Swern Oxidation**. Note that these conditions **do not** oxidize an aldehyde to the carboxylic acid.

Example VI.2.3

Draw the major product for each of the following reactions:

1° → aldehyde
2° → ketone

OH
TFAA
DMSO, –70 °C

OH
TFAA
DMSO, –70 °C

Solution VI.2.3

Both reactions employ TFAA in DMSO at –70 °C, which are Swern Oxidation conditions, so the top reaction will convert (*R*)-2-butanol to 2-butanone and the bottom reaction will convert 1-butanol to butanal:

OH
TFAA
DMSO, –70 °C
O
ketone product

OH
TFAA
DMSO, –70 °C
O
H
aldehyde product

*Lesson VI.2.4 Oxymercuration/Demercuration of Alkynes.*

As we saw in Lesson III.15 (OC1 Primer), alkynes react with $HgSO_4$ in the presence of $H_2O$ to produce ketones. The net result is addition of an H to one carbon in the C≡C bond and an OH to the other carbon to yield an enol ("*-en*" for the alkene functional group and "*-ol*" for the alcohol). The enol form of a molecule is typically less stable than the keto form, so the enol typically tautomerizes to the keto very rapidly:

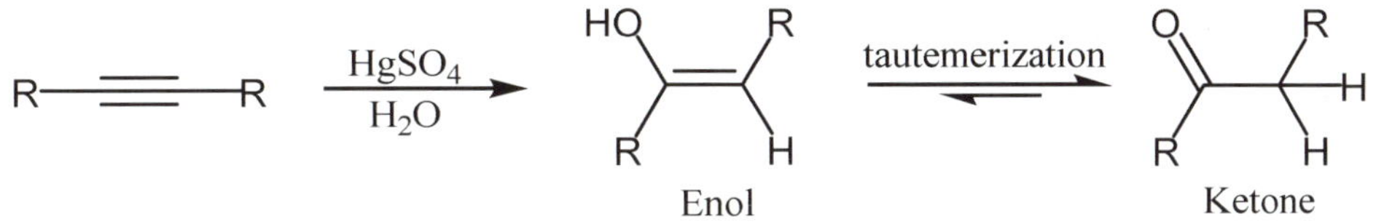

HO R
HgSO$_4$ / $H_2O$ → tautemerization ⇌
R—≡—R → Enol → Ketone

The above reaction is stereoselective (anti addition) and can be regioselective (Markovnikov). We know that alkynes can be either internal or terminal. In an internal alkyne, both carbons are disubstituted, which does not provide a driving force for regioselectivity of one site over the other. As a result, hydration of an internal alkyne that is asymmetric ($R \neq R'$) will yield a mixture of two ketone constitutional isomers. In a terminal alkyne, however, a single ketone isomer is formed because the alkyne carbon with the H substituent is substantially less electron rich than the alkyne carbon with the R substituent:

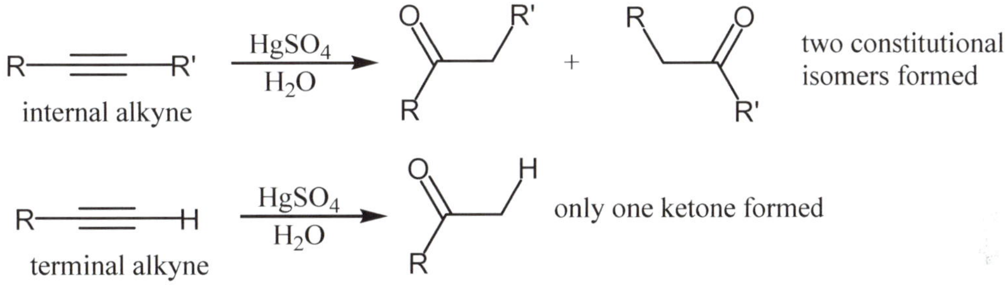

R—≡—R' (internal alkyne) → HgSO$_4$ / $H_2O$ → two constitutional isomers formed

R—≡—H (terminal alkyne) → HgSO$_4$ / $H_2O$ → only one ketone formed

Example VI.2.4

Give the product(s) of the following reaction:

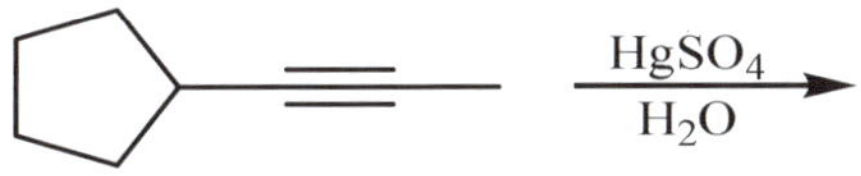

HgSO$_4$ / $H_2O$

Solution VI.2.4

Because the starting material is an internal alkyne, the oxygen lone pair in $H_2O$ can attack either carbon with equal probability, the C–OH σ-bond can form at either carbon, yielding a mixture of two enol isomers (enol I and enol II). Following tautomerization, we will obtain two ketone products (ketone I and ketone II).

$HgSO_4$, $H_2O$

enol I + enol II

ketone I + ketone II

*Lesson VI.2.5 Hydroboration/Oxidation of Alkynes.*

We learned in Lesson III.15 (OC1 Primer) that internal and terminal alkynes undergo hydroboration followed by oxidation to afford ketones and aldehydes respectively. The reaction can exhibit anti-Markovnikov regioselectivity, whereby hydroboration/oxidation of a terminal alkyne will afford an aldehyde (the oxygen ends up on the less substituted carbon). However, an internal alkyne will yield a mixture of two ketone constitutional isomers (similar to oxymercuration/demercuration):

internal alkyne — 1. $BH_3$ 2. $NaOH, H_2O_2, H_2O$ — two constitutional isomers formed

terminal alkyne — 1. $BH_3$ 2. $NaOH, H_2O_2, H_2O$ — only one aldehyde formed

Example VI.2.5

Give the major product(s) of the following reaction:

1. $BH_3$

2. $NaOH, H_2O_2, H_2O$

Solution VI.2.5

Because the starting material is a terminal alkyne, the boron will end up on the less substituted carbon, and is converted to an OH group in the second set of conditions. Only one enol isomer will form and it will tautomerize to give the aldehyde as the major product

1. $BH_3$
2. NaOH, $H_2O_2$, $H_2O$

enol

aldehyde
*isolated major product*

*Lesson VI.2.6 Ozonolysis of Alkenes*

In Lesson III.10, we learned that an alkene can react with ozone, which cleaves the C=C bond and replaces it with two C=O bonds. Recall that this reaction requires a second step involving either an oxidant (usually $H_2O_2$) or a reductant (usually $R_2S$). Reducing conditions afford ketones and aldehydes (depending on the degree of substitution of the alkene reactant), whereas oxidizing conditions afford ketones and carboxylic acids (the aldehyde is oxidized to the acid).

Example VI.2.6

Give the products of the reaction of 2-methyl-2-butene with $O_3$ followed by either an oxidant (top reaction) or reductant (lower reaction).

1. $O_3$
2. $(CH_3)_2S$

Solution VI.2.6

In the first step of this reaction, the C=C breaks and two carbonyl-containing molecules are formed. For the second step of the reaction a ketone and an aldehyde are formed:

1. $O_3$
2. $(CH_3)_2S$

ketone + aldehyde

## Lesson VI.3. Classifying Reactions of Carbonyls

### *Lesson VI.3.1 Introduction to Carbonyl Reaction Type A*

In this Primer, we divide carbonyl reactions into types A, B, C and D to help with studying and organizing the reactions when you study. These are particular classifications used in this book as a way to help you sort reactions to study them, and are not general terms for all chemists. Carbonyl reactions that we call **Type A** involve two steps:

1) nucleophilic addition of an anionic carbon or hydrogen then

2) protonation of the anionic carbonyl oxygen produced by this addition.

For the purpose of discussion in this book, we will refer to reactions that follow these steps as **carbonyl reaction type A**.

Nucleophilic addition to a carbonyl group is driven by Coulombic attraction between a nucleophile and the $C^{\delta+}$ of the C=O unit. Upon nucleophilic attack, the $\pi$-bond between the carbon and oxygen breaks, leading to formation of negative charge on the oxygen. The reaction is followed by a workup using dilute aqueous acid to form the alcohol product. The general mechanism is as follows:

**Carbonyl Reaction Type A**

R' Y :Nu *nucleophilic addition* R' Nu Y $H-OH_2^{+}$ *protonation* R' OH Nu Y Alcohol

Y = alkyl or aryl (ketones)
or H (aldehydes)

Nu: = hydride or carbanion

**The net result is of a Carbonyl Reaction Type A is:**

1. Add a nucleophile (Nu) to the carbonyl C to replace the pi bond

2. Protonate the carbonyl O.

*Lesson VI.3.2 Introduction to Carbonyl Reaction Type B*

Unlike aldehydes and ketones, carboxylic acids and their derivatives (anhydrides, acid chlorides, amides and esters) undergo **nucleophilic acyl substitution**.

Nucleophilic acyl substitution is a two-step process:

1. Nucleophilic addition to the carbonyl carbon and
2. Nucleophilic elimination of the best leaving group from what was the carbonyl carbon.

For the purposes of categorizing reactions as a way to help study them in this particular book, we will refer to the reactions that follow this pattern as being **Type B**.

Why do carboxylic acids and their derivatives exhibit fundamentally different mechanisms of reaction from what we saw for aldehydes and ketones? The difference is stems from the fact that the carbonyl carbon in carboxylic acids/derivatives is connected to a heteroatom, whereas it is connected to an alkyl/aryl or a hydrogen in ketones/aldehydes, respectively. The presence of heteroatoms in carboxylic acids and their derivatives can provide the leaving groups needed to favor nucleophilic elimination. The following figure summarizes the general reaction mechanism, which involves the formation of a **tetrahedral intermediate**:

**Carbonyl Reaction Type B**

R'' O Y Nu — *Nucleophilic Addition* → R'' O Nu Y — *Nucleophilic Elimination* → R'' O Nu + $Y^-$, HY or "other products"

*tetrahedral intermediate*

**The net result is of a Carbonyl Reaction Type B is:**

Substitute one nucleophile (Nu) for one leaving group (Y) attached to the carbonyl carbon.

*Lesson VI.3.3 Introduction to Reaction Type C*

When treating carboxylic acids and their derivatives with some nucleophiles, the nucleophilic substitution (reaction Type B) step may be followed by a nucleophilic addition then protonation (reaction Type A). For the purpose of discussion in this book, **we will call these reactions of Type C**, which follow the general net reaction:

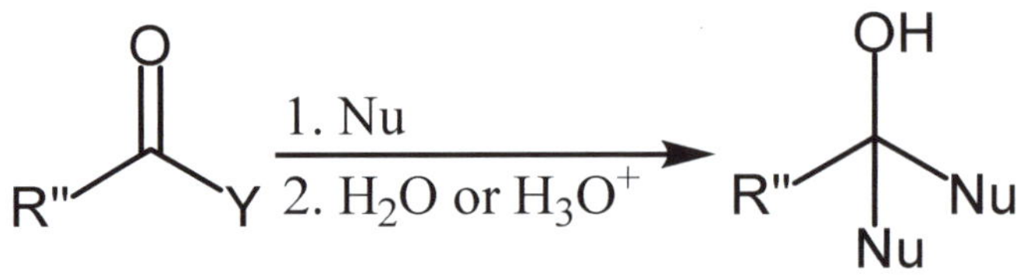

$$\text{R''C(=O)Y} \xrightarrow[\text{2. } H_2O \text{ or } H_3O^+]{\text{1. Nu}} \text{R''C(OH)(Nu)(Nu)}$$

The general mechanism for these reactions is:

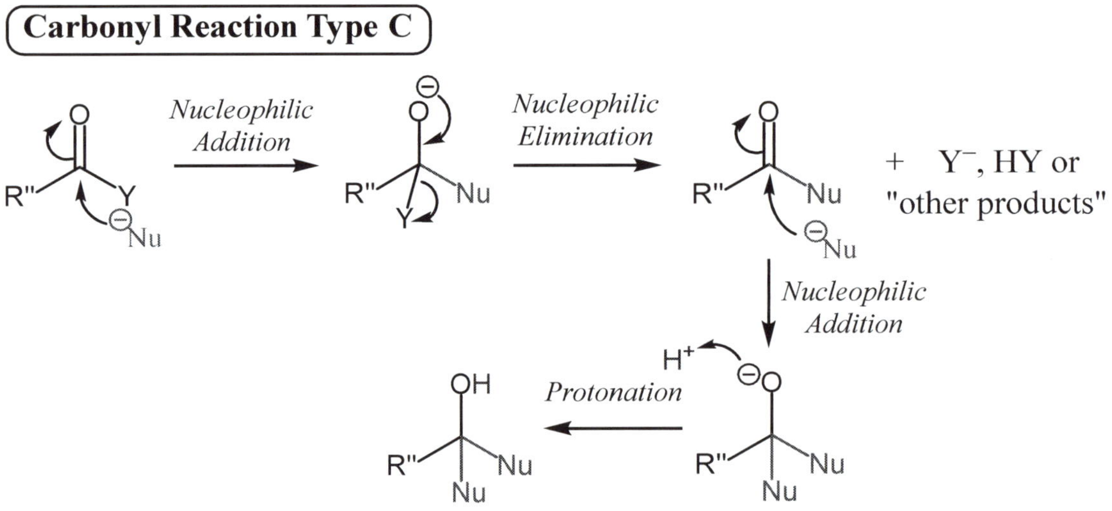

**Carbonyl Reaction Type C**

*Nucleophilic Addition* → *Nucleophilic Elimination* → + $Y^-$, HY or "other products"

*Nucleophilic Addition* ↓

← *Protonation* ($H^+$)

**The net result is of a Carbonyl Reaction Type C is:**

1. Substitute one nucleophile for one leaving group attached to the carbonyl carbon.
2. Add a nucleophile to the carbonyl C to replace the pi bond, push negative to O.
3. Protonate the carbonyl O.

*Lesson VI.3.4 Introduction to Reaction Type D*

Carbonyl reactions of Type D are reactions of carbonyl compounds in which both of the ca C=O bonds are broken and replaced with bonds from the carbonyl C to reactants. These special reacti of carbonyl compounds are very specific to the type of carbonyl reactant used as well as the reagents present. They do not have a general mechanism other than the first step, which involves a nucleophilic addition where a C—O π bonds breaks. There are four main options for replacing the two C=O bonds to C with new bonds so that the carbonyl C still has four bonds in the product:

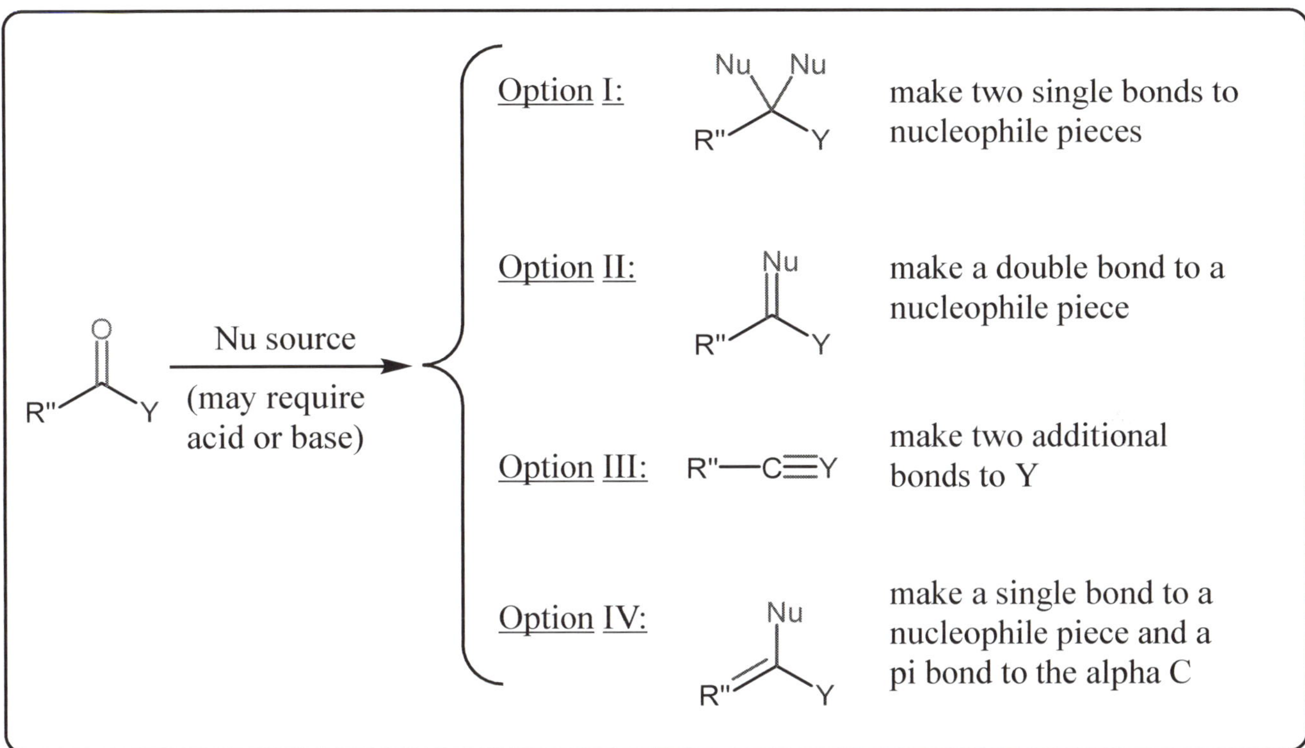


**The net result of a Carbonyl Reaction Type D is:**

1. Remove the oxygen from the carbonyl carbon.

2. Make two new bonds to the carbonyl carbon.

These are some of the more complex carbonyl reactions that we will study, but sorting them into the "Type D" category as we encounter them will remind you to replace *both* of the bonds to the carbonyl O and to get rid of the O, whereas all of the other reaction types keep an O attached to the carbonyl C in the net reaction product.

ognizing Patterns in Carbonyl Reactions

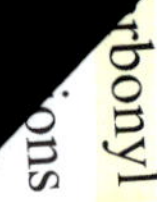

chart summarizes most of the reactions of carbonyl functional groups that we will We can start to recognize patterns in their reactivity on the basis of whether the ntial leaving group and whether the nucleophile is more reactive or less reactive. It to refer back to this summary frequently while going through the rest of the reactions

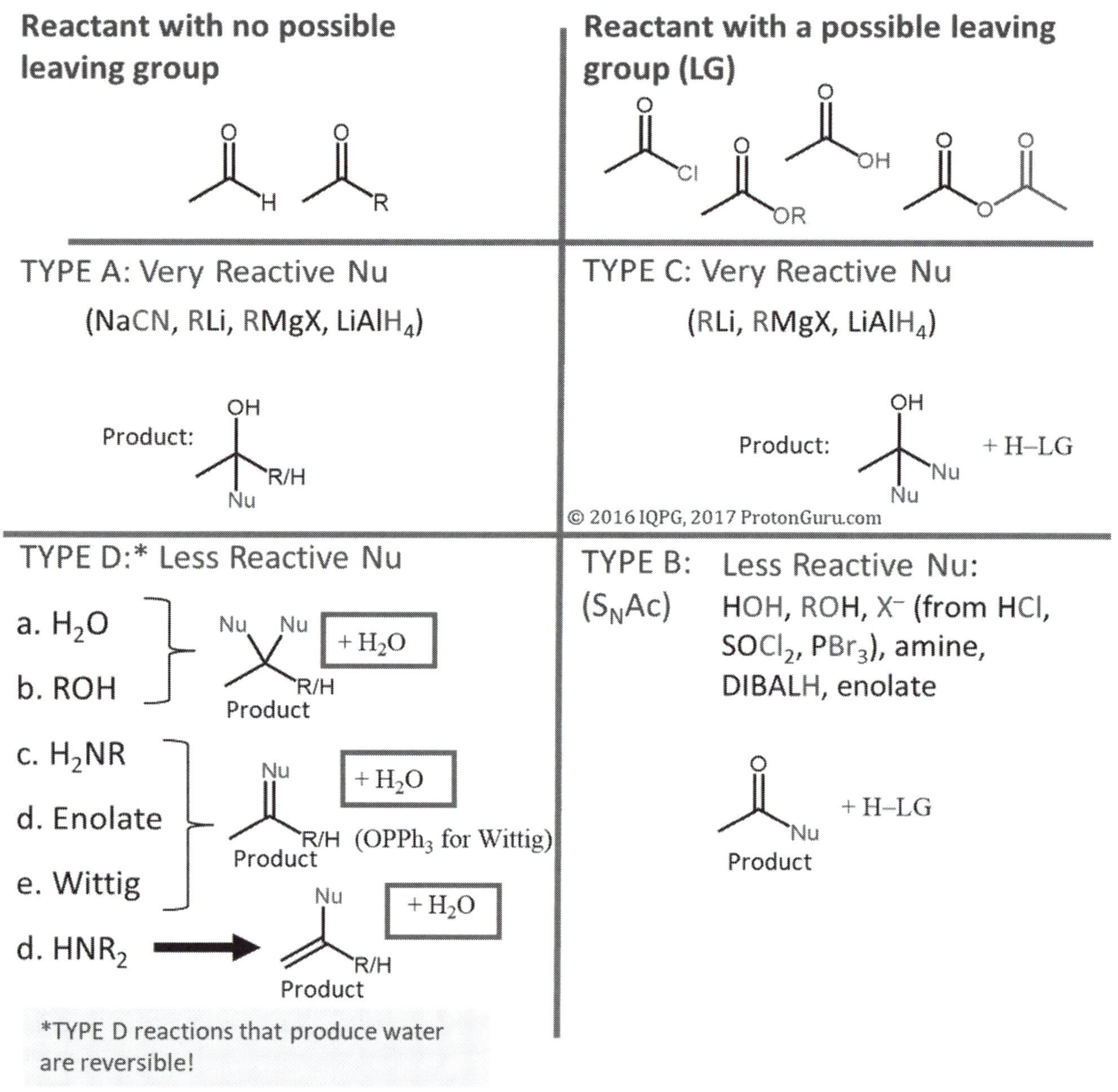

Reactant with no possible leaving group
O
H
R
Reactant with a possible leaving group (LG)
Cl
OR
OH
O
TYPE A: Very Reactive Nu
(NaCN, RLi, RMgX, LiAlH4)
Product:
OH
R/H
Nu
TYPE C: Very Reactive Nu
(RLi, RMgX, LiAlH4)
Product:
OH
Nu
Nu
+ H–LG
© 2016 IQPG, 2017 ProtonGuru.com
TYPE D:* Less Reactive Nu
a. H2O
b. ROH
Nu Nu
R/H
Product
+ H2O
c. H2NR
d. Enolate
e. Wittig
Nu
R/H (OPPh3 for Wittig)
Product
+ H2O
d. HNR2
Nu
R/H
Product
+ H2O
*TYPE D reactions that produce water are reversible!
TYPE B: (SNAc)
Less Reactive Nu:
HOH, ROH, X− (from HCl, SOCl2, PBr3), amine, DIBALH, enolate
O
Nu
+ H–LG
Product

## Lesson VI.4. Relative Rates of Nucleophilic Attack on Carbonyl Functional Groups

*Lesson VI.4.1 Carbonyl Functional Groups Have Different Reactivity to Nucleophilic Addition.*

Many of the reactions involving carbonyls that are covered in sophomore organic chemistry involve the addition of a nucleophile to the carbonyl carbon. The different carbonyl-based functional groups have different reactivity towards nucleophilic addition, and it is important to understand the origins of these differences in reactivity. For example, treating acetyl chloride with $NH_3$ (the nucleophile) will afford the amide, but treating the amide with HCl ($Cl^-$ is the nucleophile) does not produce acetyl chloride, even though the anionic $Cl^-$ is a better nucleophile than the neutral $NH_3$.

$NH_3$ → (amide); HCl → no net reaction

The difference in reactivity among the carbonyl-based functional groups arises from the electronic and steric effects of substituents at the carbonyl carbon. We learned to classify substituents based on their ability to donate or withdraw electrons and their influence on electrophilic aromatic substitution reactions (Lesson IV.11). A resonance or inductive donor group (collectively called electron-donating groups, EDGs) attached to the carbonyl carbon increases the electron density in the carbonyl unit (i.e., makes the $C^{\delta+}$ smaller), which renders the carbonyl less susceptible to nucleophilic addition. If electron withdrawing groups (EWGs, such as $-CF_3$) are present, however, the electron density in the carbonyl unit decreases (i.e., makes the $C^{\delta+}$ larger), which increases attraction towards a nucleophile. These are examples of **electronic effects**. In the rest of this lesson, we will examine how electronic effects produce the differences in reactivity among the carbonyl-based functional groups.

*Lesson VI.4.2 Explaining Differences in Reactivity using Sterics and Electronics*

We will now order each of the carbonyl-based functional groups according to their relative reactivity towards nucleophilic addition.

Acid chlorides

Among all the carbonyl-based functional groups, acid halides have the most partial positive character ($\delta+$) on the carbonyl C, and are thus the most reactive towards nucleophiles. As we will see, acid halides react with water, alcohols, amines, all organolithium, Grignard, Gilman, $NaBH_4$ and $LiAlH_4$ reagents. It is notable that Gilman reagents do not react with any of the other carbonyl-based functional groups besides acid halides.

Anhydrides

These are the second-most reactive of the carbonyl-based functional groups after acid chlorides, but their analyses are complicated by the fact that one anhydride molecule contains two carbonyl groups which can undergo nucleophilic attack, so we will omit some reactions in this book.

Aldehydes

Aldehydes feature the next-most electron deficient carbonyl carbons, due to the fact that an H atom is not as electron-donating as an alkyl group (as in the case of a ketone) or a heteroatom (such as the O in an ester or carboxylic acid or the N in an amide), thus aldehydes will be the next-most reactive towards nucleophiles of the carbonyl-based functional groups.

Ketones

Ketones are somewhat less reactive to nucleophilic addition than are aldehydes, but the difference is sufficiently small that aldehydes and ketones typically undergo the same types of reactions. Any differences in relative reactivity derives from steric or electronic effects. For example, butanal reacts with a nucleophile more rapidly than 2-butanone (see below) because the nucleophile must overcome greater steric repulsion from the two alkyl groups adjacent to the carbonyl group in a ketone vs. the one alkyl group of the aldehyde:

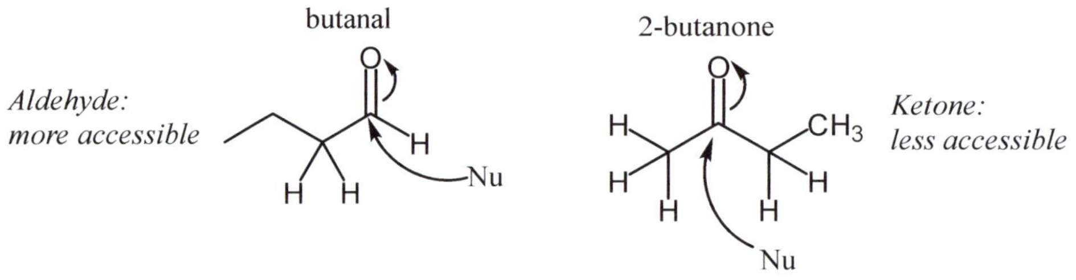

Carboxylic acids

Carboxylic acids feature substantially more electron-rich carbonyl carbons than ketones due to the fact that the carbonyl carbon is attached to a resonance donating OH group. Recall that resonance donors are better donors than are inductive donor groups (Lesson IV.11), so the δ+ on the carbonyl carbon in a carboxylic acid is smaller than in a ketone or aldehyde. As a result, carboxylic acids are even less reactive towards nucleophiles than ketones. It is important to recall, however, that the p$K_a$ for a carboxylic acid is substantially lower than the p$K_a$ for an alkane or hydrogen (e.g., p$K_a$ = 3.75 for $HCO_2H$ vs. 56 for $CH_4$ and 35 for $H_2$). The R– or H– anion from an organometallic or metal hydride reagent is sufficiently basic to deprotonate the carboxylic acid, therefore the COOH group will always be converted to $COO^-$ (carboxylate) before any nucleophilic substitution at the carbonyl carbon can occur. Keep in mind that deprotonation of the carboxylic acid makes the carbonyl group substantially more electron rich (it has an adjacent anionic O, after all) and, as a result, significantly reduces its susceptibility towards nucleophilic attack. As we will see later in the book, only the most reactive of the nucleophiles – organolithium reagents and $LiAlH_4$ – will do nucleophilic attack at the carbonyl carbon in a carboxylate ($COO^-$) unit. Amines, neutral water, alcohols, Grignard reagents and $NaBH_4$ will all lead only to deprotonation the COOH group to form a carboxylate salt and do not react further.

Esters

Esters feature more electron-rich carbonyl carbons than those in carboxylic acids due to the fact that an OR group is a better resonance donor than an OH group. The ester, however, is not susceptible to deprotonation to form carboxylates, so the reaction of an esters with a nucleophilic organometallic or metal hydride reagent is mechanistically more straightforward. Nonetheless, esters are less reactive towards nucleophiles than ketones. Although **organolithium, Grignard and $LiAlH_4$ reagents react with esters, $NaBH_4$ (which has less nucleophilic $H^{\delta-}$ than does $LiAlH_4$) does not.**

Carboxamides

Carboxamides (also known as "amides") are the most electron rich of the carbonyl-based functional groups. This is because the electronegativity of nitrogen is lower than that of oxygen, thus the N of an amide donates its electrons more effectively to the carbonyl carbon than the O of a carboxylic acid or an ester. Consequently, amides are the least reactive of the carbonyl-based functional groups to nucleophilic addition. In fact, **only $LiAlH_4$ reacts with the carboxamide functional group**. Without forcing conditions, most nucleophilic reagents that can undergo nucleophilic addition to other carbonyl functional groups will not react with carboxamides. Even very reactive nucleophiles derived from organolithium, Grignard, and $NaBH_4$ reagents cannot do nucleophilic addition to carboxamides.

## Lesson VI.5. Addition of C and H Nucleophiles to Aldehydes/Ketones

### *VI.5.1 Aldehydes and Ketones Undergo Reaction Type A*

Aldehydes and ketones undergo reaction type A with hydride or C-based nucleophiles. Recall from Lesson VI.3 that Type A reactions involve two steps:

1. nucleophilic addition of an anionic carbon or hydrogen

2. protonation of the anionic oxygen produced by this addition

The general mechanism is as follows:

***Carbonyl Reaction Type A***

*nucleophilic addition*

*protonation*

Alcohol

Y = alkyl or aryl (ketones) or H (aldehydes)

Nu: = hydride or carbanion, including alkyl, acetylide, or cyanide anions

The nucleophiles for this type of reactions bear negative charge on either a carbon ($R^-$, provided by an organometallic reagent, an acetylide or cyanide anion) or a hydrogen ($H^-$, provided by a metal hydride). The reactions exhibiting this type of behavior are summarized here in the following figure:

Addition of $NaBH_4$ or $LiAlH_4$:

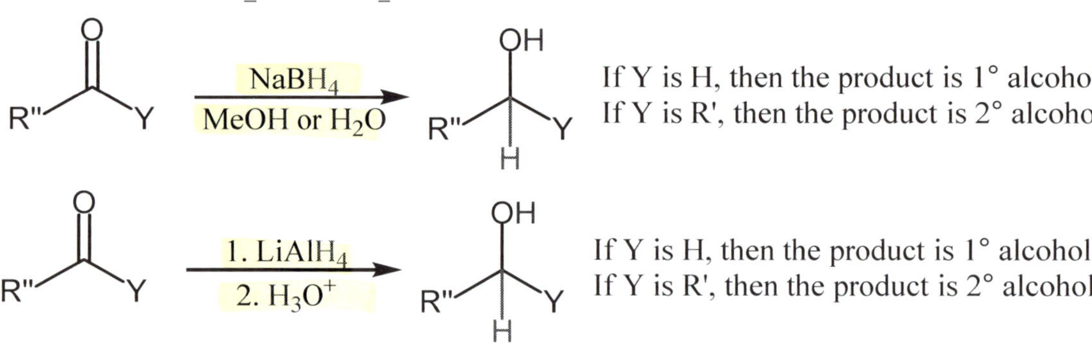

Addition of Grignard Reagent:

$R''C(=O)Y \xrightarrow[\text{2. } H_3O^+]{\text{1. RMgX}} R''C(OH)(R)Y$

(X is Cl, Br or I)

If Y is H, then the product is 2° alcohol
If Y is R', then the product is 3° alcohol

Addition of Organolithium Reagent:

$R''C(=O)Y \xrightarrow[\text{2. } H_3O^+]{\text{1. RLi}} R''C(OH)(R)Y$

If Y is H, then the product is 2° alcohol
If Y is R', then the product is 3° alcohol

Addition of Cyanide Nucleophile:

$R''C(=O)Y \xrightarrow[\text{2. } H_3O^+]{\text{1. NaCN}} R''C(OH)(CN)Y$

A product with a –OH and a –CN on the same carbon is called a **cyanohydrin**

Example VI.5.1

Draw and provide the IUPAC name for the major product for each of the following reactions:

Reaction 1 $\xrightarrow[\text{MeOH}]{NaBH_4}$

Reaction 2 $\xrightarrow[\text{2. } H_3O^+]{\text{1. PhMgBr}}$

Solution VI.5.1

Each reaction above is Type A: 1) nucleophile attack to break the C–O π-bond, then 2) protonation to yield the alcohol products. For Reaction 1, the nucleophile is a $H^-$ (from $NaBH_4$) and for Reaction 2 the nucleophile is a $Ph^-$ (from PhMgBr). Note that each product contains a chiral center (marked by *), but because each starting material is achiral, the product will be a racemate:

Reaction 1 — NaBH$_4$ / MeOH → 3-methyl-2-butanol, *racemic mixture*

Reaction 2 — 1. PhMgBr; 2. $H_3O^+$ → 3-methyl-1-phenyl-1-butanol, *racemic mixture*

## Lesson VI.6. Addition of Water or Alcohols to Aldehydes/Ketones

### *Lesson VI.6.1 Hydration of Aldehydes and Ketones*

Water itself can do nucleophilic addition to an aldehyde or ketone to form a **hydrate**. Aldehyde/ketone hydration reactions can occur in an acidic, basic, or neural medium. The mechanism of hydrate formation in neutral media is as follows:

*Nu addition* — *deprotonation* — *protonation*

Y = H, alkyl or aryl

**Hydrate**

Addition of $H_2O$ across the C=O double bond is a reversible reaction, and the equilibrium position is determined by the by the carbonyl substituents (R" and Y). The presence of EWG as substituents on the carbonyl (e.g., Y and R" are $CX_3$, where X = F or Cl) shifts the equilibrium to the right (favors the products), whereas EDG carbonyl substituents (e.g., Y and R" are alkyl or aryl groups) shift the equilibrium to the left. For example, <0.1% of acetone (Y = R" = $CH_3$) exits as the hydrate in aqueous solution, but >99.9% of formaldehyde (Y = R" = H) or hexafluoroacetone (Y = R" = $CF_3$) exists as the hydrate in aqueous solution. Note that the overall reaction is a simple replacement of the leaving group (Y) with part of the nucleophile and protonation of the carbonyl O. This is a Type A reaction (see Lesson VI.3).

### *Lesson VI.6.2 Reactions of ROH with Aldehydes and Ketones to form Acetals and Ketals*

An alcohol can likewise do nucleophilic addition to an aldehyde or ketone to form a **hemiacetal** or **hemiketal**, respectively, upon addition of one equivalent of ROH. These species can be generated in acidic, basic, or neutral media. The mechanism of hemiacetal/hemiketal formation is essentially the same as in the hydration reaction. The mechanism in neutral media is as follows:

*Nu addition* (ROH) — *deprotonation* — *protonation*

Y = H, alkyl or aryl

**Hemiacetal** (Y = H)
**Hemiketal** (Y = $C_xH_y$)

This is another example of a Type A reaction (see Lesson VI.3).

## *Lesson VI.6.3 Reactions of ROH with Aldehydes and Ketones to form Acetals and Ketals*

Treating an aldehyde or ketone with two or more equivalents of alcohol (ROH) will afford molecules called **acetals** (from aldehydes) and **ketals** (from ketones). An acetal (or ketal) features a carbon having two C—OR bonds. The reaction is catalyzed by acid.

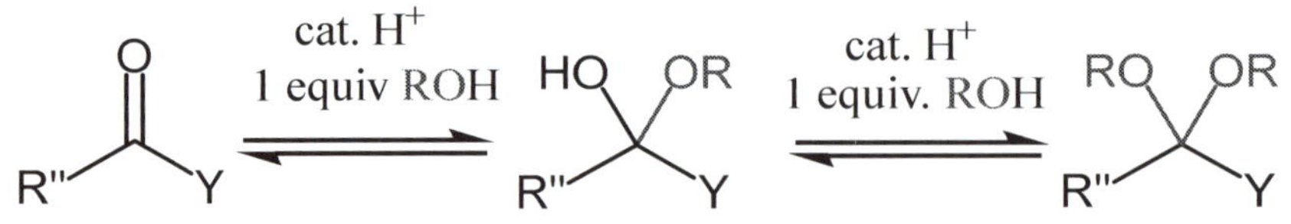

cat. H+
1 equiv ROH
cat. H+
1 equiv. ROH
O
R''
Y
HO
OR
RO


**Aldehyde** (Y = H) **Hemiacetal** (Y = H) **Acetal** (Y = H)
**Ketone** ($Y = C_xH_y$) **Hemiketal** ($Y = C_xH_y$) **Ketal** ($Y = C_xH_y$)

Acetal and ketal formation are reversible, which means they can be easily converted back to aldehydes and ketones upon reaction with water in presence of an acid catalyst (a hydrolysis reaction). If we want to push the reaction to the right, we apply LeChatelier's principle by either employing excess ROH or removing water from the reaction mixture.

The mechanism for acetal and ketal formation from the hemiacetal/hemiketal (which forms by reaction with the first equiv of ROH) is depicted below:

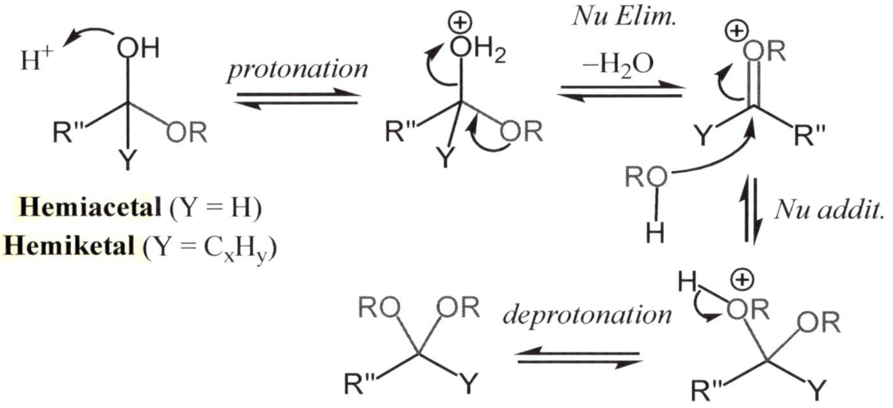

H+
OH
R''
OR
Y
protonation
OH2
Nu Elim.
–H2O
OR
Nu addit.
RO
H
deprotonation


**Hemiacetal** (Y = H)
**Hemiketal** ($Y = C_xH_y$)

**Acetal** (Y = H)
**Ketal** ($Y = C_xH_y$)

Note that the net reaction is replacement of *both* carbonyl C=O bonds with new single bonds to the –OR groups, so this is a Type D reaction (see Lesson VI.3). Note also that every step in the formation of hemiacetal/hemiketal and acetal/ketal formation is reversible. Furthermore, water is a product of acetal/ketal formation. This means that we need to distill the water away as it forms to push the equilibrium to the ketal/acetal product. Conversely, if an excess of water is added the ketal/acetal reverts to the carbonyl:

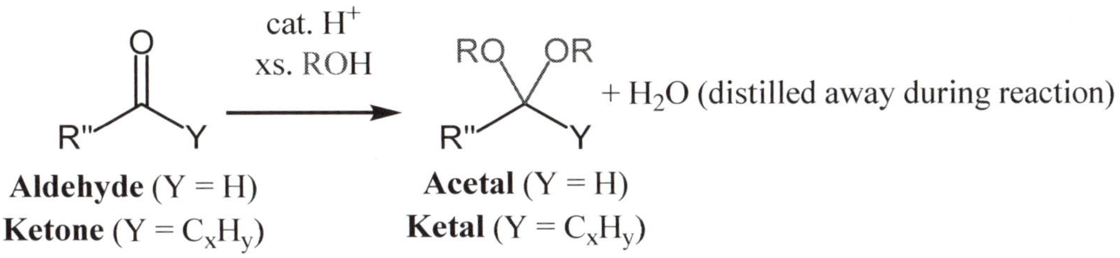

cat. $H^+$
xs. ROH

RO OR

+ $H_2O$ (distilled away during reaction)

R'' Y → R'' Y

**Aldehyde** (Y = H)
**Ketone** (Y = $C_xH_y$)

**Acetal** (Y = H)
**Ketal** (Y = $C_xH_y$)

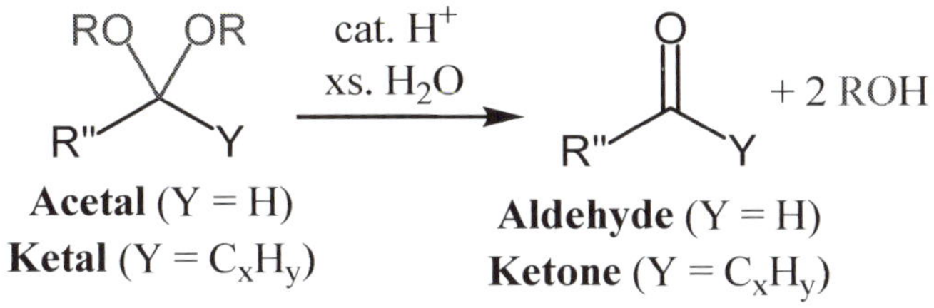

RO OR

cat. $H^+$
xs. $H_2O$

+ 2 ROH

R'' Y → R'' Y

**Acetal** (Y = H)
**Ketal** (Y = $C_xH_y$)

**Aldehyde** (Y = H)
**Ketone** (Y = $C_xH_y$)

*Lesson VI.6.4 Acetals and Ketals as Protecting Groups*

Because they can be formed reversibly, acetals and ketals are sometimes referred to as **protecting groups**. This is because whereas an aldehyde/ketone is reactive to nucleophilic attack under neutral conditions, an acetal/ketal is not because an acetal/ketal does not have a good leaving group. With those facts in mind, let us consider a case where a protecting group could be useful. Say we want to do the following reaction:

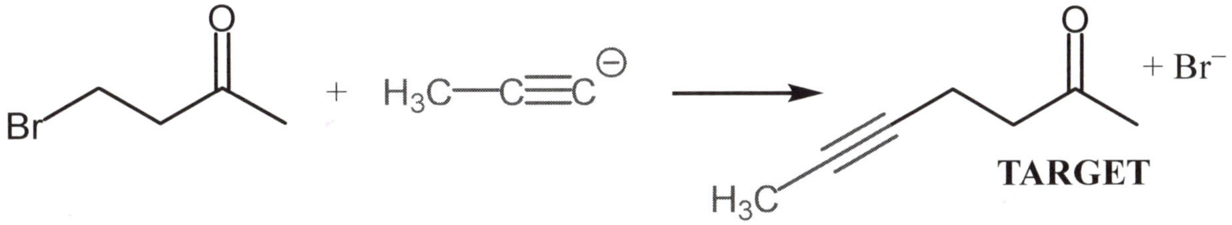

Br + $H_3C—C{\equiv}C^{\ominus}$ → + $Br^-$

$H_3C$ **TARGET**

The yield of this reaction would be low, because acetylide nucleophile could add either to the carbonyl C or displace the Br leaving group. A better approach would be as follows:

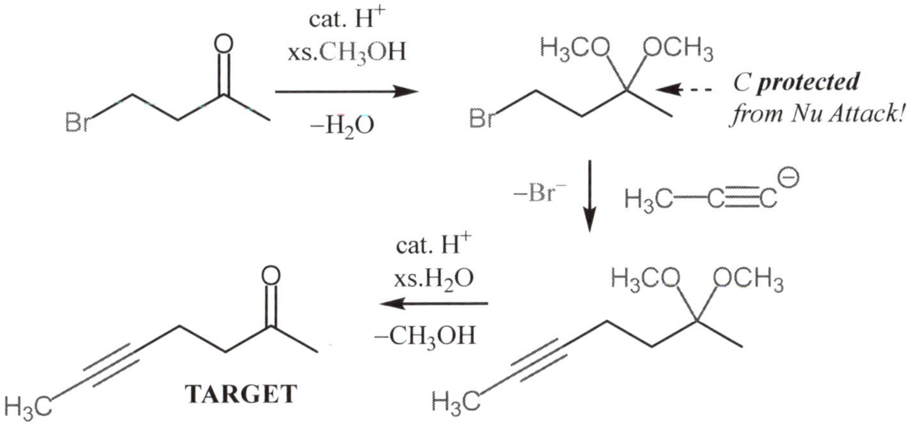

cat. $H^+$
xs.$CH_3OH$
$-H_2O$

Br → Br

$H_3CO$ $OCH_3$

*C protected from Nu Attack!*

$-Br^-$ ↓ $H_3C—C{\equiv}C^{\ominus}$

cat. $H^+$
xs.$H_2O$
$-CH_3OH$

$H_3CO$ $OCH_3$

$H_3C$ **TARGET** ← $H_3C$

By protecting one of the two sites of possible nucleophilic attack before adding the nucleophile, we now have a plan where each of the steps will give a very good yield of product on the way to our ultimate product.

Example VI.6.1

Provide the product of this reaction:

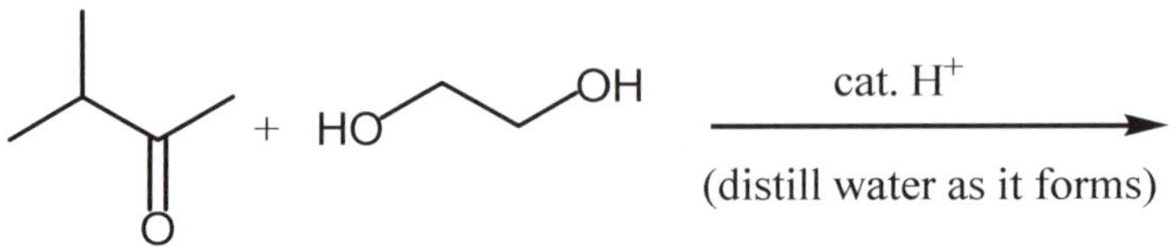

+ HO OH    cat. $H^+$    (distill water as it forms)

Solution VI.6.1

We identify this reaction as one between a ketone and an alcohol, with an acid catalyst provided. The product should be a ketal. The unique thing about this alcohol, though, is that the two O-nucleophiles that will end up added to the carbonyl C are also attached to each other by a bridging $-CH_2CH_2-$ unit:

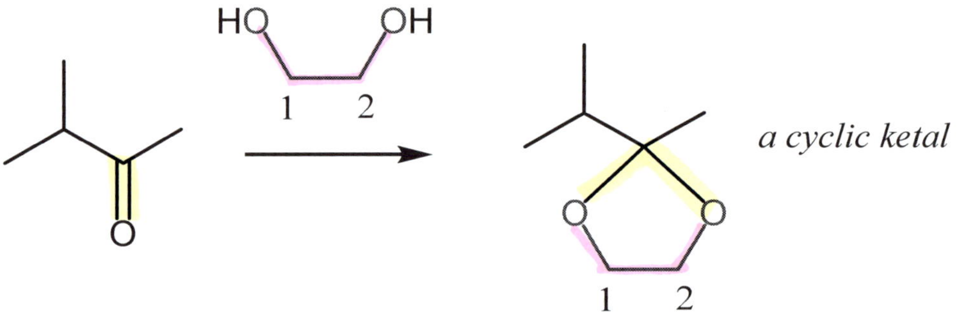

HO OH
1 2

1 2

*a cyclic ketal*

## Lesson VI.7. Addition of Amine Derivatives to Aldehydes/Ketones

*Lesson VI.7.1 Reactions of $RNH_2$ and $NH_3$ with Aldehydes and Ketones to form Imines*

Reacting an aldehyde and ketone with ammonia ($NH_3$) or a 1° amine will lead to replacement of the C=O of the reactant with a C=N bond in the product. This is therefore a Type D reaction (Lesson VI.3). A functional group with a C=N group is called an **imine** (sometimes called a Schiff base). The reaction takes place in acidic or basic media and the reaction is reversible, so the extent of formation of the product is determined by the structure of the amine and the aldehyde or ketone.

The general mechanism of this reaction is very similar to the acetal formation, with a difference that after the nucleophilic elimination step occurs the second hydrogen on the nitrogen is abstracted to yield the final imine product:

*protonation*

Y=H: Aldehdye
Y=R': Ketone

$N(R)H_2$

*Nu Addit.*

*deprotonation*

*protonation*

$-H_2O$
*Nu Elim.*

**Iminium Ion**

*deprotonation*

**Imine**
(R = alkyl, aryl or H)

Note that the imine can exhibit different configurational isomers about the C=N bond. The major product is generally the one in which the bulkiest groups are positioned farther apart from one another:

cat. $H^+$
(distill $H_2O$ away as it forms)

*minor product* *major product*

As with acetals/ketals, imines form reversibly, and so imines revert back to the amine/carbonyl starting materials upon reaction with excess water.

*Lesson VI.7.2 Reductive Amination: Forming an Amine from an Imine*

Imine formation can be combined with a reduction step using $H_2$/Pd, $LiAlH_4$, $NaBH_4$ or $NaBH_3(CN)$ as the reductant:

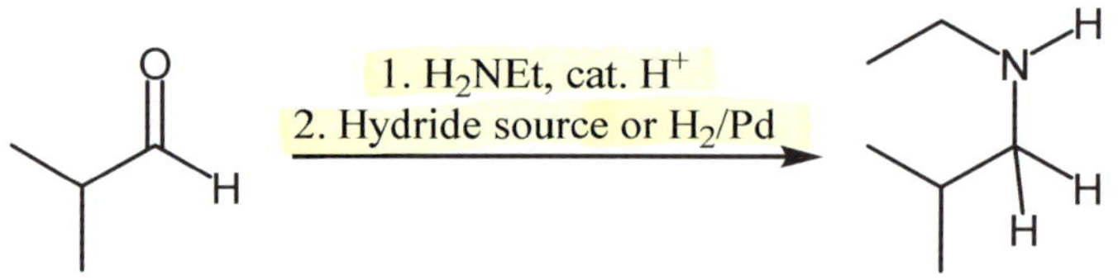

1. $H_2NEt$, cat. $H^+$
2. Hydride source or $H_2$/Pd

This to facilitate formation of a new amine via a reaction mechanism called **reductive amination**. So how does this reaction work mechanistically? Well, we just saw the mechanism for formation of the imine from the aldehyde and $H_2NEt$, and this portion of the reaction proceeds by essentially the same mechanism here as well. Once that imine is formed, however, a hydride nucleophile from the hydride source can undergo nucleophilic addition to the polar N=C bond in the same way that a nucleophile can add to a polar O=C bond to give the final amine:

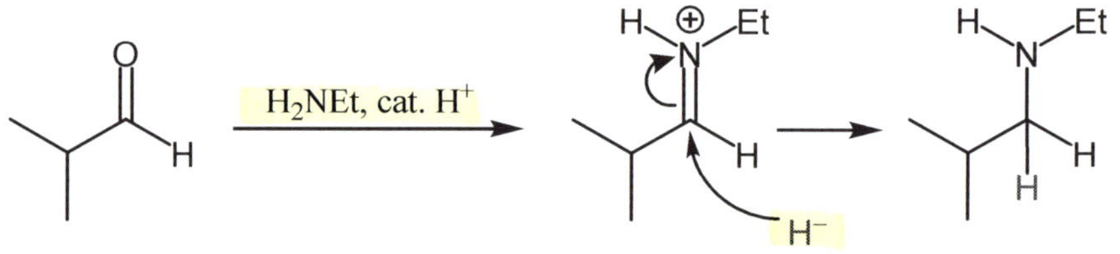

$H_2NEt$, cat. $H^+$

Note that if the least reactive of the hydride sources, NaBH3(CN), is used as the reductant, we can actually mix all of the components together in a single step rather than having to do the reaction in two steps. In such cases, the reductive amination could be written as:

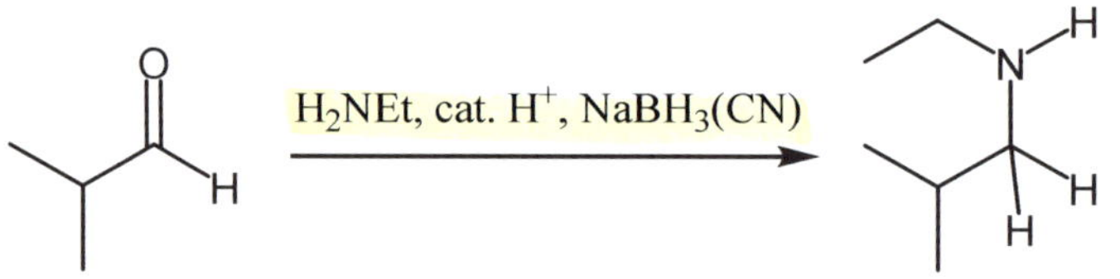

$H_2NEt$, cat. $H^+$, $NaBH_3(CN)$

*Lesson VI.7.3 Reactions of $R_2NH$ with Aldehydes and Ketones, Formation of Enamines*

Reacting an aldehyde or ketone with a secondary amine will give an **enamine** – an alkene that has an amine on the alkene C=C bond.

Mechanistically, enamine formation starts off similarly to the preparation of an imine, up to the point at with the iminium ion forms. When a secondary amine is used, however, the iminium intermediate cannot be deprotonated at the N, because the N has no H to lose. So, when base is added, it will deprotonate the most acidic site on the molecule, the C next to the C=N unit (called the $\alpha$-carbon). This leads to the neutral enamine:

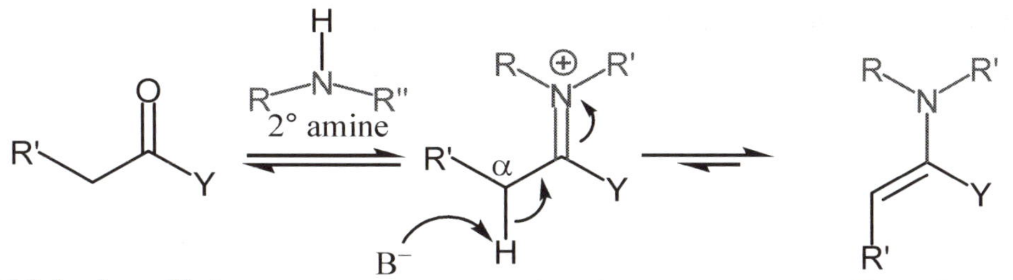

Aldehyde or Ketone ⇌ ($2°$ amine) Iminium Ion ⇌ Enamine

Example VI.7.1

Give the products of the following reactions

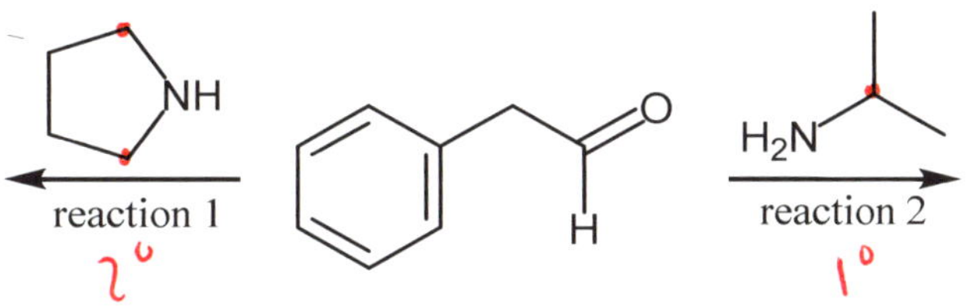

reaction 1 ← phenylacetaldehyde → reaction 2

Solution VI.7.1

Both reaction 1 and 2 are addition reactions of an amine to phenylacetaldehyde: reaction 1 employs a 2° amine, whereas reaction 2 employs a 1° amine. We know that a 2° amine will convert a carbonyl group into an enamine. The enamine reaction requires deprotonation of an α site (labeled below). We also know that a 1° amine will convert a carbonyl group into an imine. In both cases, it is important to indicate as the major product the isomer that is least sterically-encumbered about the newly-formed double bond:

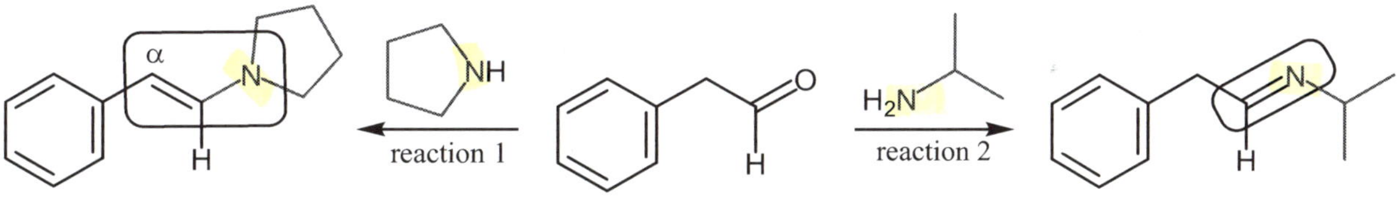

Enamine ← reaction 1 — reaction 2 → Imine

*Lesson VI.7.4 Reductions of Aldehydes and Ketones to Alkanes*

The carbonyl group of an aldehyde or ketone can be fully reduced to an alkane via two reactions: (1) Wolff–Kishner Reduction and (2) Clemmensen Reduction.

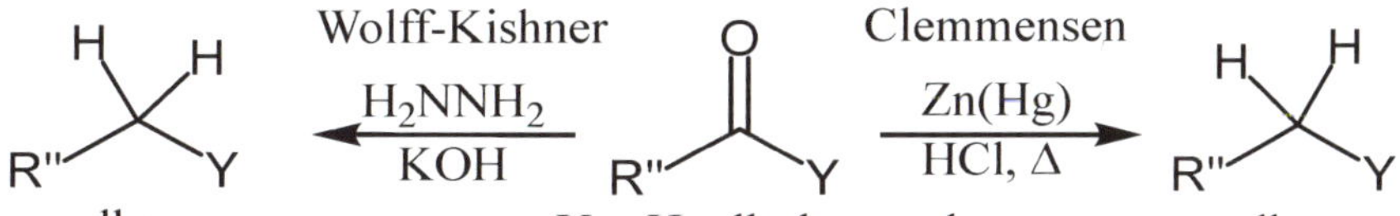

alkane ← Wolff-Kishner ($H_2NNH_2$, KOH) — Clemmensen (Zn(Hg), HCl, Δ) → alkane

Y = H, alkyl or aryl

In the **Wolff-Kishner Reduction**, the carbonyl group of an aldehyde or ketone reacts with hydrazine ($H_2NNH_2$) to give hydrazone via a Type D reaction mechanism analogous to the mechanism of imine formation shown earlier in this lesson:

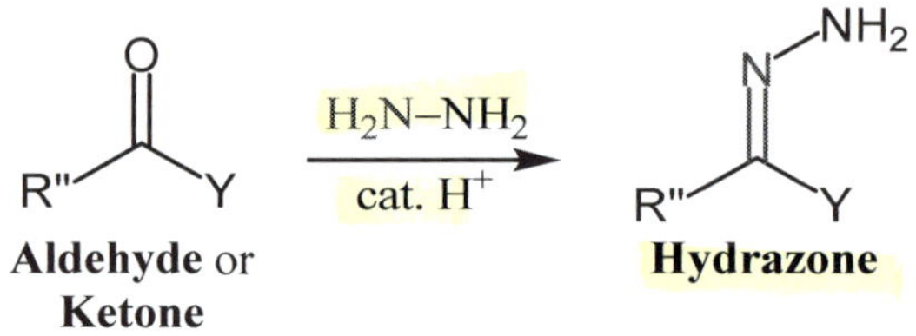

R''C(=O)Y —($H_2N$–$NH_2$, cat. $H^+$)→ R''C(=N–$NH_2$)Y

**Aldehyde** or **Ketone** → **Hydrazone**

The second mechanistic part of the Wolff-Kishner Reduction involves treating the hydrazone with aqueous base will promote the formation of the alkane and nitrogen gas as byproduct, a process driven forward by the very thermodynamically favorable release of $N_2$ gas:

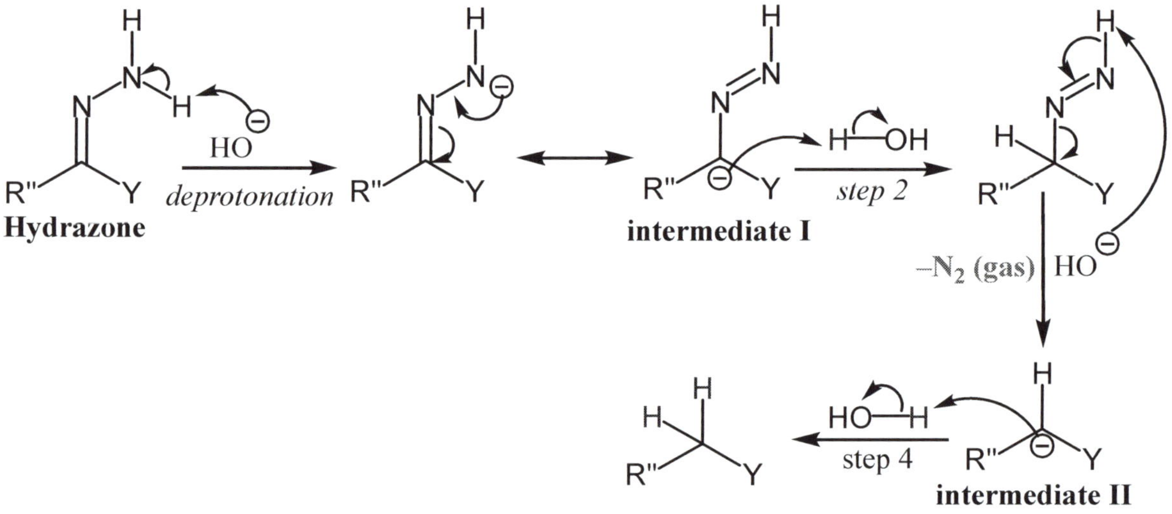

**Hydrazone** —(HO⁻, *deprotonation*)→ ⟷ **intermediate I** —(H–OH, *step 2*)→ —($-N_2$ (gas), HO⁻)→ **intermediate II** —(HO–H, step 4)→ R''CH₂Y

The Clemmensen Reduction of an aldehyde/ketone will lead to the same alkane product that is produced by the Wolff-Kishner Reduction, but the mechanism of the Clemmensen reduction involves removal of the O by the Zn. This is a mechanism better covered in an inorganic chemistry text, and so is not covered here.

## Lesson VI.8. The Wittig Reaction

*Lesson VI.8.1 Reactions of Phosphonium Ylides with Aldehydes and Ketones*

The **Wittig reaction** is a Type D reaction (Lesson VI.3) in which the =O of a C=O group is replaced with $=CR_2$, to give an alkene. The reagents used in this reaction, known as Wittig reagents, are typically **triphenylphosphonium ylides** and they have the structure of $Ph_3P{=}C(R)R'$ where R is mostly H (but can be alkyl or aryl) and R' can be H or alky or aryl:

**Aldehyde** or **Ketone** — **Phosphonium Ylide** ($R_2\overset{\ominus}{C}-\overset{\oplus}{P}Ph_3$) → **Alkene** + $O{=}Ph_3P$

Phosphonium ylides are prepared by deprotonation of an alkyl triphenylphosphonium halide with *n*BuLi or some other very strong base (e.g., $NaNH_2$). The alkyltriphenylphosphonium halide starting material can be synthesized via the $S_N2$ reaction of alkyl halide with triphenylphosphine. The phosphonium salt can then be deprotonated by a basic reagent like *n*BuLi to give the phosphonium Ylide (often simply called the Wittig Reagent):

$Ph_3P$ + H–C(X)(R)(R') —$S_N2$→ **Phosphonium salt** ($Ph_3\overset{\oplus}{P}$–CH(R)R' $\overset{\ominus}{X}$)

*deprotonation* / *n*BuLi ↓

**Phosphonium Ylide**: [$Ph_3P{=}C(R)R'$ ⟷ $Ph_3\overset{\oplus}{P}-\overset{\ominus}{C}(R)R'$]

The mechanism for the Wittig reaction is presented in the figure below. The reaction begins when the ylide undergoes [2+2]-cycloaddition with the C=O bond to form an intermediate called an oxaphosphetane. The intermediate then collapses in a [2+2]-cycloreversion (i.e., the reverse of a [2+2]-

cycloaddition) to release triphenylphosphine oxide and the alkene product. A key driving force for this step, in addition to relieving the ring strain, is the formation of the very strong P–O bond.

[2+2] cycloaddition

Oxaphosphetane

[2+2] cycloreversion

Alkene

$Ph_3P{=}O$

For alkene products in which multiple stereoisomers are possible (e.g., *cis*/*trans* or *Z*/*E*), **the more sterically encumbered isomer will generally be favored as the major product**. This is initially counterintuitive, but the more sterically encumbered alkene actually comes from the *less* sterically-encumbered and thus more stable oxaphosphetane intermediate. The phosphorus atom with the three large phenyl rings is much larger than any typical R groups on the carbon atoms in the ring, so placing the $PPh_3$ *cis*- to the smallest R group is the most important for minimizing ring strain. As a result, the R groups tend to end up *cis*- to one another. In the example below, the oxaphosphetane is drawn as a puckered ring like we saw when we learned about the typical shapes for different ring sizes in Lesson I.16 to better show the real shape of the molecule. We can see that the very voluminous $PPh_3$ unit being on the same side of the 4-membered ring as the small H in the oxaphosphetane is the most favorable conformation, but this conformation places the two R groups *cis* to one another. We can see then that the most stable conformer of the oxaphosphetane thus leads to the *Z*-alkene as the major product!

Oxaphosphetane

[2+2] cycloreversion

***Z*-Alkene**

$+ Ph_3P{=}O$

Example VI.8.1

Give the products of the following reaction:

Solution VI.8.1

We first recognize the unique ylide nucleophile and ascertain that this is a Wittig reaction. We know that the Wittig reaction replaces the carbonyl O with a double bond to make an alkene. We also know that alkenes can have different configurations (*E*- or *Z*-). So, we have two possible stereoisomers for the product:

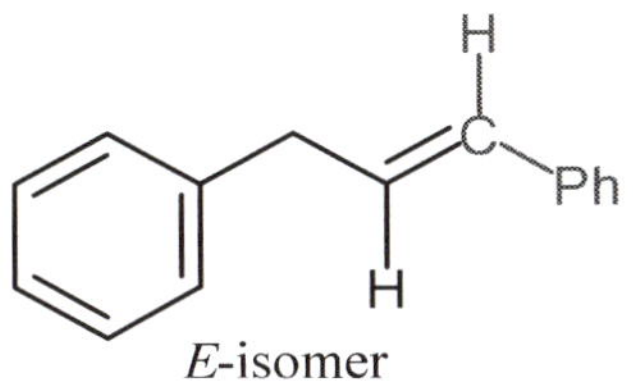

*E*-isomer
(minor product)

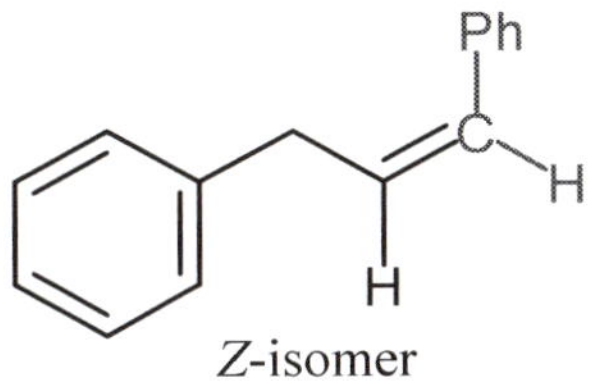

*Z*-isomer
(major product)

We predict that the *Z*-isomer will be the major product because it is derived from the most stable oxaphosphetane intermediate.

Z-isomer = same side

E-isomer = opposite side

## Lesson VI.9. Nucleophilic Acyl Substitution of Acid Chlorides and Anhydrides

*Lesson VI.9.1 Introduction to Nucleophilic Acyl Substitution*

Unlike aldehydes and ketones, carboxylic acids and their derivatives (anhydrides, acid chlorides, amides and esters) undergo **nucleophilic acyl substitution**, sometimes abbreviated **$S_NAc$**. Nucleophilic acyl substitution is a two-step process:

1) Nucleophilic addition to the carbonyl carbon

2) Nucleophilic elimination of the best leaving group from what was the carbonyl carbon.

For the purposes of categorizing reactions as a way to help study them in this particular book, we will refer to the reactions that follow this pattern as being **Type B**.

***Carbonyl Reaction Type B*** *NOT for aldehydes or ketones*



**The net result is replacing the bond to Y with a new bond to a nucleophile component**. The identity of Y, HY and "other products" in the figure above depend on the specific reaction and the starting functional group. The reason that ketones/aldehydes do not do nucleophilic acyl substitution it because in a ketone/aldehyde, the "Y" group is always $C_xH_y$ or H. These are terrible leaving groups, and cannot even be protonated to become good leaving groups. Without any good leaving group, it is impossible for aldehydes or ketones to lead to a favorable nucleophilic elimination step. All of the other carbonyl functional groups we have introduced can undergo nucleophilic acyl substitution reactions, as we will see in the next group of lessons.

*Lesson VI.9.2 Nucleophilic Acyl Substitution of Acid Chlorides.*

As we saw in Lesson VI.4, acid chlorides are the most reactive carboxylic acid derivatives. The Cl of the acid chloride is an outstanding leaving group, so acid chlorides very readily undergo $S_NAc$ reactions. Acid chlorides can be converted to anhydrides, esters, amides, and carboxylic acids through the general nucleophilic acyl substitution mechanism. If a neutral nucleophile is used it will need to be deprotonated to give a neutral product in the end, whereas no deprotonation is needed if you use a nucleophile that is already anionic:

Anionic nucleophile (e.g. anions produced from NaOC(O)R, NaOH, NaOR, etc.)

This is a nucleophilic acyl substitution (Type B), where step 1 is nucleophilic addition and step 2 is nucleophilic elimination:

**General Mechanism:**

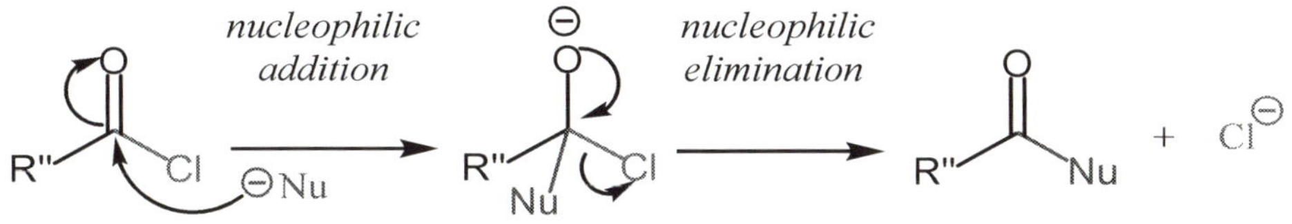

**Specific Examples:**

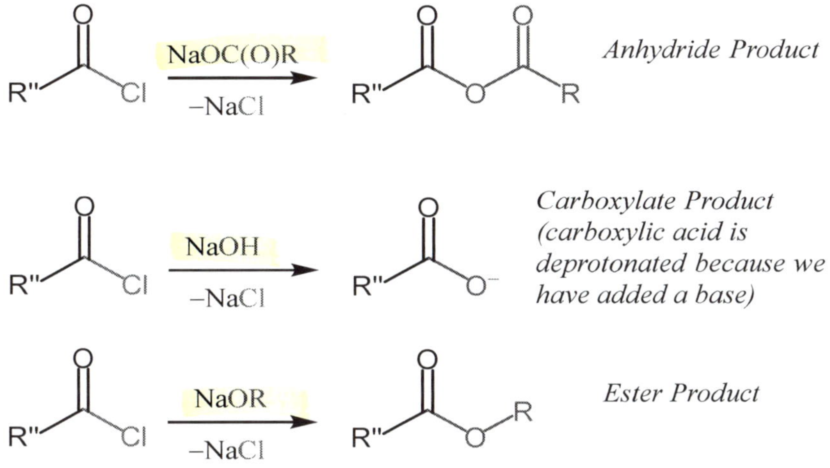

Neutral nucleophile (e.g. ROH, $HNR_2$, $RNH_2$, $NH_3$ or $H_2O$)

This mechanism generally requires the presence of base catalyst. If the NuH is itself a base (NuH = $HNR_2$, $RNH_2$ or $NH_3$), however, then we just use two equivalents of NuH (one to be the nucleophile, one to be the base). This mechanism consists of nucleophilic addition first as usual, but a deprotonation step is required before nucleophilic elimination can occur:

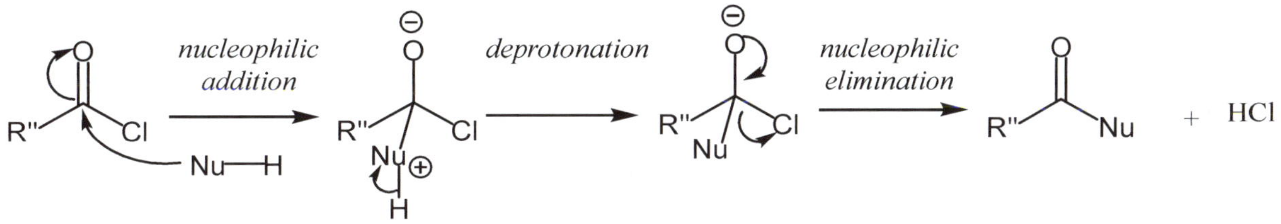

Some specific examples of $S_NAC$ of acid chlorides with neutral nucleophiles are provided here:

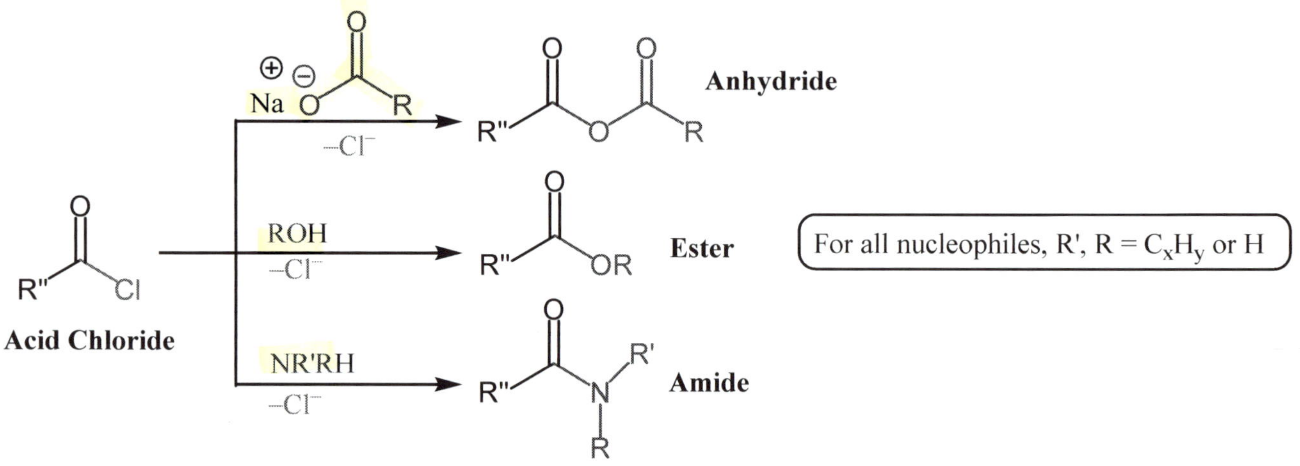

*Lesson VI.9.3 Nucleophilic Acyl Substitution of Anhydrides*

The second-most reactive carboxylic acid derivatives are acid anhydrides, and they can be converted to esters, amides and carboxylic acids. Anhydrides, however, cannot be converted to acid chlorides, since $Cl^-$ is a better leaving group than $RC(O)O^-$. The general mechanism for the nucleophilic acyl substitution of anhydrides follows that we saw for acid chlorides with a neutral nucleophile:

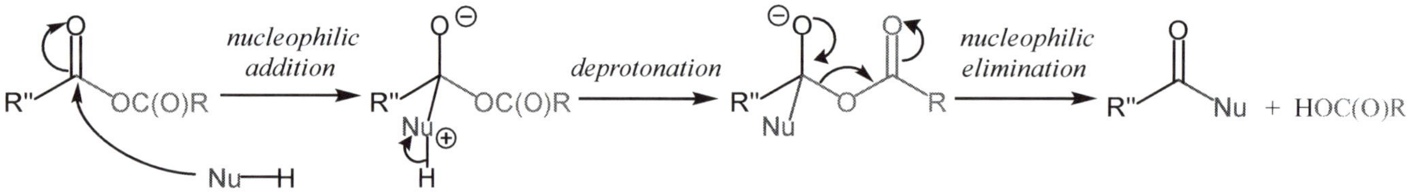

In these cases, Nu–H can be $H_2O$, ROH, a 1° amine, or a 2° amine. The following chart shows the nucleophilic acyl substitution reactions of anhydrides that give rise to carboxylic acids, esters, and amides.

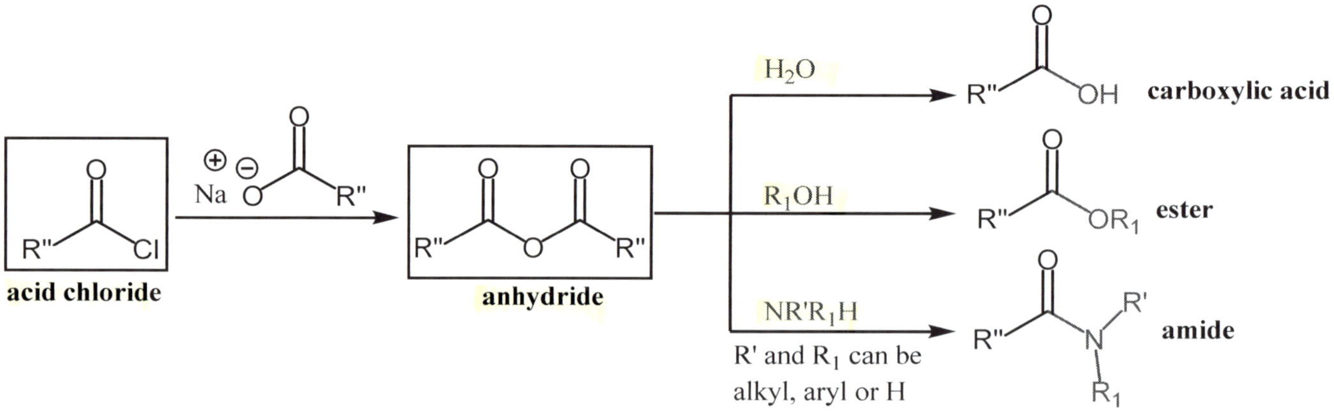

Example VI.9.1

Provide the missing reactant for the transformation shown. Would the proposed product be the major or minor product upon reaction of this acid anhydride with the missing reactant?

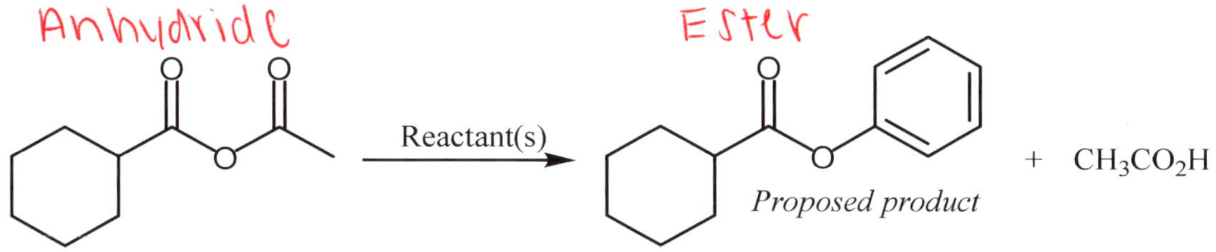

Anhydride     Ester

Reactant(s) → *Proposed product* + $CH_3CO_2H$

Solution VI.9.1

The above reaction is a nucleophilic acyl substitution where an anhydride is converted to an ester. If we circle the nucleophile in the reaction (as shown below) we can see it is a phenoxide (Ph—O$^-$), so the missing reactant (NuH) should be PhOH with catalytic base.

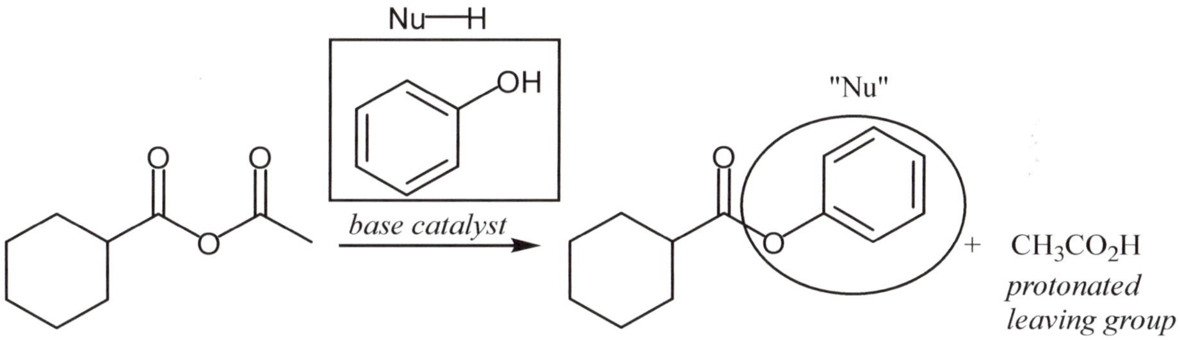

Nu—H

OH

*base catalyst*

"Nu"

+ $CH_3CO_2H$ *protonated leaving group*

In this particular example, the two sides of the anhydride are different. One of the anhydride carbonyl carbons has a methyl substituent, whereas the other has a cyclohexyl substituent. Generally, we would predict that the nucleophile would attack the less hindered side. We would predict that the major product would result from nucleophilic attack on the acetate side and that the proposed product in the question would actually be the minor product:

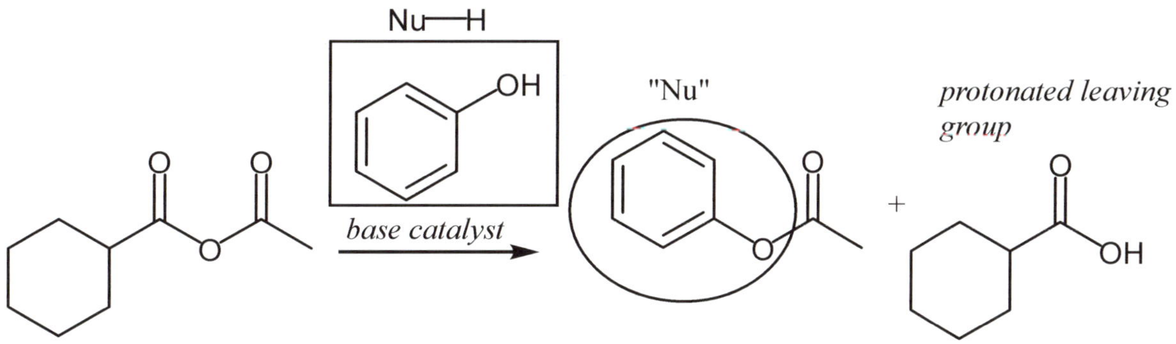

Nu—H

OH

*base catalyst*

"Nu"

*protonated leaving group*

+

OH

## Lesson VI.10. $S_NAc$ of Carboxylic Acids to form Acid Chlorides

Carboxylic acids react with phosphorus trichloride ($PCl_3$) or thionyl chloride ($SOCl_2$) to give acid chlorides. In these cases, the –OH is substituted for a –Cl:

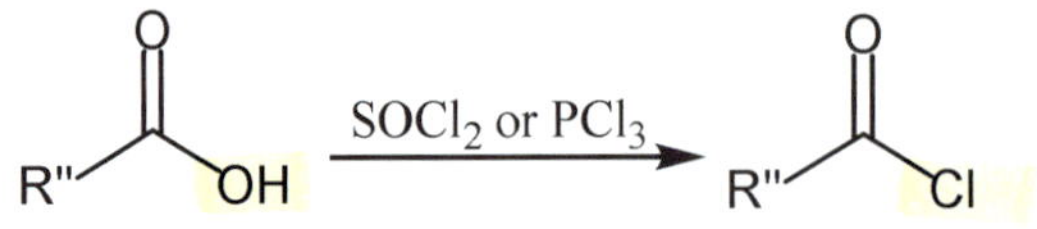

$SOCl_2$ or $PCl_3$

This nucleophilic acyl substitution is initiated by the reaction of the carbonyl oxygen with the sulfur of $SOCl_2$ or the phosphorus of the $PCl_3$, generating the chloride ion that will act as a nucleophile:

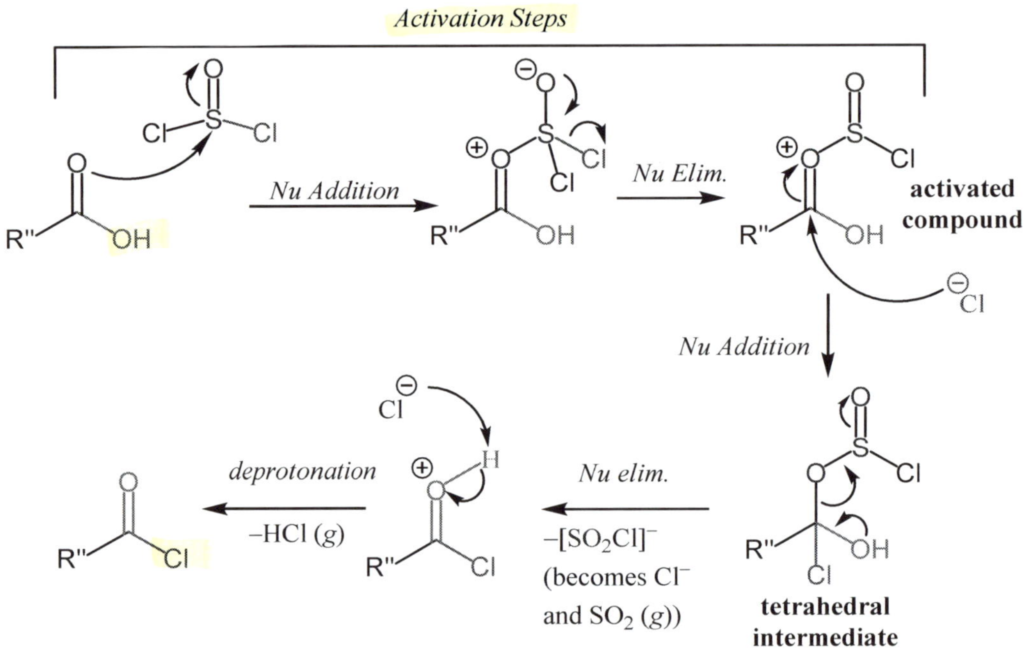

*Activation Steps*

*Nu Addition*

*Nu Elim.*

**activated compound**

*Nu Addition*

**tetrahedral intermediate**

*Nu elim.*

$-[SO_2Cl]^-$ (becomes $Cl^-$ and $SO_2$ (*g*))

*deprotonation*

$-HCl$ (*g*)

After initial activation, the usual steps of the $S_NAc$ (Type B) reaction occur. The production of sulfur dioxide and HCl as side products makes the purification of this reaction rather easy, because both of these side products are gases that bubble out of the reaction vessel and can be collected. Sometimes pyridine is added to neutralize the HCl as it forms instead of letting it escape as a gas.

The reactions of carboxylic acids with $PCl_3$ proceed through a very similar mechanism to that observed for reaction with $SOCl_2$. The carbonyl O again acts as a nucleophile to activate it for nucleophilic acyl substitution:

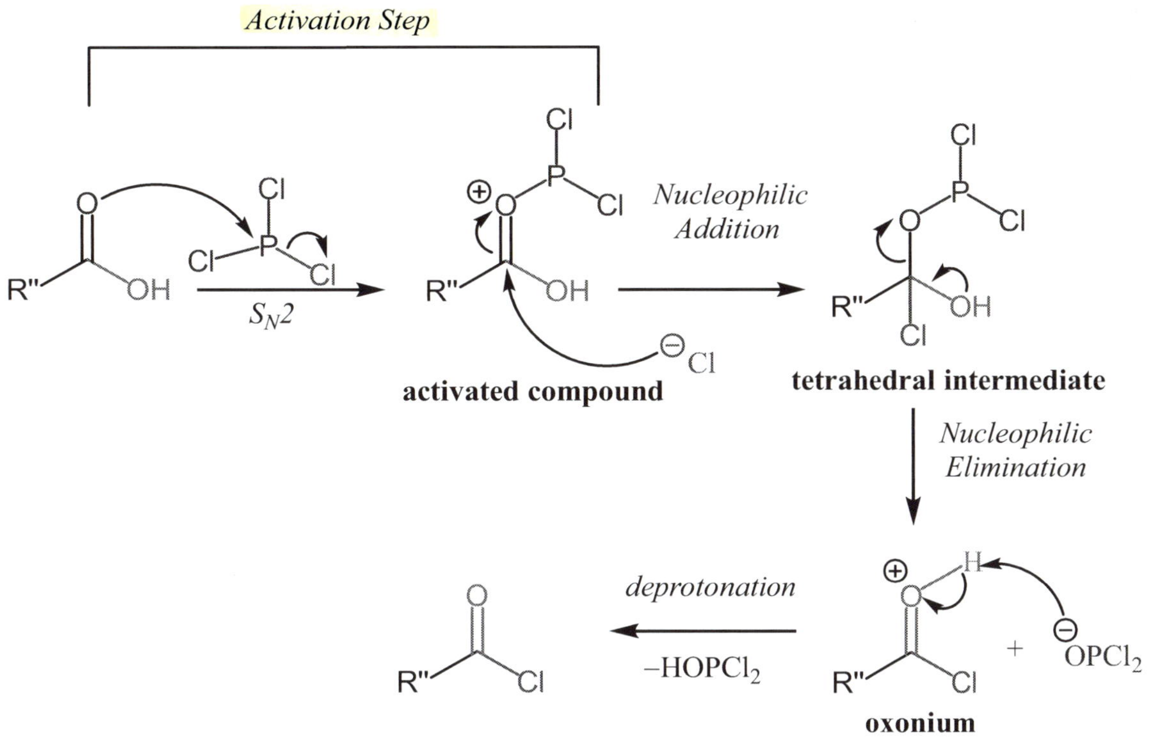

*Activation Step*

Nucleophilic Addition

activated compound

tetrahedral intermediate

Nucleophilic Elimination

deprotonation

$-HOPCl_2$

oxonium

In this case, the carbonyl O undergoes an $S_N2$ reaction with $PCl_3$ in the activation step, and then the nucleophilic acyl substitution takes place. Step 4 is needed to deprotonate the oxonium to yield a neutral product. A key difference from a practical standpoint is that the $HOPCl_2$ produced in step 4 is not a gas like the $HCl/SO_2$ byproducts of reaction with $SOCl_2$, which makes the purification steps harder.

---

Example VI.10.1

Give the final product of the following two-step sequence:

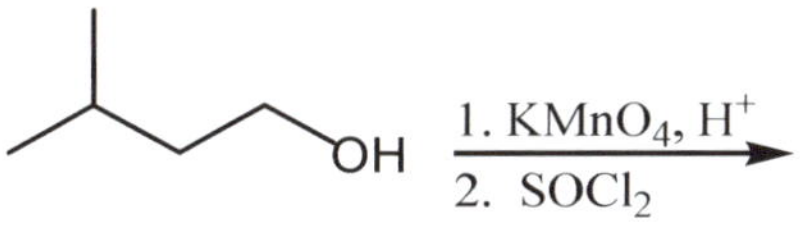

1. $KMnO_4$, $H^+$
2. $SOCl_2$

Solution VI.10.1

The first reaction is an oxidation reaction with a strong oxidant ($KMnO_4$), which will produce a carboxylic acid from the given primary alcohol. The second reaction is a simple nucleophilic acyl substitution where the –OH will be substituted with –Cl

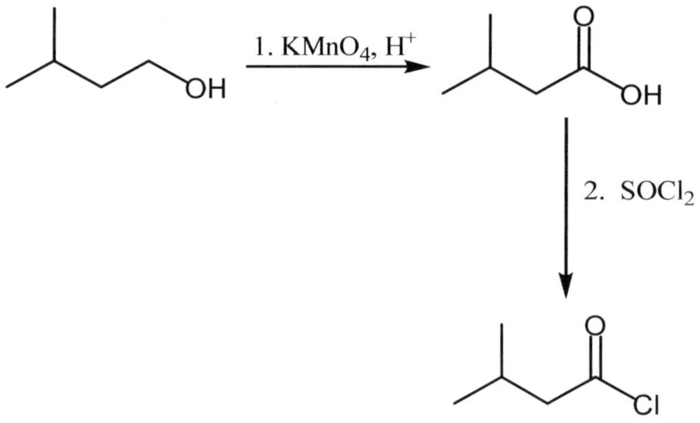

OH

1. $KMnO_4$, $H^+$

O

OH

2. $SOCl_2$

O

Cl

## Lesson VI.11. SNAc Reaction of Oxygen Nucleophiles with Carboxylic Acids and Esters

*Lesson VI.11.1 Nucleophilic Acyl Substitution of Carboxylic Acids to form Esters*

The new reactions in this Lesson are SNAc (or Type B as described in Lesson VI.3) reactions in which we replace the leaving group with a group supplied by the nucleophile.

Carboxylic acids react with alcohols in the presence of an acid catalyst (such as $H_2SO_4$) and heat to produce esters (and water as a byproduct), this reaction is known as **Fischer Esterification**. The general reaction is the following:

ROH
heat, acid catalyst
R''COOH ⇌ R''COOR + $H_2O$

The following figure shows a reasonable arrow-pushing mechanism for Fischer esterification:

*protonation*
*nucleophilic addition*
ROH
*deprotonation*
:B
*protonation*
$H^{\oplus}$
**tetrahedral intermdiate**
*nucleophilic elimination*
$-H_2O$
:B
*deprotonation*
**ester**

Although this mechanism looks complicated, you will see that the key steps are nucleophilic addition of the nucleophile and nucleophilic elimination, the usual nucleophilic acyl substitution steps. All of the other extra steps are just moving protons on and off as necessary.

Note that **each step of the reaction is reversible**. In fact, an ester can react with water and hydrolyze to give back the carboxylic acid, a useful reaction called **ester hydrolysis:**

ROH
heat, acid catalyst
R''COOH ⇌ R''COOR + $H_2O$

To ensure the production of the ester in high yields we will apply our knowledge of equilibrium processes in terms of LeChatelier's principle. Specifically, we can use a large excess of the alcohol reagent so that its consumption pushes the equilibrium towards ester formation. Alternatively, we can distill the water away as it forms, again to push the reaction towards ester formation.

Example VI.11.1

Isoamyl acetate (structure below) is an ester found in banana oil and is responsible for the familiar banana aroma. Give the carboxylic acid and alcohol used in the formation of this ester.

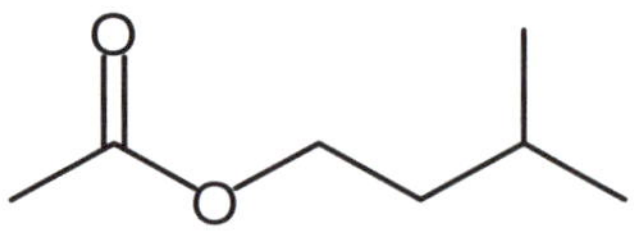

Solution VI.11.1

In order to find the reactants of this ester we put the general reaction and try to find both R" and R of the carboxylic acid and alcohol respectively. As the figure shows the carboxylic acid used is the acetic acid and isoamyl alcohol (3-methyl-1-butanol) is the alcohol used.

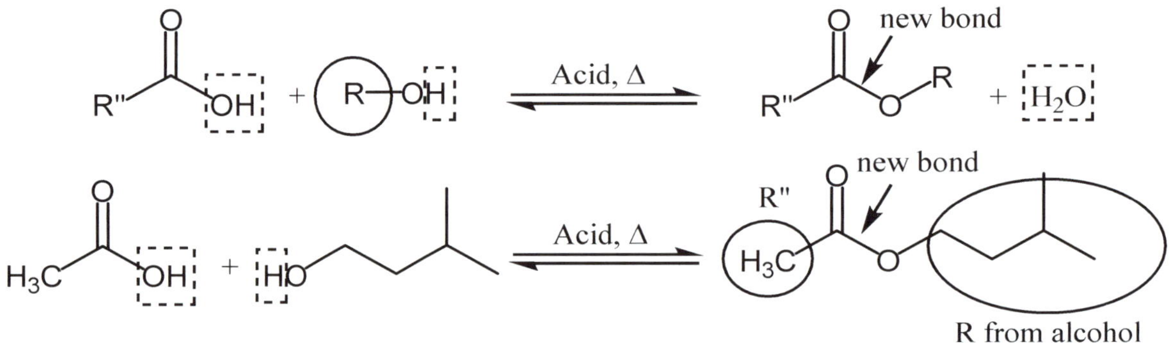

R"C(=O)OH + R–OH ⇌ (Acid, Δ) R"C(=O)O–R + $H_2O$ (new bond)

$H_3C$C(=O)OH + HO–(isoamyl) ⇌ (Acid, Δ) $H_3C$C(=O)O–(isoamyl) (new bond; R"; R from alcohol)

Example VI.11.2

Provide the missing reactants and products:

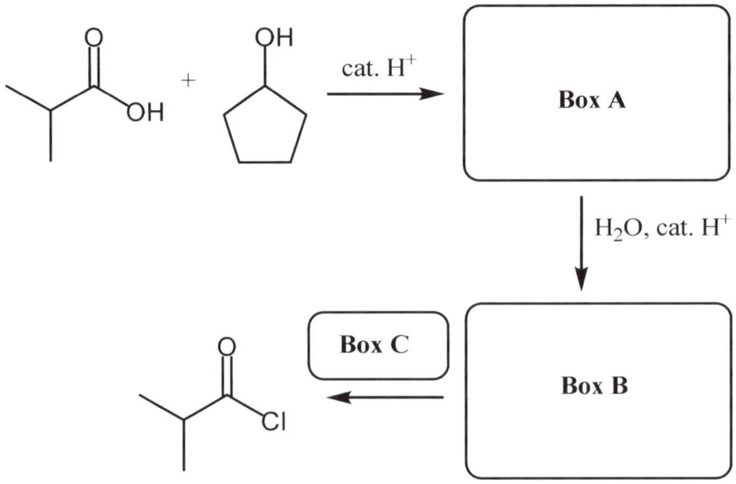

(isobutyric acid) + (cyclopentanol) → cat. $H^+$ → Box A

Box A → $H_2O$, cat. $H^+$ → Box B

Box B → Box C → (isobutyryl chloride, Cl)

Solution VI.11.2

The product in Box A will be an ester, formed via Fischer Esterification. Reaction of the Ester with water/acid catalyst takes us back to the carboxylic acid in Box B. The missing reagents in Box C must facilitate conversion of the carboxylic acid into an acid chloride. Either $SOCl_2$ or $PCl_3$ will accomplish this. The completed scheme would be:



*Lesson VI.11.2 Hydrolysis and Transesterification*

Esters cannot be converted to acid chlorides or anhydrides because this would require nucleophilic elimination of $RO^-$, which is less stable (a worse leaving group) than $Cl^-$ or $RC(O)O^-$. We have just seen that esters can, however, undergo acid-catalyzed hydrolysis (the reverse of Fischer Esterification). Esters can likewise be converted to other esters through **transesterification reactions**.

Like hydrolysis, the conversion of an ester into a different ester is a reversible reaction, which means that driving the equilibrium to favors the products can be achieved by applying LeChatelier's principle (e.g., adding more reactants, removing the products, etc.). Transesterification reactions follow two general mechanisms depending whether they are **base-catalyzed** or **acid-catalyzed** reactions.

Base-catalyzed transesterification/hydrolysis:

In this reaction, the base first deprotonates the nucleophile (not shown below). A standard $S_NAc$ reaction follows:

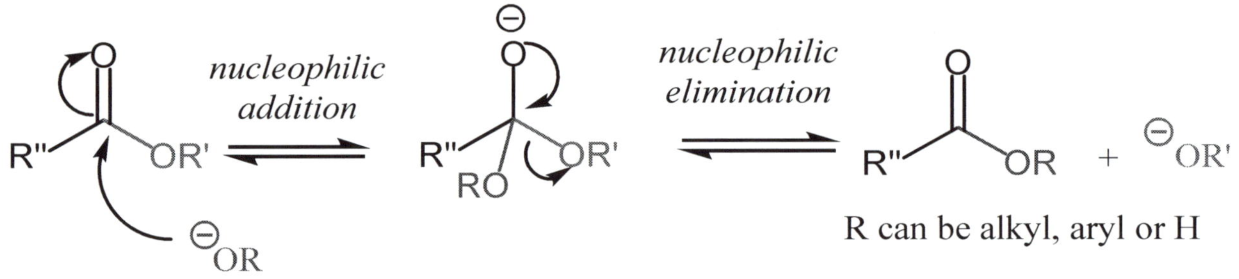

Note that if the initially-formed product is a carboxylic acid, it will be deprotonated in the basic reaction solution, so that a **carboxylate is the isolated product** when R = H:

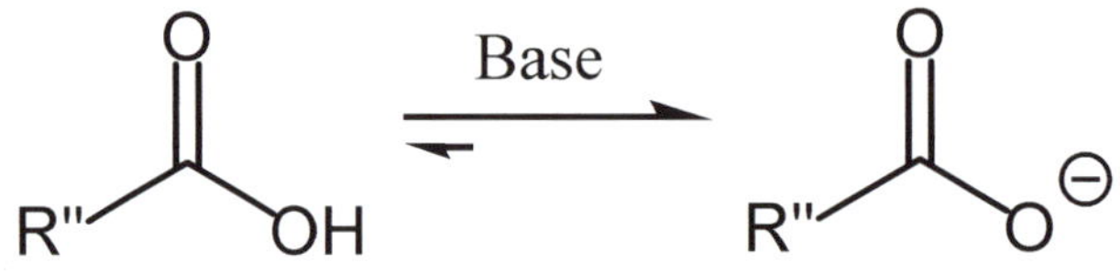

With the strong bases typically used in these reactions, the equilibrium lies overwhelmingly to the right, producing the carboxylate as the major product. One specific base-catalyzed ester hydrolysis reaction is called **saponification**. Saponification is the reaction of triglycerides to form fatty acids:

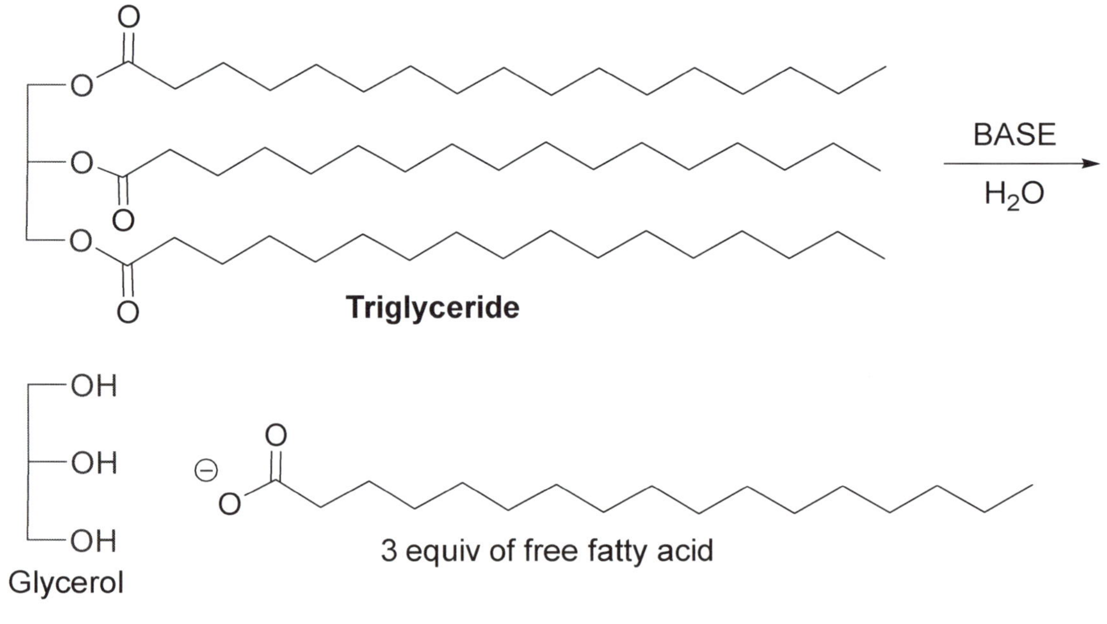

Acid-catalyzed transesterification/hydrolysis:

This reaction mechanism is essentially the same as that of Fischer Esterification, but starting with an ester instead of a carboxylic acid. The mechanism is shown below. Importantly, each step of the acid-catalyzed transesterification or acid-catalyzed hydrolysis is reversible. In order to drive the reaction in the forward direction, an excess of the nucleophile (ROH) is generally used:

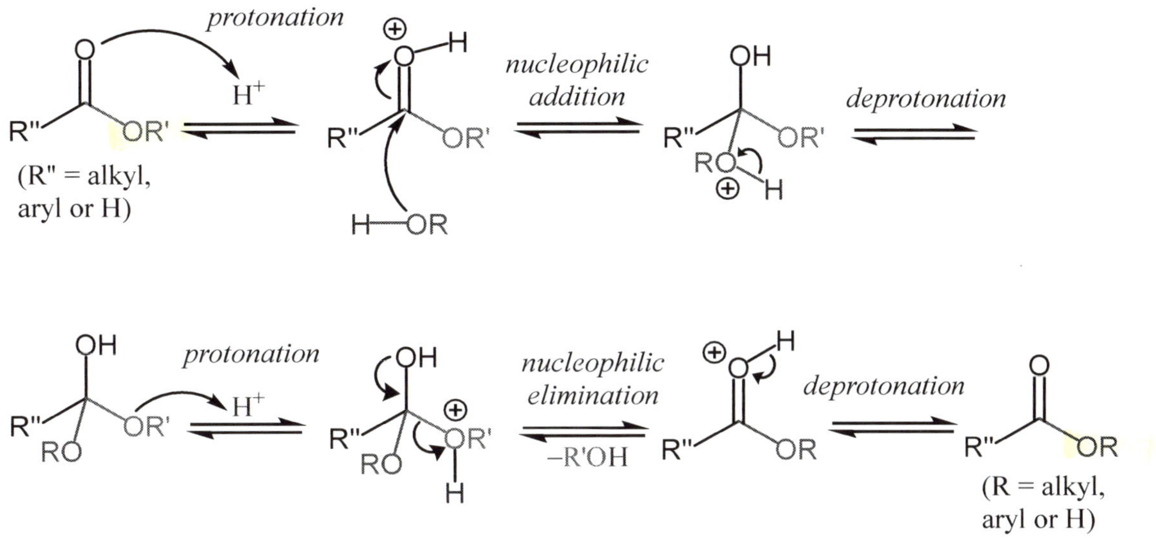

protonation

nucleophilic addition

deprotonation

(R" = alkyl, aryl or H)

protonation

nucleophilic elimination

−R'OH

deprotonation

(R = alkyl, aryl or H)

## Lesson VI.12. Amide Formation, Amide Hydrolysis and the Gabriel Synthesis

### *Lesson VI.12.1 Ammonolysis and Amidation*

Esters can be converted to amides upon reaction with 1° amines, 2° amines, or $NH_3$ under base- or acid-catalyzed conditions via mechanisms similar to the transesterification mechanisms we saw in Lesson VI.11

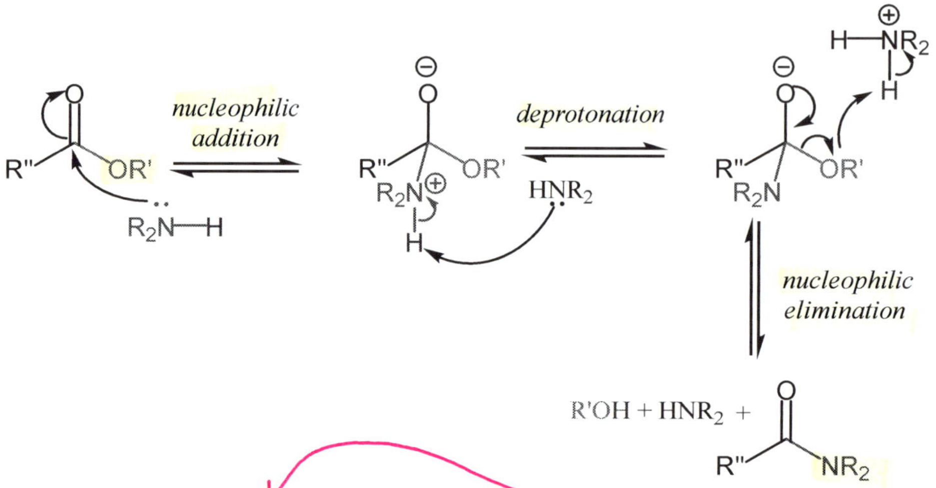

*nucleophilic addition*

*deprotonation*

*nucleophilic elimination*

R'OH + $HNR_2$ +

When ammonia is used as the nucleophile, this reaction is called **ammonolysis**. When a 1° or 2° amine is used, this reaction is called **amidation**. These are all $S_NAc$ (Type B) reactions.

Example VI.12.1

Give the missing reactant in this nucleophilic acyl substitution

Reactant

+ $CH_3OH$

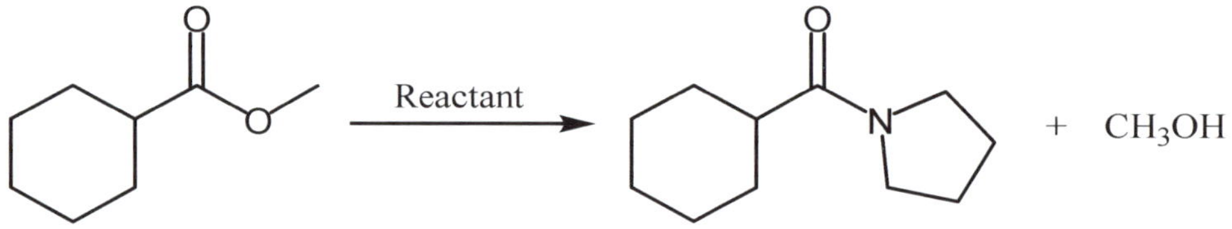

Solution VI.12.1

The above reaction is a nucleophilic acyl substitution where an ester is converted to an amide. If we circle the nucleophile in the reaction (as shown below) we can see it is a pyrrolidine anion so the missing reactant is NuH (amine)

Reactant

Nu

+ $CH_3OH$
leaving group

*VI.12.2 Amide Hydrolysis is a $S_NAc$ Reaction*

Amides are the least reactive of the carboxylic acid derivatives, and they can only undergo hydrolysis reactions to give carboxylic acids under harsh reaction conditions. Hydrolysis of amides can take place either in very strongly acidic conditions (e.g. addition of aqueous HI) or very strong basic conditions (e.g. addition of concentrated aqueous NaOH). Furthermore, the reactions require heating to higher temperatures than are used for hydrolysis of esters.

The general mechanism in either acidic or basic media is similar to the corresponding mechanism for ester hydrolysis (Lesson VI.11), with one exception: in an acidic medium, the amine ($HNR_2$) released by nucleophilic elimination is a sufficiently strong base that it will deprotonate the cationic oxygen in step 6, to afford the ammonium $[H_2NR_2]^+$.

$H^+$

*protonation*

*nucleophilic addition*

*deprotonation*

(R is alkyl, or aryl or H)

*protonation*

*Nu elim.*

$-NR_2H$

$NR_2H$

*deprotonation*

+ $\overset{\oplus}{N}R_2H_2$

### *VI.12.3 Gabriel Synthesis*

One very useful synthetic application that involves amide hydrolysis is the **Gabriel synthesis**. The Gabriel synthesis is an excellent way to make primary amines. If you try to make a primary amine by simple $S_N2$ reaction of a primary alkyl halide with ammonia, you will get some of the primary amine, but a variety of side products will also form:

Br
$NH_3$
$NH_2$
H
N
N
N
⊕

The secondary and tertiary amines, in addition to the tetraethylammonium salt, can all form as well as the targeted primary amine because the initially-formed amines can add to the ethylbromide. The Gabriel synthesis uses a N that is between two carbonyl carbons, in a molecule called **phthalimide**, as a sort of protected N to which only one R group can add:

1. NaOH
2. Primary RX
3. $H_3O^+$, $H_2O$
OH
OH
+ $[H_3NR]^+$
base
+ $H_2NR$
**Primary Amine**
**Phthalimide**

Note that if a base is added as a workup step the ammonium salt is converted to the target primary amine and the carboxylic acid is deprotonated to give the dicarboxylate. This makes separation easy, because the carboxylate is soluble in water, whereas the amine can be extracted into an organic solvent. The mechanism of steps 1 and 2 in the above scheme are a simple deprotonation to make the N into a good nucleophile, followed by an $S_N2$ reaction. These two steps are illustrated here for reaction with ethyl bromide:

$^-OH$
$-H_2O$
$-Br^-$
Br
**Phthalimide**
***N*-Ethyl Phthalimide**

The *N*-ethylphthalimide then undergoes amide hydrolysis to give the ethylammonium salt and the carboxylic acid.

Example VI.12.2

Provide a reasonable synthetic route to prepare this target amine:

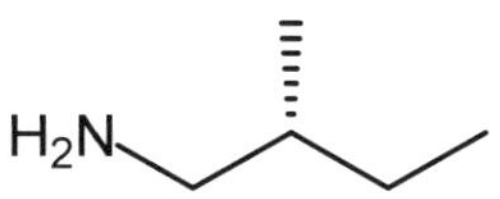

$H_2N$

Solution VI.12.2

The target is a primary amine, so the Gabriel synthesis is a good choice for its synthesis. We will need phthalimide and the appropriate alkyl halide. In this case, we need (*R*)-1-bromo-2-methylbutane:

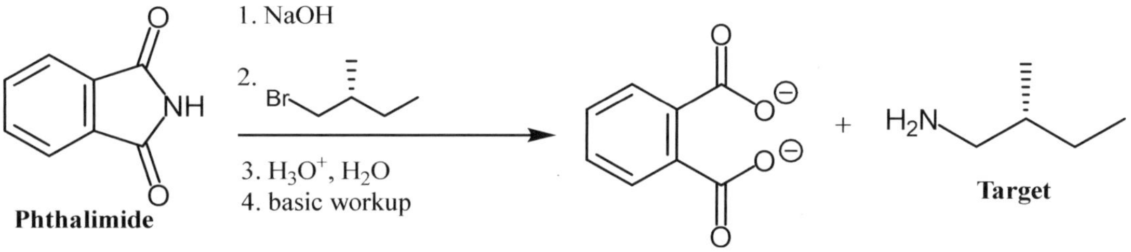

1. NaOH
2. Br
3. $H_3O^+$, $H_2O$
4. basic workup

NH

**Phthalimide**

$H_2N$

**Target**

## Lesson VI.13. Reaction of Carboxylic Acid Derivatives with H and C Nucleophiles

### *Lesson VI.13.1 Reduction of Esters or Acid Chlorides to form Aldehydes or Primary Alcohols*

When an ester reacts with a hydride ($H^-$) as nucleophile the first step is nucleophilic acyl substitution of the –OR for–H:

R''C(=O)OR' + H⁻ —*nucleophilic addition*→ R''C(O⁻)(H)OR' —*nucleophilic elimination*→ R''C(=O)H + RO⁻

If the source of the hydride for this reduction reaction is diisobutylaluminum hydride (DIBAL-H), a specialized metal hydride that we have not yet encountered in this book, the reaction stops here. So, **DIBAL-H is a great reagent for conversion of esters to aldehydes**. The reaction is usually done in toluene at low temperature. What about the other hydride reagents that we have already seen in this book, namely $NaBH_4$ and $LiAlH_4$? Well, **$NaBH_4$ is not reactive enough to react with an ester**. On the other hand, $LiAlH_4$ is extremely reactive, and continues to react with the aldehyde after the $S_NAc$ portion of the reaction has taken place. This should not be surprising, as we have already seen the mechanism for reaction of $LiAlH_4$ with an aldehyde in Lesson VI.5. Because $LiAlH_4$ reduces the ester/carboxylic acid all the way to an alcohol in a single reaction, **$LiAlH_4$ is a great reagent to convert esters into primary alcohols:**

R''C(=O)OR' —$H^-$, *$S_NAc$*, $-RO^-$→ R''C(=O)H —$H^-$, *Nu Addit.*, *then protonation (from acidic workup)*→ R''CH₂OH

So, to summarize the reactivity of hydride sources with esters:

R''C(=O)OR' —1. $NaBH_4$; 2. $H_3O^+$→ *NO REACTION*

R''C(=O)OR' —DIBAL-H; –40 °C→ R''C(=O)H **Aldehyde**

R''C(=O)OR' —1. $LiAlH_4$; 2. $H_3O^+$→ R''CH₂OH **Primary Alcohol**

The reaction of $LiAlH_4$ and an ester to form a primary alcohol is a **Type C** reaction. In type C reactions, two new bonds to the nucleophile form to the carbonyl carbon as the carbonyl carbon loses the π-bond to O and the leaving group. So, Type C reactions are $S_NAc$ followed by nucleophilic addition/protonation (Lesson VI.3).

Recall from Lesson VI.4 that acid chlorides are much more reactive towards nucleophilic attack than are esters. For this reason, even $NaBH_4$ can be used to convert an acid chloride to a primary alcohol. Of course, the much more reactive $LiAlH_4$ would also accomplish this reaction:

R''–C(=O)–Cl —(1. $NaBH_4$ or $LiAlH_4$; 2. $H_3O^+$)→ R''–CH(H)–OH **Primary Alcohol**

*Lesson VI.13.2 Reaction of Amides with $LiAlH_4$, Formation of Amines*

Reacting amides with strong reducing agents (hydride donors) like $LiAlH_4$ will reduce the C=O to two C–H bonds and an amine is produced. This reaction is commonly used to make primary, secondary, and tertiary amines. The mechanism is given below:

*Nu addition* — *Nu elimination* — *Nu addition*

R''–C(=O)–$NR_2$ —($H^⊖$)→ R''–CH(–O–$AlH_2$)–$NR_2$ —(–$LiOAlH2$)→ R''–CH=$N^⊕R_2$ —($H^⊖$)→ R''–$CH_2$–$NR_2$ **Amine**

R = H, alkyl or aryl groups

The reaction is followed by an acidic workup to separate the product from the byproducts.

*Lesson VI.13.3 Reactions with RMgX (X=Cl, Br, I) or RLi*

Treating esters and acid chlorides with Grignard reagents (RMgX, where X is Cl, Br, or I) or organolithium reagents (RLi) will give a tertiary alcohol with two new σ C–C bonds. The reaction mechanism is same as for reaction of the corresponding carbonyls with $LiAlH_4$, but with $R^-$ as the nucleophile instead of hydride. The general mechanism is as follows:

Nucleophile

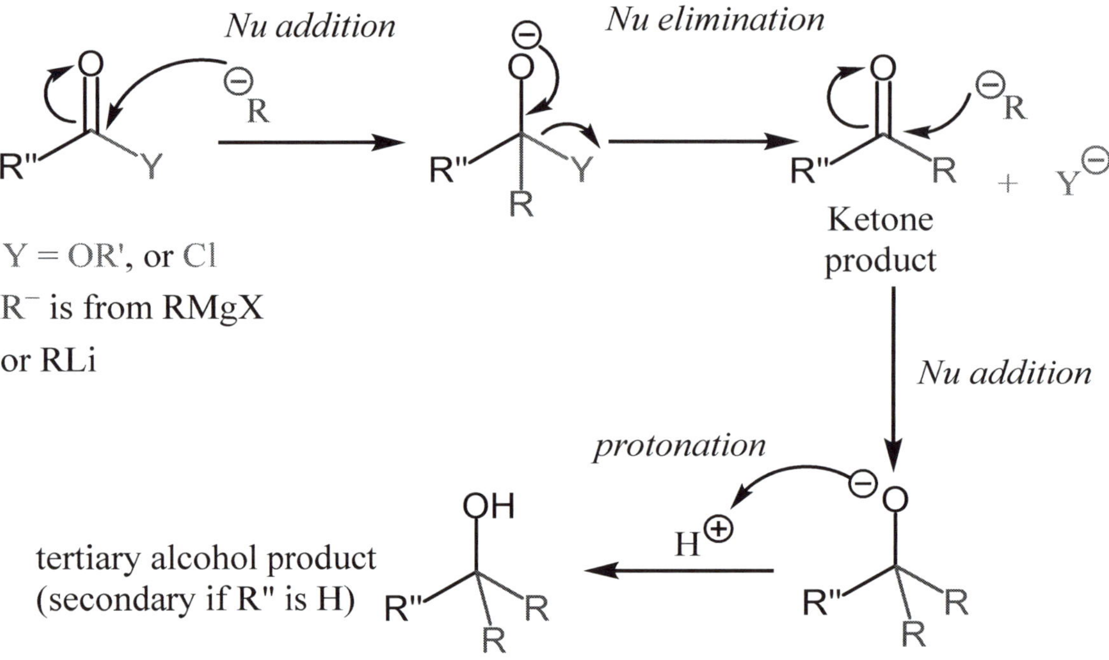


The reactivity of RMgX and RLi diverge in the case of carboxylic acids. Keep in mind that Grignard and organolithium reagents are still strong bases, so these reagents will first deprotonate a carboxylic acid to generate a carboxylate ($RCO_2^-$) before any nucleophilic addition is able to occur. This negative charge makes the carbonyl group less susceptible to nucleophilic addition in a carboxylate relative to a carboxylic acid. Consequently, the $R^-$ in RMgX is not sufficiently nucleophilic to add to the carboxylate, so reaction stops at the magnesium carboxylate salt:

R—MgX → R—H + R"CO–O–MgX

In contrast, the organolithium reagent will convert the carboxylic acid into the alcohol:

1. xs RLi
2. $H_3O^+$

**Primary Alcohol**

## Lesson VI.14. Preparation and Reaction of Nitriles

### *Lesson VI.14.1 Preparation of Nitriles*

Nitriles (R"–C≡N) are important functional groups used is many important industrial processes. These functional groups can be prepared by the dehydration reaction of primary amides (R"C(O)NH$_2$) with P$_2$O$_5$, or by the S$_N$2 reaction of alkyl halides (or sulfonate esters) with cyanide salts (NaCN, KCN or LiCN). The reaction mechanism for reaction with P$_2$O$_5$ is beyond the scope of this book.

R"C(O)NH$_2$ —($P_2O_5$, $-H_2O$)→ R"—C≡N (**nitrile**) ←($^{\ominus}$CN)— R"—LG (LG = X, OTs, OTf or OMs)

### *Lesson VI.14.2 Hydrolysis of Nitriles, Formation of Carboxylic Acids and Amino Acids*

Hydrolysis of nitriles (reaction with H$_2$O) in presence of acid or base catalysts will give carboxylic acids. The mechanism of this hydrolysis is similar to hydrolysis od carbonyls. The C≡N is ultimately substituted with one C=O and one C—OH. The mechanism of the acid-catalyzed hydrolysis is given in the figure below. The last part of the overall transformation is the amide hydrolysis we saw in Lesson VI.12

R"—C≡N (**nitrile**) —H$^+$→ R"—C≡N$^{\oplus}$—H (**I**) —H$_2$O→ **II** —H$_2$O→ **III** —H$^+$→ **IV** ↔ **V** —H$_2$O→ **amide** —(*amide hydrolysis*)→ **carboxylic acid**

One especially useful reaction sequence employing nitrile hydrolysis as one of the steps is the **Strecker amino acid synthesis**. The Strecker amino acid synthesis starts with an aldehyde that is transformed first into an imine (Step 1 in the following scheme). A cyanide anion then undergoes nucleophilic addition to the imine (Step 2). The final step is hydrolysis of the nitrile functional group and adjustment of the pH to give an **amino acid** as the final product.

**General structure of an amino acid:**

carboxylic acid functional group

The R group is usually the only thing that is different for different amino acids.

amine functional group

**Strecker amino acid synthesis:**

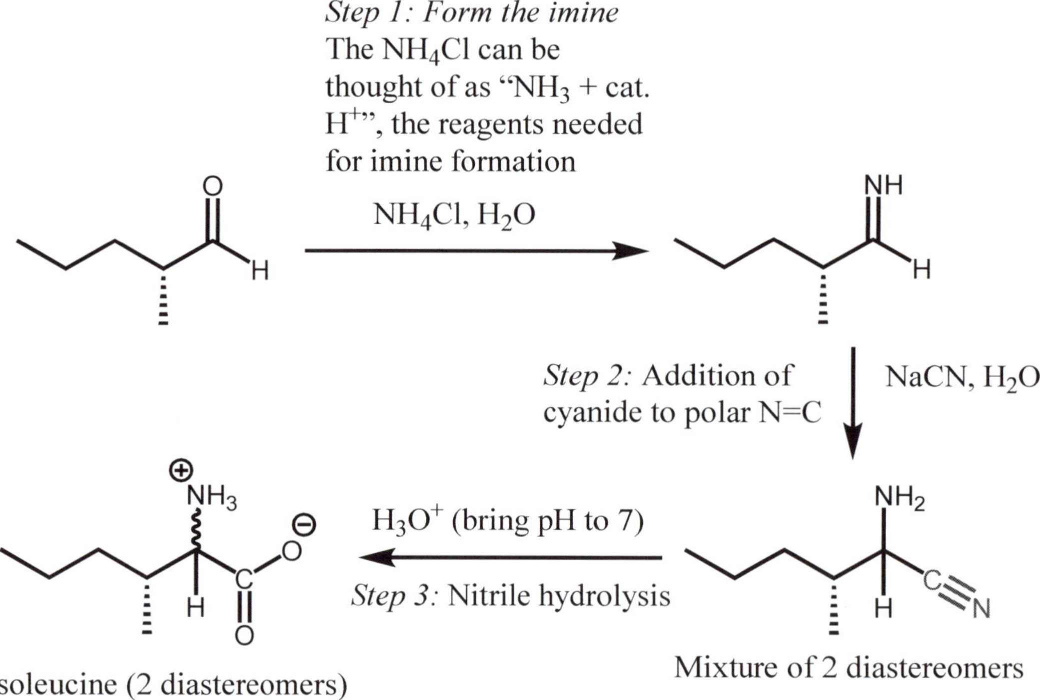


*Lesson VI.14.3 Reaction of Nitriles with RLi, RMgX, or $LiAlH_4$ to Form Primary Amines*

Reaction of nitriles with RLi, RMgX (Grignard reagents) or $LiAlH_4$ followed by acid workup will produce primary amines according to the following mechanism:

$R^-$ is from RLi or RMgX

M = Li or MgX

$H_3O^+$ workup, pH adjust

**primary amine**

$R''—C≡N \xrightarrow{H^{\ominus}} R''CH=N–M \rightarrow R''–CH_2–N^{\ominus}–M \xrightarrow[\text{workup, pH adjust}]{H_3O^+} R''–CH_2–NH_2$ **primary amine**

M = Al or Li from $LiAlH_4$

Example VI.14.1

Give the major products of the following sequence of reactions

Br $\xrightarrow{NaCN} \xrightarrow{H_3O^+} \xrightarrow{SOCl_2} \xrightarrow{MeNH_2} \xrightarrow[2.\ H_3O^+]{1.\ LiAlH_4}$

Solution VI.14.1

The first reaction is an $S_N2$ reaction where the C—Br σ bond is substituted with C—CN. Treating the nitrile with $H_3O^+$ will produce carboxylic acid, note that we added a carbon to the main chain. $SOCl_2$ will react with carboxylic acid to give acid chloride product, which upon aminolysis (addition of $CH_3NH_2$) will produce the amide product. The amide will then react with $LiAlH_4$ to give the amine as final product.

Br $\xrightarrow[ACN]{NaCN}$ CN $\xrightarrow{H_3O^+}$ OH, O $\xrightarrow{SOCl_2}$ Cl, O $\xrightarrow{MeNH_2}$ H–N, O $\xrightarrow[2.\ H_3O^+]{1.\ LiAlH_4}$ H–N

## Lesson VI.15. Preparation of Enolates and Alkylation

### *Lesson VI.15.1 Introduction to Enolates*

Enolates are molecules formed by treating carbonyl compounds with strong bases. The strong base removes a proton form the carbon adjacent to the carbonyl carbon – called the α-carbon (read "alpha carbon"). Enolates are characterized by negatively-charged oxygen next to a C–C double bond. Enolates are stabilized by delocalization of electrons (resonance forms). They are reactive nucleophiles, as we will see as we study the various reactions they can facilitate.

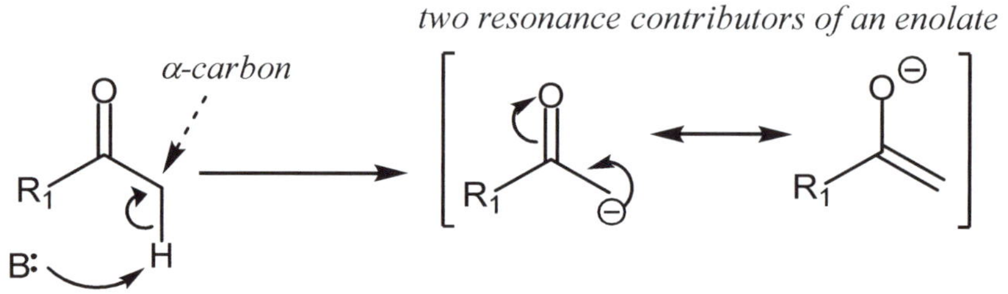

*two resonance contributors of an enolate*

*α-carbon*

$R_1$ = -H, alkyl, aryl, alkenyl, -OR, or -SR

### *Lesson VI.15.2 Enolates of Ketones and Aldehydes*

When a ketone or an aldehyde is treated with a strong base, the α-hydrogen will be removed to leave a negative charge on the α-carbon. The carbanion is resonance-stabilized by delocalization of charge onto the O as well. The stability of the enolate depends on how many resonance contributors can be formed (more resonance contributors = more stable). Treating phenyl acetaldehyde with base, for example, will lead to formation of a very stable enolate due to the resonance contributors possible by delocalizing the charge onto the adjacent benzene group:

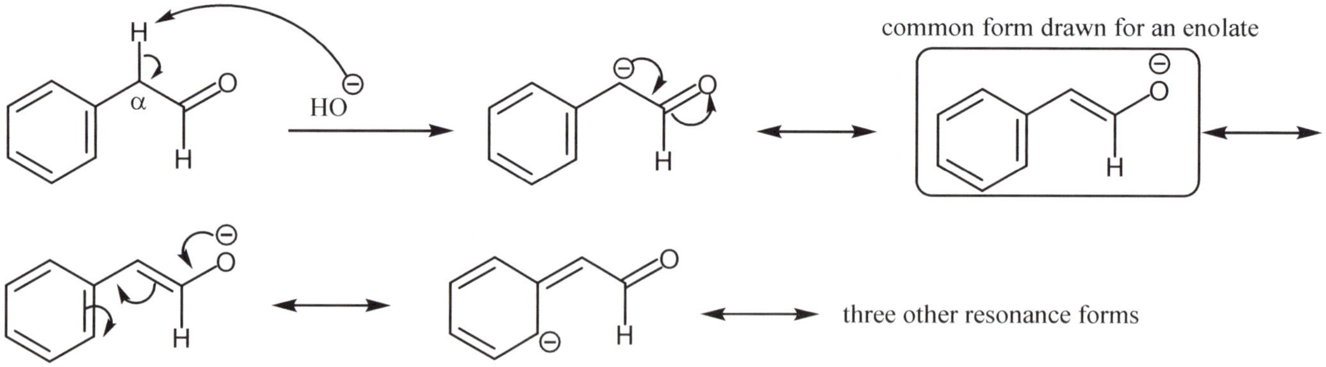

common form drawn for an enolate

three other resonance forms

When a compound has an $sp^3$-carbon between two carbonyl groups, deprotonation of the "doubly α" site will lead to a more resonance-stabilized species, and this site will therefore be deprotonated selectively over the "singly α" sites, as illustrated for 2,4-pentadione:

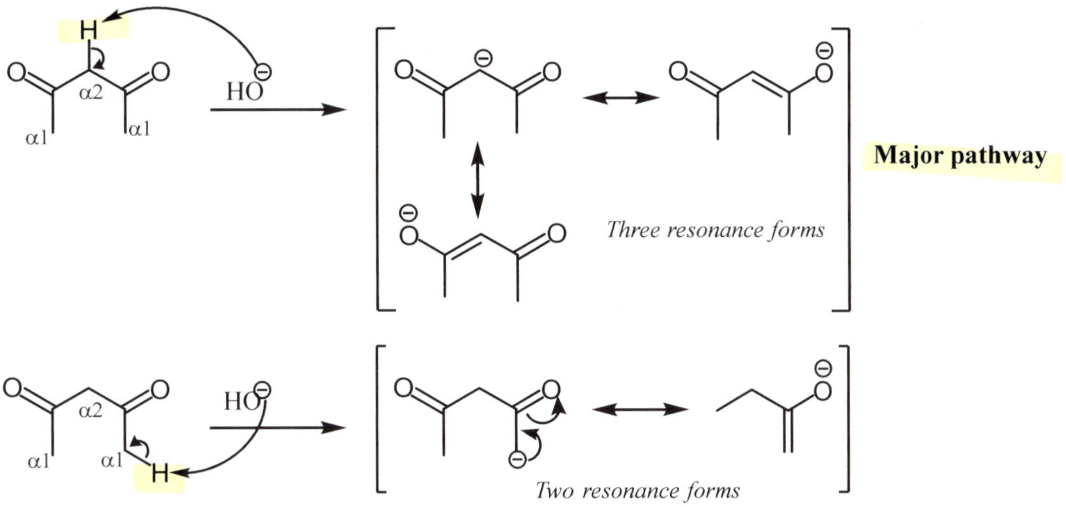

**Major pathway**

*Three resonance forms*

*Two resonance forms*

Some molecules will contain two inequivalent α-carbon positions beside the same carbonyl group, whereby deprotonation at either position affords an anion with the same extent of resonance stabilization. In these cases, we must consider inductive effects to determine which enolate anion is more stable. Consider the case of 2-butanone:

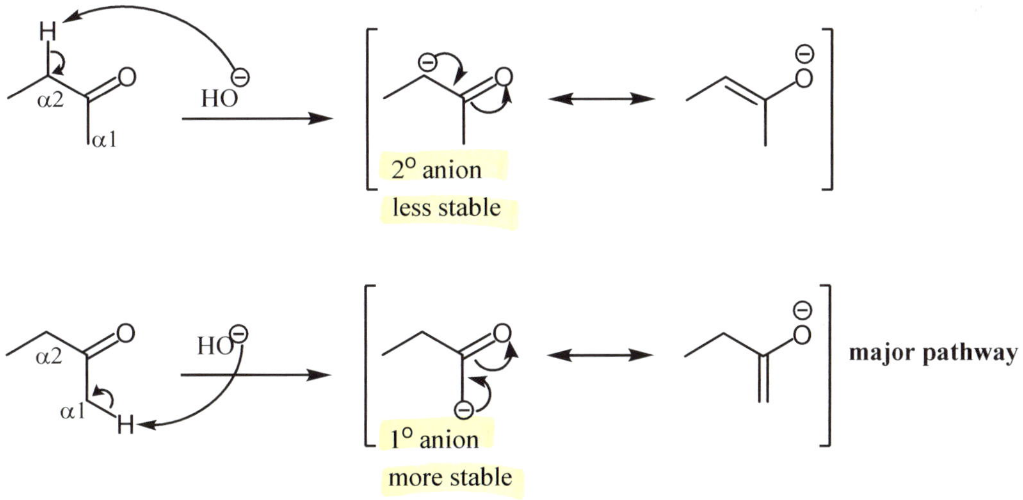

2° anion
less stable

**major pathway**

1° anion
more stable

Anions are more stable on less substituted carbon atoms due to less repulsion by adjacent σ-bonds (an example of a repulsive inductive effect, see Lesson I.10 in OC1 Primer). This makes the deprotonation of α1 more thermodynamically favorable, and therefore yields the major enolate in this case. Note that this inductive effect is **secondary** to resonance effects in determining the predominant deprotonation site.

If same # of resonance contributors → inductive effect

1° anion > 2° anion > 3° anion

*Lesson VI.15.3 Enolates of Esters*

An ester enolate can be generated by deprotonation of an ester with a base like an alkoxide ($^{-}$OR). Note that the alkoxide base used should contain the same alkyl group as the ester O-substituent to prevent undesired nucleophilic addition or substitution reactions. For example, to prepare the ester enolate of ethyl butanoate (see below) we should use ethoxide as the base (added as NaOEt salt). In this reaction, the predominant net pathway is one in which the alkoxide will act as a base rather than as a nucleophile and it will deprotonate the most acidic hydrogen in the ester, similar to what we saw for aldehydes and ketones:

Et ⁻OEt H + EtOH Et Et

*Lesson VI.15.4 Enolates as Nucleophiles for $S_N2$ Reaction: α-Alkylation*

Once an enolate forms, it can be used as a nucleophile in an $S_N2$ reaction:

same thing

Br

The net result of this reaction is adding an alkyl group to the α-carbon of a carbonyl. For this reason, this reaction is often called **α-alkylation**. In the next several lessons, we will see other reactions that employ the enolate as a nucleophile.

It is also important to note that some people draw the enolate in its other resonance form for mechanisms. If we do this to represent the exact same $S_N2$ reaction we saw above, it would look like this:

Br

# Lesson VI.16. Alpha-Halogenation and Haloform Reactions

## *VI.16.1 Alpha Halogenation*

When a carbonyl species reacts with $X_2$ (X = Cl or Br), the result is **α-halogenation**:

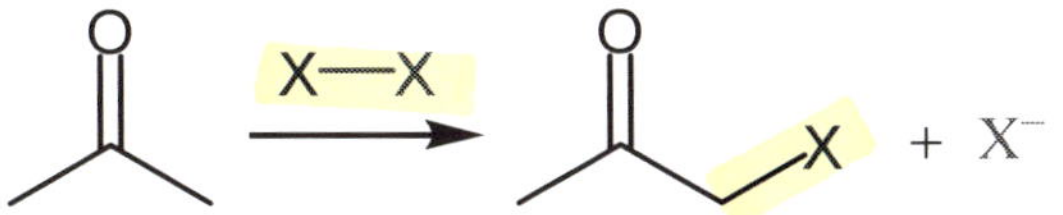

X—X
+ $X^-$

To figure out how the α-halogenation reaction works, we can think back to two other halogenation reactions we have seen. We have seen (Lessons III.6 and III.12 in OC1 Primer) that in alkene or alkyne halogenation, a π-bond can act as a nucleophile in a reaction with $X_2$ (X = $Cl_2$ or $Br_2$):

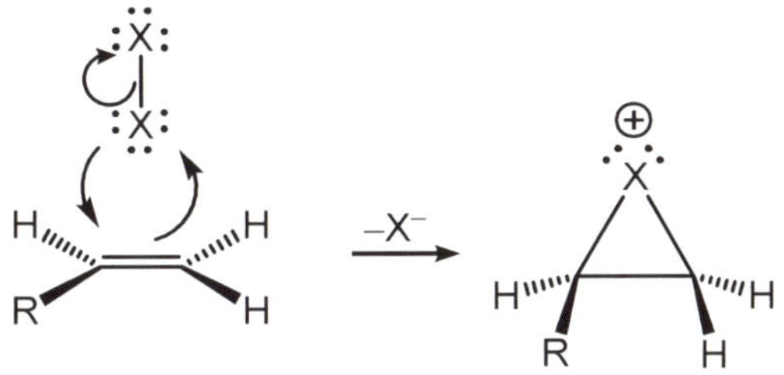

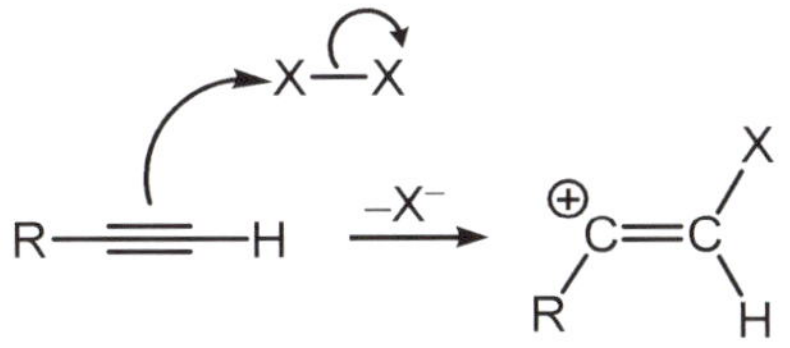

$-X^-$

$-X^-$

A reaction analogous to the alkyne reaction can be drawn for the reaction of a carbonyl species with $X_2$. When representing halogenation of the carbonyl, it is easiest to envision the reaction by considering the minor enol tautomer as the reactant, so that we can portray the π-bond attacking the halogen just like we saw for alkynes:

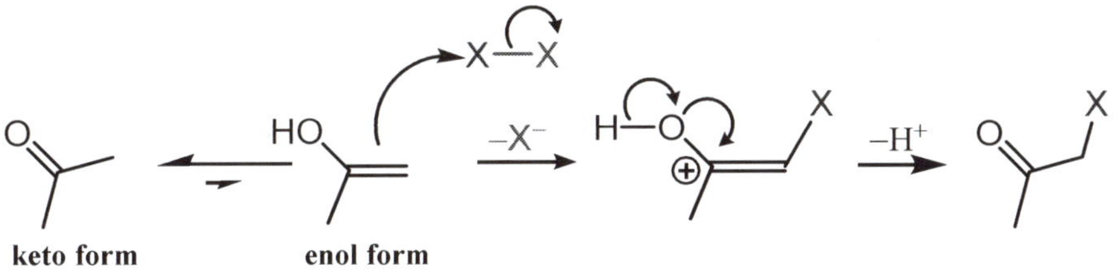

$-X^-$ $-H^+$

**keto form** **enol form**

Remember that the keto-enol equilibrium is occurring constantly for a carbonyl species in solution, so if it is being consumed by reaction, eventually all of the carbonyl can be consumed to give the halogenated product. The $H^+$ that comes off of the O and the $X^-$ can go together to make HX as well.

*VI.16.2 The Haloform Reaction*

When a ketone having a methyl group on one side reacts with $X_2$ (X = Cl, Br or I) in the presence of NaOH, the result is formation of a haloform and a carboxylate. A haloform is a molecule with the formula $CHX_3$. The specific molecule is called chloroform, bromoform or iodoform when X = Cl, Br or I, respectively. This is called the **haloform reaction:**

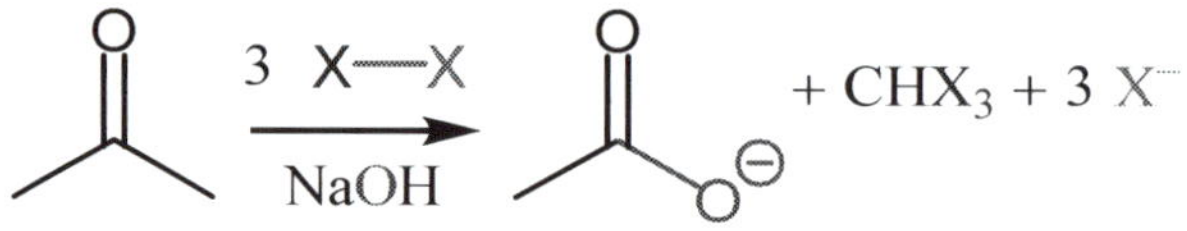

3 X—X, NaOH; + $CHX_3$ + 3 $X^-$

What is a reasonable mechanism for this reaction? Well, because we have a base and a carbonyl, we can envision that the first step will be formation of the enolate. Then, the enolate can act as a nucleophile, leading to α-halogenation. The base can then deprotonate the α-carbon again, so that a second halogen can add to the α-carbon. This deprotonation-halogenation step happens a third time, and then there are no more α-hydrogens to remove from that site, and they have all been substituted for X:

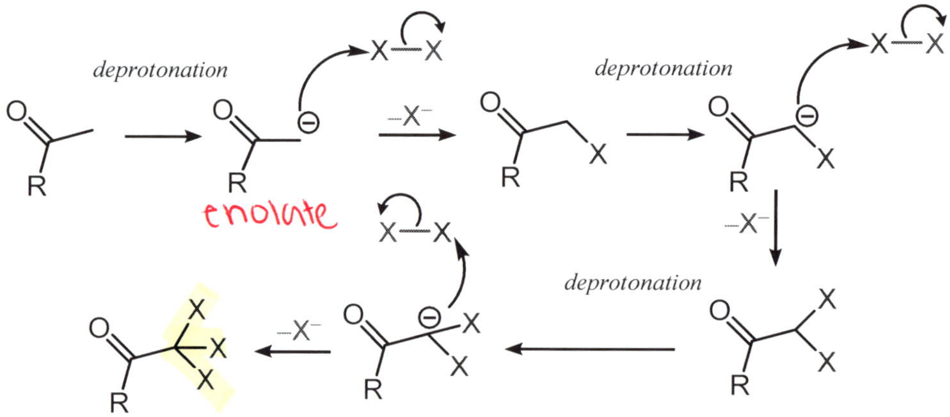

deprotonation; $-X^-$; deprotonation; $-X^-$; deprotonation; $-X^-$; enolate

At this point, the hydroxide will act as a nucleophile for the $S_NAc$ reaction. Generally, a carbanion would be too poor a leaving group for this reaction, but in this case the strong inductive stabilization afforded by the three attached halogens allows this to happen to some extent. As the carbanion and carboxylic acid form, the carbanion is rapidly protonated by the carboxylic acid, driving the equilibrium to the right and forming the final products:

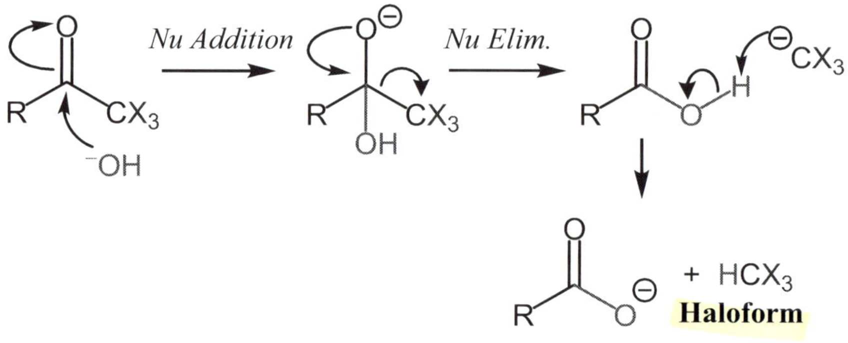

*Nu Addition*; *Nu Elim.*; + $HCX_3$ **Haloform**

## Lesson VI.17. Aldol Addition and Condensation

### *VI.17.1 Aldol Addition*

Another example of a Type A reaction is **aldol addition**, so named for the **ald**ehyde and alcoh**ol** functional groups it installs in the product for some such reactions. In the aldol reaction, an enolate is generated and functions as a nucleophile that attacks the electrophilic carbonyl carbon, followed by protonation to give a β-hydroxycarbonyl product. A strong base is required to form the enolate. An example of the aldol reaction is:

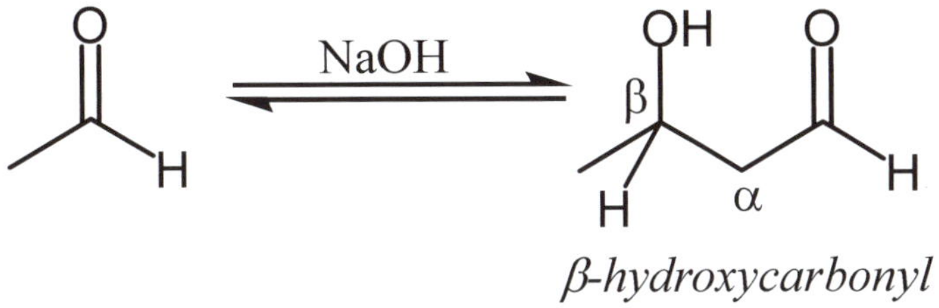

The mechanism for aldol addition is depicted in the following figure:

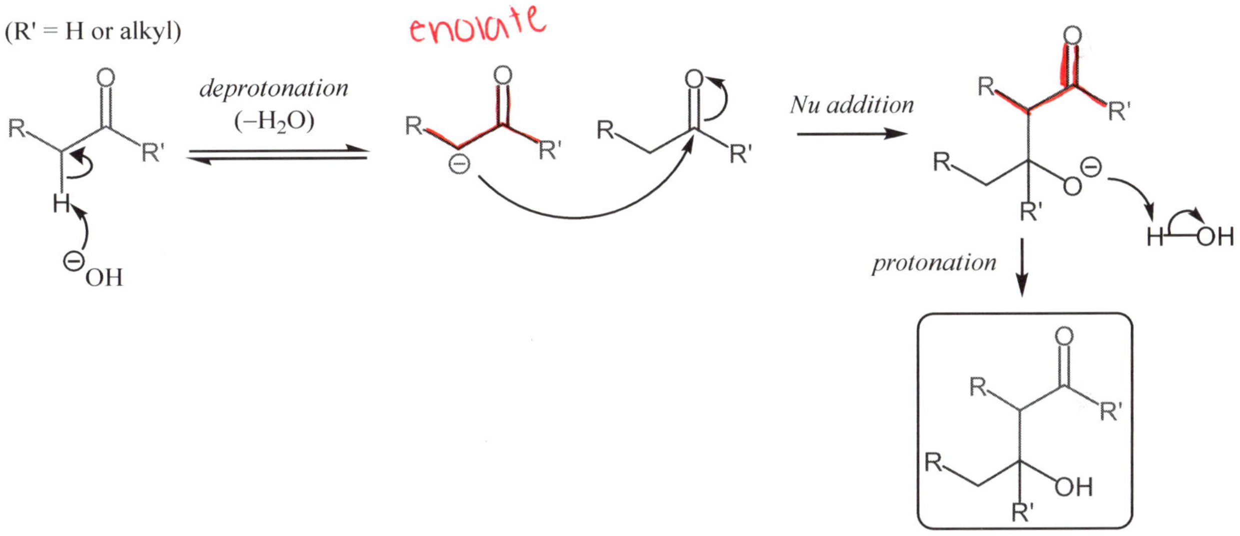

**Aldol addition product**

The aldol addition reaction is actually just a simple nucleophilic addition followed by protonation (Type A reaction) that is common for ketones and aldehydes. The only difference between this Type A reaction and those we have seen previously is that we are using the enolate as the nucleophile, so the enolate must be generated from some of the starting carbonyl itself!

Example VI.17.1

Draw the major addition product of the following reaction

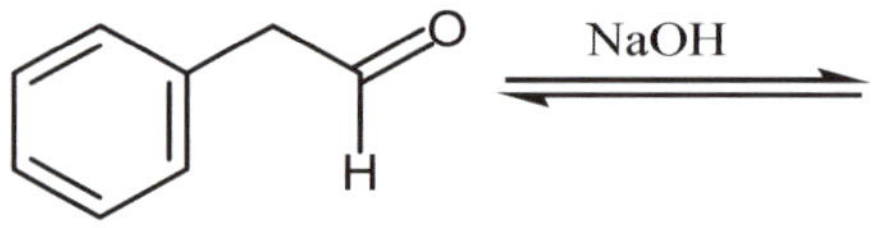

NaOH

Solution VI.17.1

This is an aldol reaction, which is a Type A nucleophilic addition, wherein the base ($^{-}$OH) deprotonates one molecule of the reactant aldehyde to form the nucleophilic enolate, which then adds to the carbonyl carbon and breaks the C–O $\pi$-bond. Subsequent protonation affords the alcohol product. Furthermore, because the product has a chiral center but the reactant is achiral, the product will be a **racemic mixture** of both the *R* and *S* stereoisomers.

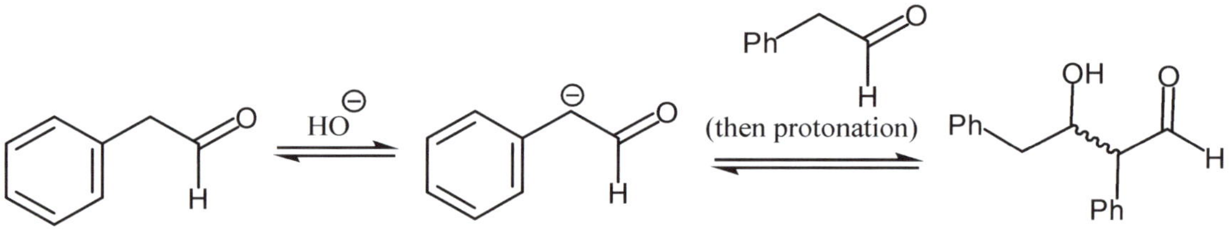

HO⊖ (then protonation) Ph= phenyl ring

*Lesson VI.17.2 Aldol Condensation*

If the aldol addition product is heated in the presence of a base, further reaction will take place that leads to elimination of another molecule of water. The water often form condensation on the reaction vessel, so this additional reaction to form the elimination product is called **aldol condensation**.

The mechanism for the elimination is shown in the figure below:

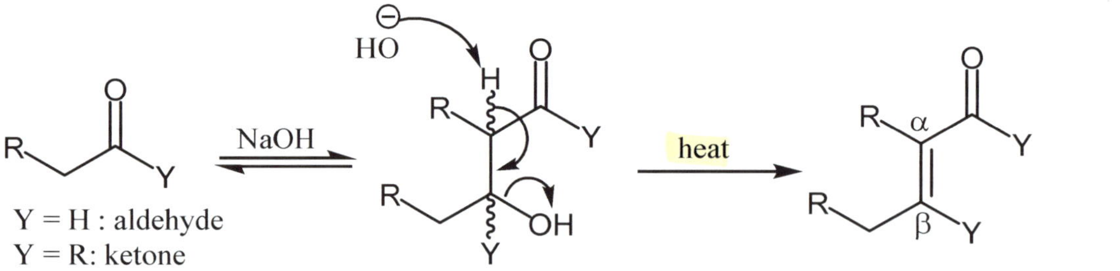

NaOH heat

Y = H : aldehyde
Y = R: ketone

**Aldol addition product** **Aldol condensation product**

When an aldol condensation reaction has the potential to produce two geometrical isomers (for example an *E*- versus a *Z*-alkene) the more stable (thermodynamically favored) product is the major product.

## Lesson VI.18. Claisen Condensation

*Lesson VI.18.1 Claisen Condensation*

**Claisen condensation** is the nucleophilic acyl substitution of esters, in which an ester enolate (Lesson VI.15) acts as the nucleophile. This leads to β-keto esters as products:

NaOCH$_3$ + HOR

The mechanism involves making the enolate and then doing the usual S$_N$Ac sequence:

**ester enolate**

*ester enolate formation*

*Nu addition*

*Nu elimination*

H$^+$ workup

**β-keto ester**

**β-keto ester**

The reaction generally produces a chiral center at the α carbon of the Claisen product, yielding a racemic mixture. The intramolecular version of the Claisen condensation is called the and **Dieckmann Condensation**.

Example VI.18.1

Give the product of the following Claisen condensation reaction

NaOEt

Solution VI.18.1

The ethoxide ion will produce the ester enolate, which will react with the carbonyl in another ester molecule as a nucleophile, to produce the β-keto ester. Since we will have a chiral center at the α-carbon then we will have a racemic mixture of the two enantiomers:

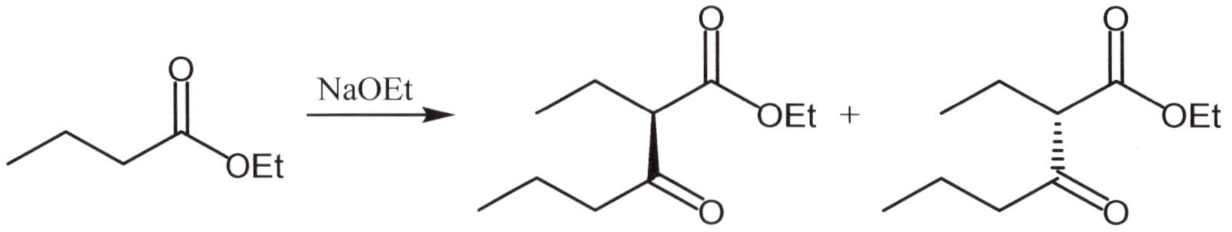

NaOEt

Example VI.18.2

Give the product of the following reaction:

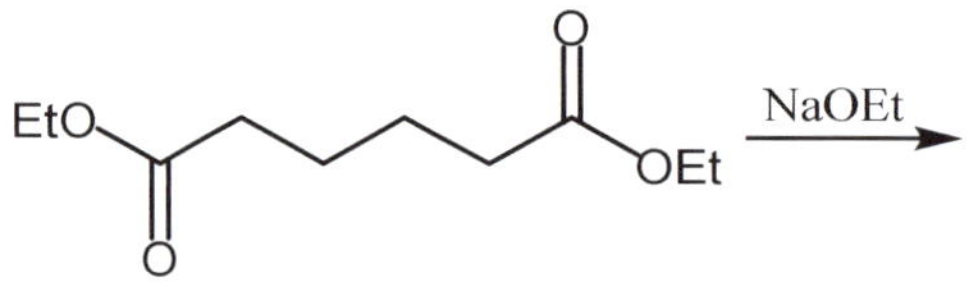

NaOEt

Solution VI.18.2

Both esters are attached to the same chain, so this will be a Dieckmann Condensation. We identify the α-carbon that will be the nucleophile once deprotonated, and simply attach it in place of the OEt on the other end (the usual $S_NAc$ reaction):

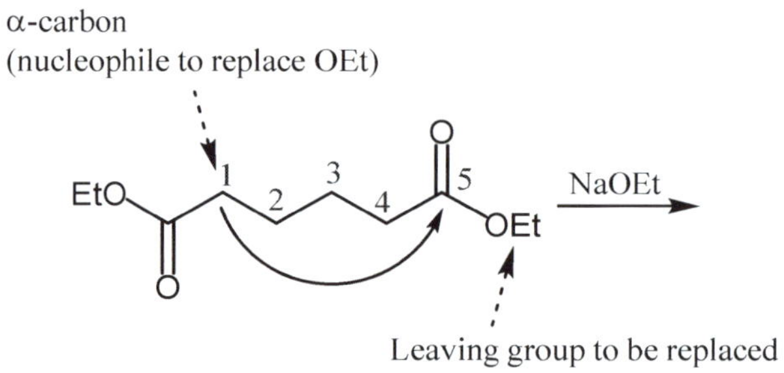

α-carbon
(nucleophile to replace OEt)

NaOEt

Leaving group to be replaced

We can see that the ring we will form has five carbon atoms in it. It is helpful to number the carbons that will form the ring to be certain, as in the structure above. The final product then is:

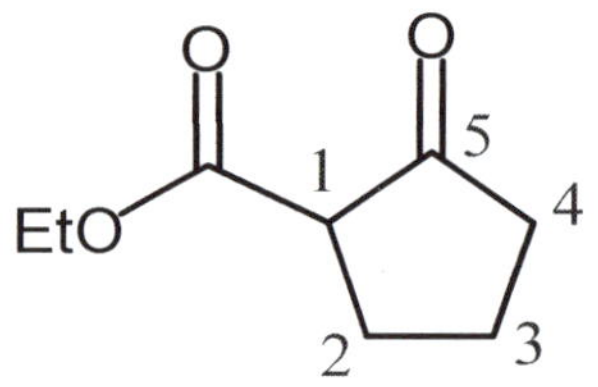

## Lesson VI.19. Decarboxylation and Synthetic Applications

### *Lesson VI.19.1 Decarboxylation of 3-oxo Carboxylic Acids*

Carbonyls are typically relatively stable upon heating to moderate temperatures, but 3-oxo carboxylic acids decompose to release $CO_2$ if they are heated to above 120 °C. This is a process called **decarboxylation**. An arrow-pushing mechanism for the process is:

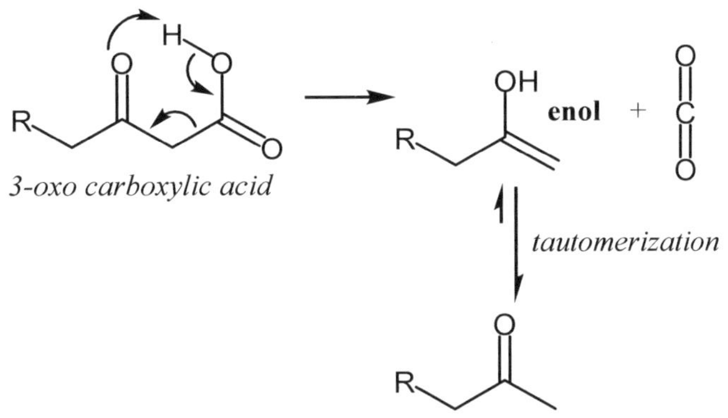

3-oxo carboxylic acid

enol + 

tautomerization

Decarboxylation is specific to 3-oxo carboxylic acids (or carboxylates at higher temperature) and is a useful way to remove a carbon from a molecule. Some especially useful synthetic procedures that employ decarboxylation as a step are discussed in the remainder of this lesson.

### *Lesson VI.19.2 Acetoacetic Ester Synthesis*

The **acetoacetic ester synthesis** involves several steps, all of which we have seen throughout this book. The net reaction is:

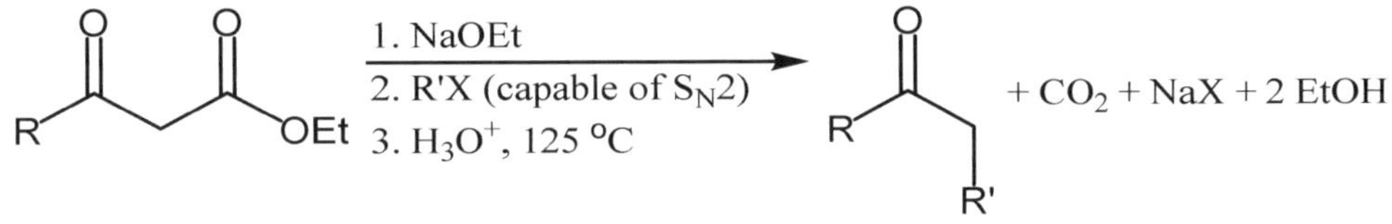

1. NaOEt
2. R'X (capable of $S_N2$)
3. $H_3O^+$, 125 °C

$+ CO_2 + NaX + 2\ EtOH$

Let us walk through all of the steps (see the scheme on the following page). First, the added base will deprotonate the doubly-α carbon between the two carbonyl units to form an enolate nucleophile. The enolate nucleophile then participates in an $S_N2$ reaction with the alkyl halide (R'X) that is added in step 2. Once this reaction is complete, the compound is heated in the presence of acid and water ($H_3O^+$). We saw in Lesson VI.11 that these conditions lead to acid-catalyzed ester hydrolysis, converting the ester into a carboxylic acid. Upon further heating at elevated temperature, this species will undergo decarboxylation to yield a ketone. The **acetoacetic ester synthesis is a convenient synthetic route to ketones** with a variety of chain compositions:

*deprotonation* $-HOEt$ ... $S_N2$ $-Br^-$ ... *ester hydrolysis* $H_3O^+$ ... *decarboxylation* $-CO_2$

**ketone** **3-oxo carboxylic acid**

*Lesson VI.19.3 Malonic Ester Synthesis*

The malonic ester synthesis is another important industrial process that makes use of several reactions that we have seen. The net reaction is:

1. NaOEt
2. R'X (capable of $S_N2$)
3. $H_3O^+$, 125 °C

$+ CO_2 + NaX + 3\ EtOH$

The steps are fundamentally the same as for the acetoacetic ester synthesis. First, the added base will deprotonate the doubly-α carbon between the two carbonyl units to form an enolate nucleophile. The enolate nucleophile then participates in an $S_N2$ reaction. The $S_N2$ product is heated in the presence of acid and water ($H_3O^+$), causing ester hydrolysis. In this case, both of the ester units are converted into carboxylic acids, giving a 3-oxo carboxylic acid. Upon further heating at elevated temperature, this species will undergo decarboxylation to yield a carboxylic acid.

The **malonic ester synthesis is a convenient synthetic route to carboxylic acids** with a variety of chain compositions. The complete mechanism is given as:

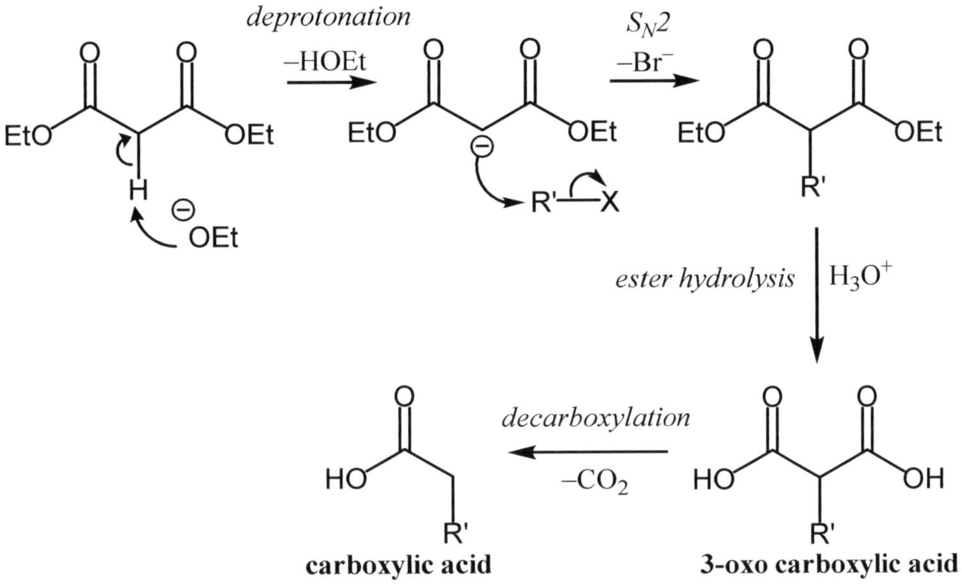

*deprotonation*

$-HOEt$

*$S_N2$*

$-Br^-$

*ester hydrolysis* $H_3O^+$

*decarboxylation*

$-CO_2$

EtO, OEt, H, $^\ominus$OEt, R'—X, R'

HO, OH

**carboxylic acid**

**3-oxo carboxylic acid**

*Lesson VI.19.4 Gabriel Malonic Ester Synthesis of Amino Acids*

We saw the Gabriel synthesis of primary amines in Lesson VI.12. Amino acids are one specific type of (typically) primary amines. We have seen the general form of amino acids and one route to prepare amino acids (the Strecker synthesis) in Lesson VI.14. Another route to amino acids is the **Gabriel Malonic Ester Synthesis of Amino Acids**. The first step of the Gabriel Malonic Ester Synthesis is an $S_N2$ reaction using a phthalimide anion as the nucleophile (Step 1 in the following scheme). Following addition of the phthalimide fragment, base is used to deprotonate the α-carbon for $S_N2$ reaction to add the R group (Step 2; just like in the typical malonic ester synthesis). When NaOH and $H_2O$ are added (Step 3 in the scheme) , the ester groups undergo hydrolysis (again just like in the typical malonic ester synthesis), but the amide bond of the phthalimide unit are also hydrolyzed. After decarboxylation, the amino acid can be isolated:

**Gabriel Malonic Ester Synthesis of Amino Acids**:

phthalimide

*Step 1* $S_N2$

*Step 2* 1. Base 2. **R**–Br ($S_N2$)

*Step 3* NaOH, $H_2O$

*Step 4* $H^+$, $H_2O$, Δ

+ $CO_2$

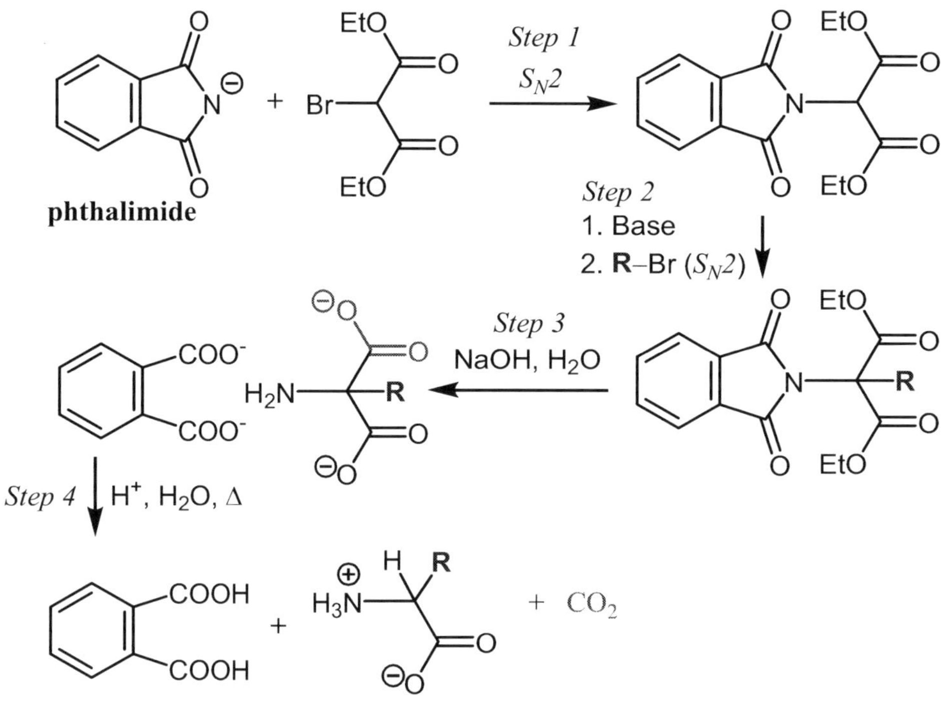

## Lesson VI.20. Addition of Nucleophiles to α,β-Unsaturated Carbonyls

### *Lesson VI.20.1 Electrophilic Sites in α,β-Unsaturated Carbonyls*

In Lesson VI.17, we learned how to prepare α,β-unsaturated carbonyls via aldol condensation. We also saw that two conjugated C=C bonds can undergo either 1,2- or 1,4-electrophilic addition reactions (Lesson IV.2). In this lesson, we will investigate how the double bonds in an α,β-unsaturated carbonyl can undergo nucleophilic addition. To understand the reactivity of α,β-unsaturated carbonyls with respect to nucleophilic addition, we must first identify which sites in an α,β-unsaturated carbonyl are electrophilic by examining the possible resonance contributors and corresponding distribution of charges within the resonance hybrid:

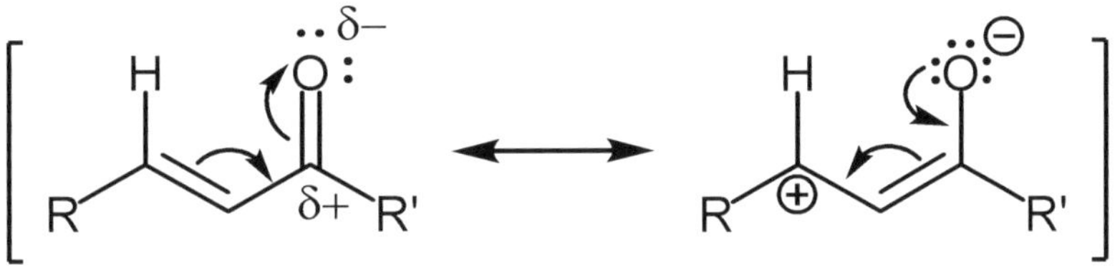

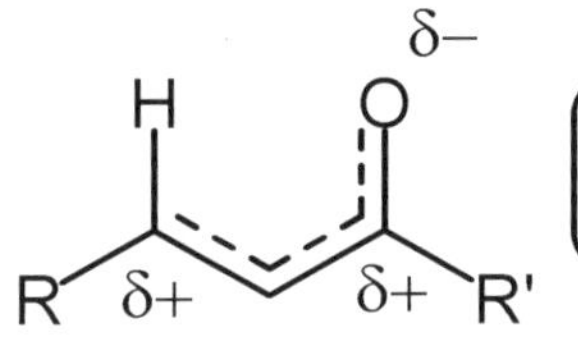

*resonance hybrid*
*(better represents the true structure)*

There are two electrophilic sites: the carbonyl carbon and the β-carbon. So, just as conjugated dienes can undergo 1,2- (direct) or 1,4- (conjugate) *electrophilic* addition, **α,β-unsaturated carbonyls can undergo 1,2- (direct) or 1,4- (conjugate) *nucleophilic* addition**:

1,2-addition (Direct Addition)

*nucleophilic addition* → *protonation* →

**Isolated Product I**

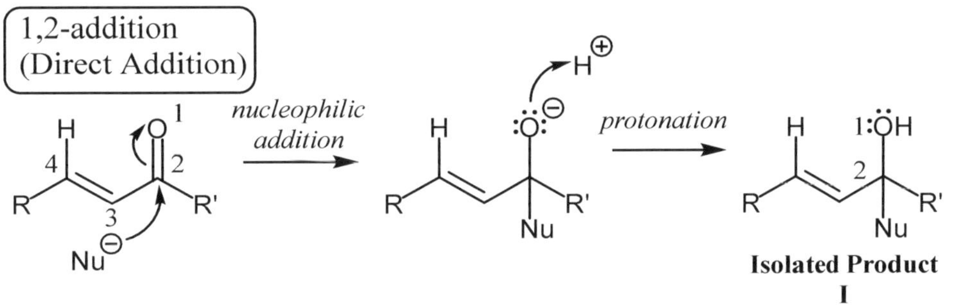

The isolated product of direct addition to an α,β-unsaturated carbonyl is the same as for a ketone that is not unsaturated: it results from nucleophilic addition at the carbonyl carbon (atom 2) followed by protonation at the carbonyl oxygen (atom 1). We have previously classified this as carbonyl reaction type A in this book.

1,4-addition
(Conjugate Addition)

nucleophilic addition → Resonance-Stabilized Anion → protonation → enol → tautomerization → Isolated Product II

Conjugate addition features nucleophilic addition to the β-carbon (atom 4), then protonation of the carbonyl oxygen (atom 1). Because the H and Nu have added to atoms 1 and 4, this is also called 1,4-addition. The isolated product, however, is not this 1,4-addition species itself, because we know from Lesson III.14 (OC1 Primer, reviewed in Lesson VI.15) that an enol spontaneously tautomerizes to the keto form. Conjugate addition to an α,β-unsaturated carbonyl occurs therefore affords a product like **II**.

*Lesson VI.20.2 Nucleophile Stability Dictates Regioselectivity*

Whether 1,2-addition or 1,4-addition leads to the major product depends on how good a leaving group the nucleophile is. To understand why this is, we need to consider the thermodynamics of the nucleophilic addition step.

We first consider the case in which the **nucleophile is an unstable anion** (i.e. a strong base; $CH_3O^-$ or less stable for this case). In this case, the nucleophile is a **bad leaving group**, so after it does nucleophilic addition, the step is **irreversible** regardless of whether the nucleophile adds to the 2-C or the 4-C of the α,β-unsaturated carbonyl. In general, a reaction that consumes a very unstable anion will favor the product side. In the current case, the anion resulting from 1,4-addition is resonance-stabilized, whereas the anion resulting from 1,2-addition is not. Finally, because there is greater δ– charge at the carbonyl C than at the β-carbon, the nucleophilic addition to the carbonyl C will be faster, meaning that the energy of activation for 1,4-addition ($E_{a1,4}$) will be higher than the energy of activation for 1,2-addition ($E_{a1,2}$). This information can be most clearly displayed on a qualitative **reaction coordinate diagram for nucleophilic addition of a strongly basic nucleophile**:

**1,2-Addition Dominates**

> **Both reactions are Irreversible:**
> **Reaction is Under Kinetic Control**
> *The faster reaction leads to the major product*
> *Occurs when Nu is a poor leaving group (strong base)*

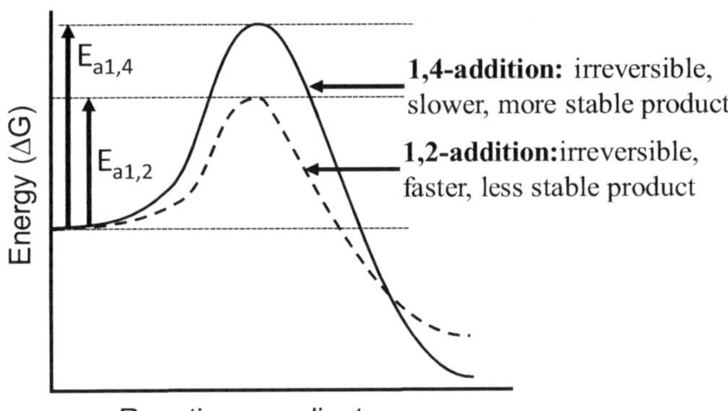


When two competing pathways are both irreversible, the faster process dominates, and the reaction is under kinetic control (Lesson IV.2). So, **when the nucleophile is a strong base, 1,2-addition is the major pathway**.

Next, let us consider the case in which the **nucleophile is a more stable anion** (a weak base, weaker than $HO^-$ for this case). In this case, the nucleophile is stable enough to serve as a leaving group, so addition is **reversible** and therefore it is possible to reach a state of equilibrium. In an equilibrium process, the reaction is under thermodynamic control. As we discussed above, 1,4-addition produces a more stable anion. Here is a qualitative reaction coordinate diagram for nucleophilic addition of a weakly basic nucleophile:

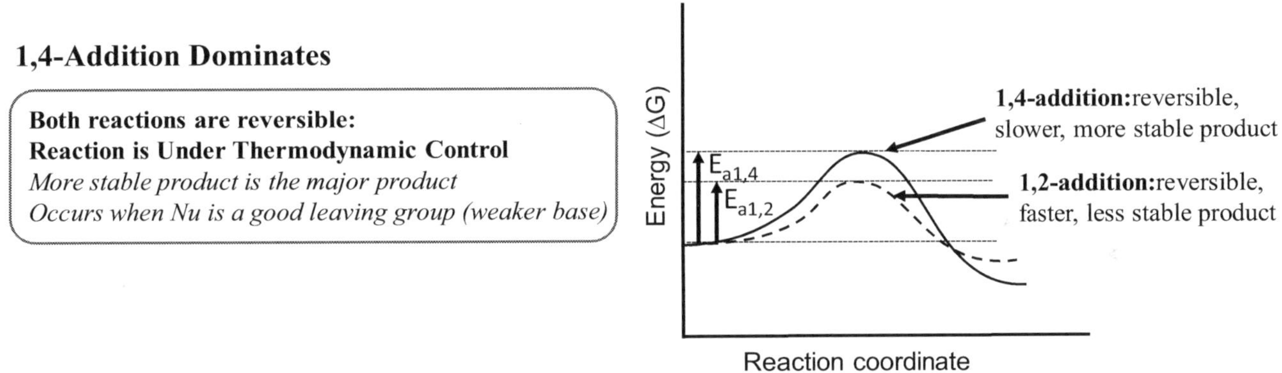


**So, when the nucleophile is a weak base, 1,2-addition is the major pathway.**

*Lesson VI.20.3 Addition of Organometallic Reagents to α,β-Unsaturated Carbonyls*

In our study of organometallic reagents (Part V), we saw that Grignard reagents and organolithium reagents are significantly more reactive (which is the same as saying they are less stable) than are Gilman reagents. We also learned that Grignard reagents and organolithium reagents are very strong bases. For this reason, **RMgX and RLi do 1,2-addition (direct addition) to α,β-unsaturated carbonyls**:

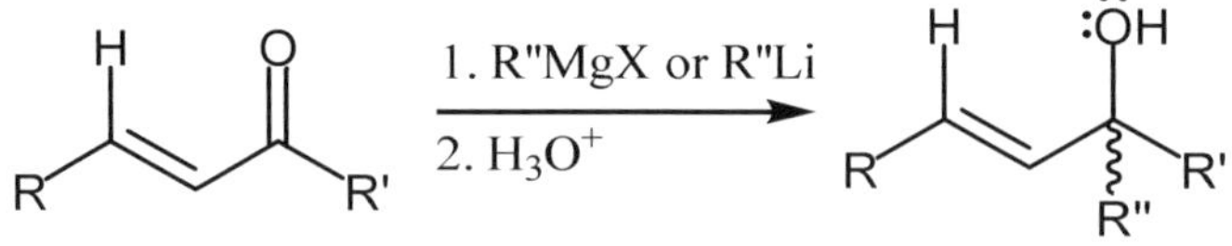

The direct addition is further driven by the strong interaction of the Li or Mg metal with the oxygen atom in the anionic intermediate. The Gilman reagents were developed as a less reactive (more stable) alternative to Grignard and organolithium reagents. They are stable enough that they **$LiCuR''_2$ undergo 1,4-addition (conjugate addition) to α,β-unsaturated carbonyls**:

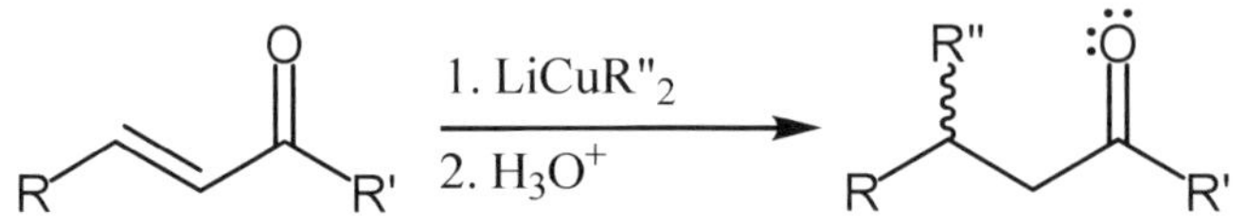

This process is also driven by the stronger interaction of the Cu with the C=C than with the C=O. The reasons for this are beyond the scope of this text.

*Lesson VI.20.4 Michael Addition of Enolates to α,β-Unsaturated Carbonyls*

We saw in Lesson VI.15 that enolates can be generated by deprotonation at a site α- to a carbonyl. These enolates are stabilized by resonance, making them more stable than a typical alkoxide. For this reason, **enolates undergo 1,4-addition (conjugate addition) to α,β-unsaturated carbonyls**:

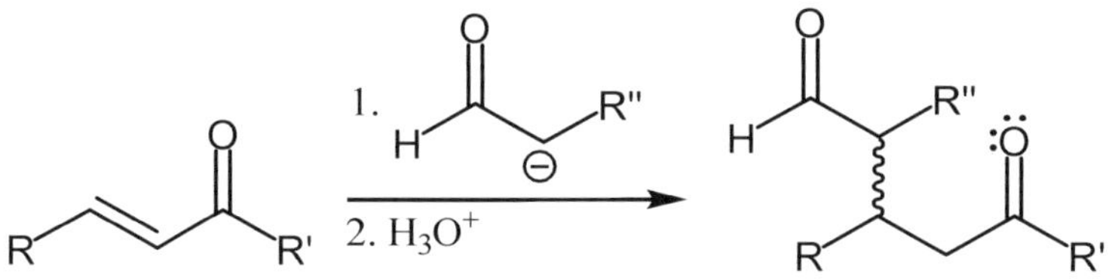

The conjugate addition of an enolate to an α,β-unsaturated carbonyl is called **Michael addition**. If we attempt to conduct a Michael addition by combining an α,β-unsaturated carbonyl and a ketone/aldehyde with a base (to generate the enolate), we will encounter undesired side reactions, because a ketone/aldehyde can also undergo aldol condensation under these conditions:

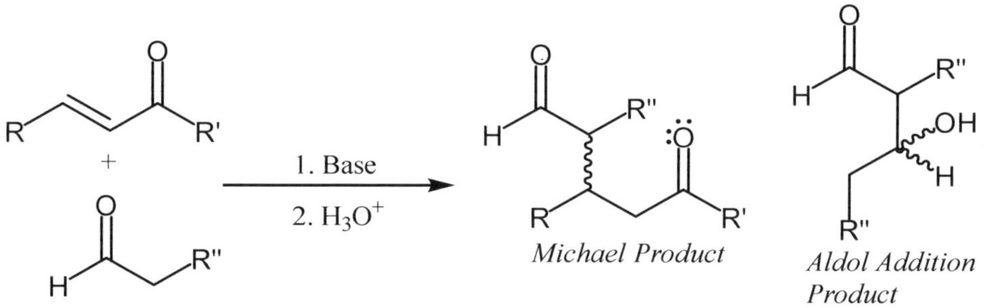

*Michael Product*

*Aldol Addition Product*

A better nucleophile for a Michael addition is a 1,3-dicarbonyl compound that will form a more stable enolate:

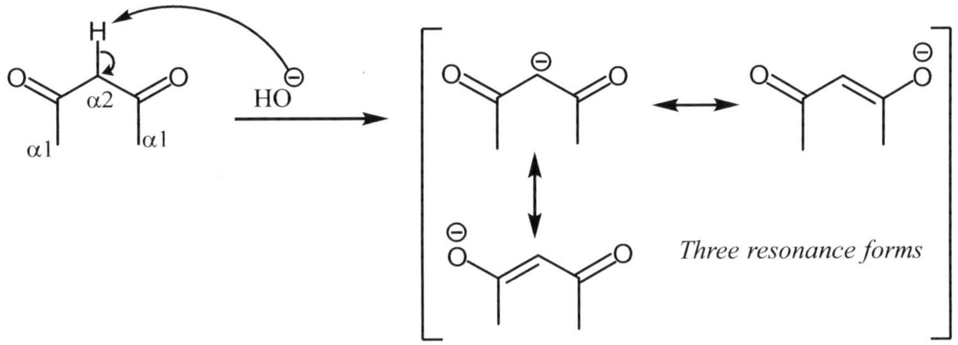

Here is an example of a Michael addition reaction using an enolate derived from a 1,3-diketone:

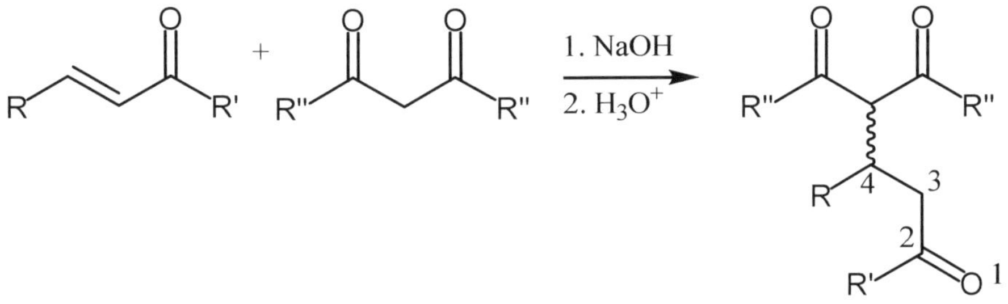

*Lesson VI.20.5 Robinson Annulation*

The Michael addition reaction always produces a product that has two carbonyl groups in it. We know that two carbonyl compounds can react with one another by the **aldol condensation** reaction (Lesson VI.17). We can now employ a Michael addition reaction followed by intramolecular aldol condensation in a useful sequence called the **Robinson Annulation** reaction. Consider the example below. First, the enolate resulting from deprotonation of cyclohexanone undergoes Michael addition to the α,β-unsaturated carbonyl. Then, the α-carbon labelled 1 in the scheme is used as the nucleophile for the aldol condensation to form the double bond to the carbon labelled 6 in the scheme:

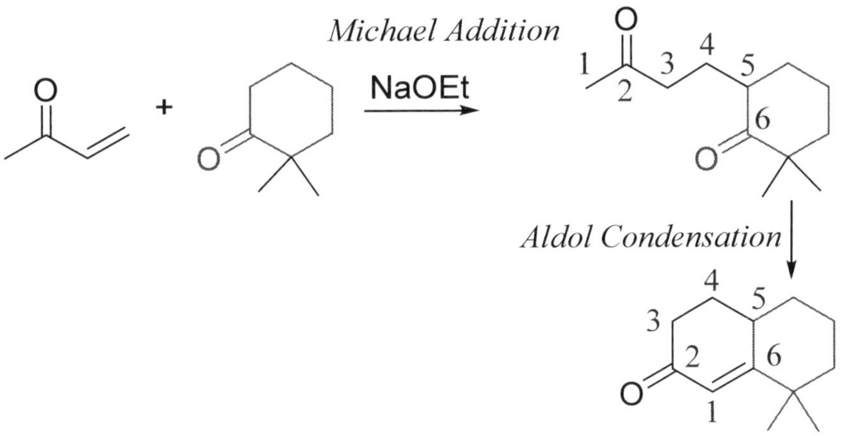

## Part VII: Methods for Determining the Structure of Organic Compounds

Lesson VII.1. Interaction of Ultraviolet and Visible Light with Molecules

Lesson VII.2. UV–Visible Spectroscopy

Lesson VII.3. Interaction of Infrared Light with Molecules

Lesson VII.4. Infrared Spectroscopy

Lesson VII.5. Introduction to Nuclear Magnetic Resonance

Lesson VII.6. Carbon-13 Nuclear Magnetic Resonance Spectrometry

Lesson VII.7. Proton Nuclear Magnetic Resonance Spectrometry

Lesson VII.8. Introduction to Mass Spectrometry

Lesson VII.9. Introduction to Fragmentation Mechanisms

Lesson VII.10. Fragmentation of Alkanes

Lesson VII.11. Fragmentation of Heteroatom-Containing Aliphatics

Lesson VII.12. Fragmentation of Carbonyl-Containing Molecules

Lesson VII.13. Fragmentation of Arenes

Lesson VII.14. Fragmentation of Alkenes

Lesson VII.15. Fragmentation of Other Unsaturated Molecules

## Lesson VII.1. Interaction of Ultraviolet and Visible Light with Molecules

*Lesson VII.1.1 The Electromagnetic Spectrum*

The electromagnetic spectrum stretches from small wavelength gamma rays to long wavelength radio waves. Different wavelengths of light interact with matter in different ways. The region of the electromagnetic spectrum with which we are most familiar is the visible region because light in this range is detectable by our unaided eyes. At longer wavelengths than visible light is **infrared (IR) light**, which we can feel as heat. At shorter wavelengths than visible light is **ultraviolet (UV) light**, which we (unintentionally) detect as sunburned skin and other damage to our tissues. In the next few lessons, we will learn how we can gain structural information about molecules by studying how they interact with electromagnetism in the UV, visible, IR, and radio frequency ranges of the electromagnetic spectrum. These regions of the electromagnetic spectrum are shown in terms of relative energies below:

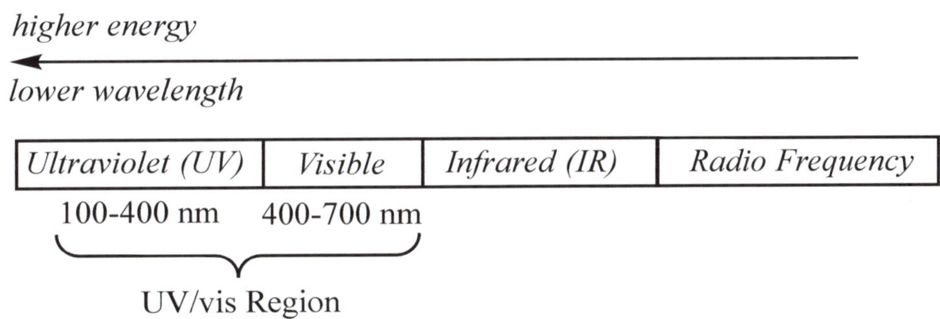


In this lesson, we will focus on how UV and visible light, collectively abbreviated **UV–visible (UV/vis) light**, interact with organic molecules. The UV/vis part of the spectrum we will consider spans a wavelength range from ~100–700 nm.

*Lesson VII.1.2 UV and Visible Light Cause Electronic Transitions in Molecules*

Each covalent bond in an organic compound consists of two electrons shared between the nuclei joined by the bond. When a molecule absorbs UV/vis radiation of an appropriate energy, it causes one of the electrons to undergo an **electronic transition** to a higher-energy orbital that. In terms of energy, the **highest occupied molecular orbital** (abbreviated HOMO) is commonly the orbital holding the electron that is promoted to higher energy upon absorption of light energy. The **lowest unoccupied molecular orbital** is abbreviated LUMO. Here, "occupied" and "unoccupied" refer to whether or not the orbitals contain electrons in them before energy absorption. Energy absorption (in the form of a photon) can promote an electron from the HOMO to the LUMO. The orbital that contains a pair of bonding electrons is called a **bonding orbital**. If the bonding orbital holds electrons in a σ-bond, the orbital is given the symbol σ, whereas if the bonding orbital holds electrons in a π-bond, it is given the symbol π. Upon absorption of an appropriate energy photon of UV/vis light, a σ-bonding electron will generally be promoted to a σ-**antibonding orbital**, given the symbol σ*:

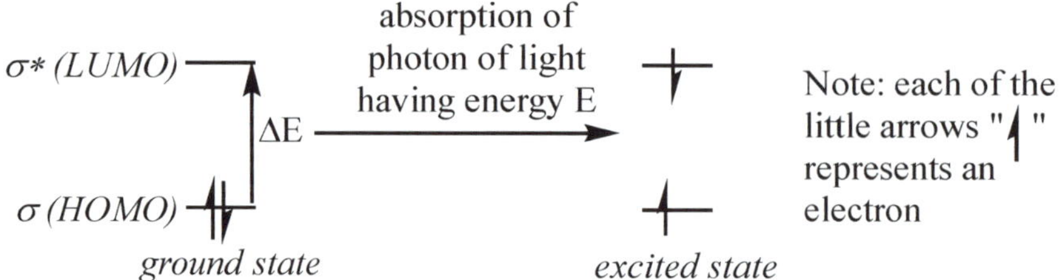


This type of transition is called a **σ→σ* transition** (read "sigma to sigma star transition"). Similarly, a π-bonding electron generally gets promoted to a π-antibonding orbital, given the symbol π*:

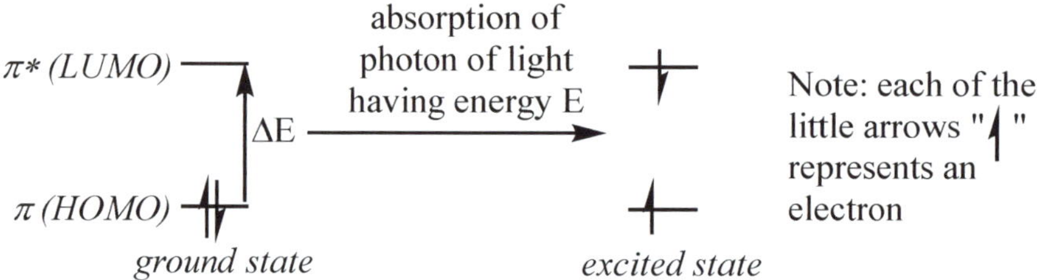


This type of transition is called a **π→π* transition** (read "pi to pi star transition").

Regardless of it is a σ→σ* transition or a π→π* transition, the energy of the photon must be identical to the energy gap between the electronic energy levels (ΔE in the figures above) for the photon to be absorbed and cause the electronic transition to occur. Longer wavelength light – towards the visible end of the UV/vis range – can only cause transitions requiring less energy. The lower energy transitions in organic molecules are usually π→π* transitions. By measuring the wavelength of UV/vis light that is absorbed by the sample, some knowledge about the bonding in the sample molecule can be inferred. The types of information we can gain by measuring absorption of UV-vis light are detailed in Lesson VII.2.

<u>Example VII.1.1</u>

Which of these molecules will have a lower-energy electronic transition? What type of transition would the HOMO–LUMO transition represent in each?

**I** **II**

Solution VII.1.1

The lower-energy transitions in organic molecules tend to be π–π* transitions, so molecule **I**, which has a π-bond, would be expected to have a lower energy (longer wavelength) transition than would molecule **II**. The transition observed in the UV-vis spectrum of molecule **I** will be a π–π* transition, while the transition observed for molecule **II** would be a σ–σ* transition.

*Lesson VII.1.3 Comparing σ→σ* to π→π* Transitions, and the Effect of π-Conjugation*

Because σ-bonding electrons lie nearer the nuclei than π-bonding electrons, they are held more tightly by Coulombic attraction to the nucleus than are π-bonding electrons. It therefore takes more energy ($\Delta E_2$ in the figure below) to pull an electron out of a σ orbital to promote it to a σ*-orbital than it does to promote a π-bonding electron to a π*-orbital (requiring $\Delta E_1$ in the figure below). A qualitative diagram showing both the σ→σ* and the π→π* transitions in a molecule that has both types of bonds will consequently look like this:

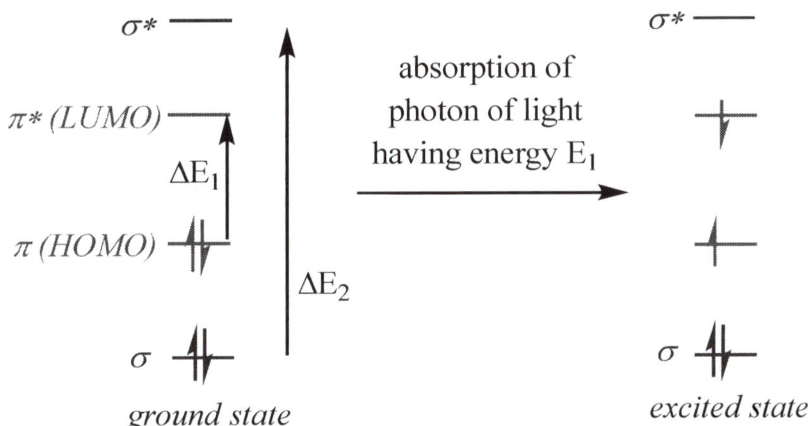


What if a molecule has more than one π-bond? The answer depends on whether the π-bonds in the molecule are isolated or conjugated (see Lesson IV.1). If the two π-bonds are isolated, they will not interact with one another, so the π-orbitals lie at similar energies. This results in an electronic transition at about the same energy as if there were only one π-bond. However, two photons will be needed per molecule to cause both to promote both the π→π* transitions (i.e., a molecule with two isolated π-bonds will absorb twice as much light as a molecule with only one π bond). The amount of light absorbed per mole of a sample is called the **molar absorptivity** or **molar extinction coefficient**.

We also know that conjugated π-bonds are more stable than isolated π-bonds. It takes less energy to cause an electronic transition of an electron in a conjugated π-system than it does to cause the transition in an isolated π-bond. One reason for this is that the single electron that is left in the bonding orbital – which we can think of as a radical – has resonance stabilization, so it takes less energy to form than it would take to form the radical resulting from an isolated π-bond that lacks resonance stabilization. In fact, **the longer the π conjugated system, the lower the energy of the photon needed to promote the π→π* transition.** The energy of UV-vis light absorbed and the number of photons absorbed at that energy allow us to assess the relative number of π-bonds in a compound and the extent to which these π-bonds are conjugated.

Example VII.1.2

Which of these molecules will have the lowest-energy electronic transition? Which would have the highest molar absorptivity?

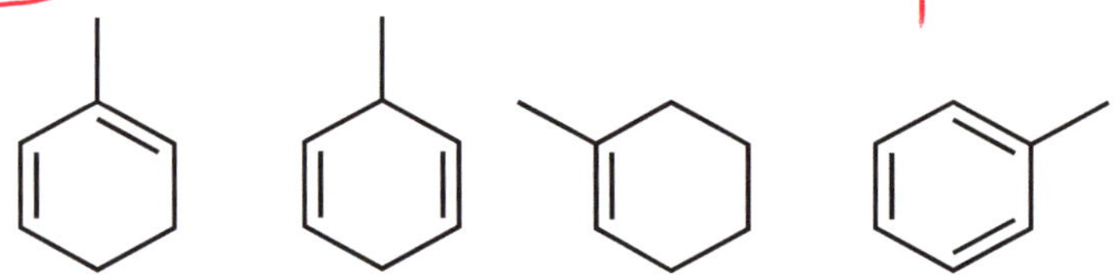

**I** **II** **III** **IV**

Solution VII.1.2

The lowest-energy transition will occur for the molecule with the most-extended π-conjugated system. In molecules **II** and **III** have only isolated alkenes. Molecule **I** has a π-conjugated system of two π-bonds, whereas molecule **IV** features a π-conjugated system of three π bonds. This analysis suggests that molecule **IV** will have the lowest-energy (highest wavelength) band in its UV-vis spectrum. Molecule **IV** is also expected to have a higher molar absorptivity because it has the greatest number of π bonds per molecule as well.

## Lesson VII.2. UV–Visible Spectroscopy

*Lesson VII.2.1 The UV/vis Spectrum*

In Lesson VII.1, we learned that an electronic transition occurs when a molecule absorbs UV-vis light of an appropriate energy. In the current lesson, we will learn how scientists have developed a technique, called **UV/vis spectroscopy**, that correlates the absorption of light at a given wavelength with molecular properties, intermolecular forces, and chemical reactions. The operating principle of a typical UV-vis spectrometer is illustrated as follows:

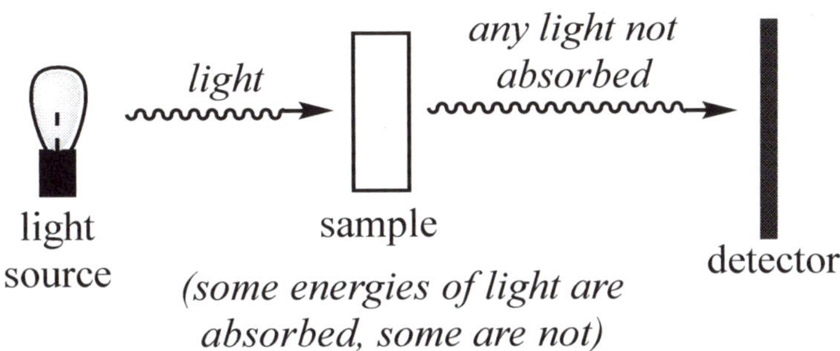


The light source generates a beam of light, which is directed through the sample. The source provides one specific wavelength at a time, scanning over the range of wavelengths that is set by the experimentalist. The detector, located on the opposite side of the sample relative to the source, detects the intensity of the light that passes through the sample relative to the amount of light emitted by the source. If the sample absorbs light at a particular wavelength, the detector will detect a decrease in light coming through at that wavelength. The computer system then provides the user with a plot of how much light was absorbed versus the wavelength at which it was absorbed. A sample that has electrons in bonds that undergo an electronic transition at 220 nm, for example, might produce a UV-vis spectrum like this:

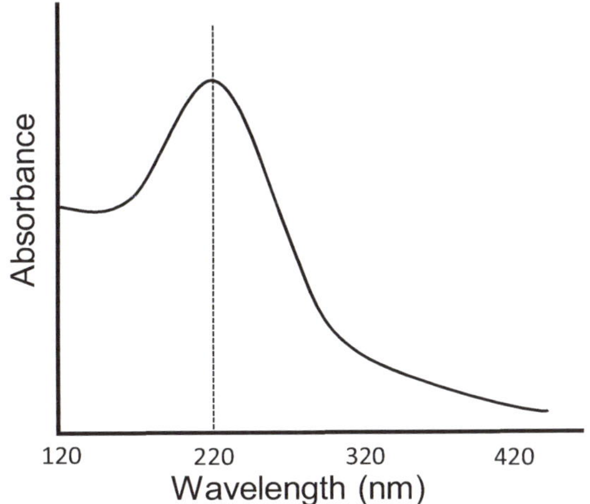

A few things should be noted at this point. If we increase the **concentration** (*c*) of the compound in the sample cell that is absorbing light at 220 nm, then more light will be absorbed. If we increase the size of the sample cell (the **pathlength**, *b*), then then more light will be absorbed. For a certain concentration and a certain pathlength, the amount of light absorbed will be a constant for a given electronic transition. This constant that relates the absorbance to the concentration and the pathlength is called the **absorptivity**, or the **extinction coefficient**, and it is given the symbol ε. The **Beer-Lambert Law** provides an equation relating the absorbance (*A*), pathlength (*b*), concentration (*c*) and extinction coefficient (ε):

$$A = \varepsilon bc$$

The UV-vis spectra shown below illustrate the Beer-Lambert Law:

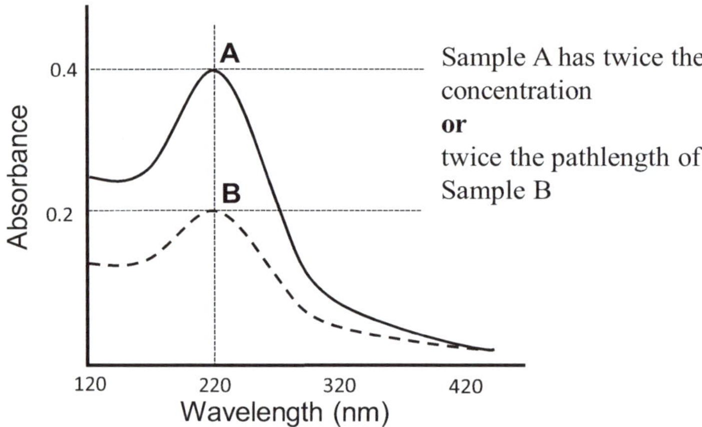


Usually, a sample cell with a pathlength of 1 cm is used, which simplifies the relationship between absorbance and concentration. The well-defined way that absorbance relates to concentration for a given pathlength allows us to measure the concentration of a molecule as a function of time with very high accuracy and sensitivity. As a result, **UV-vis spectroscopy is an excellent way to quantify reaction rates.** Let us consider, as an example, a sample that contains 1 mole each of reactants **A** and **B**, where **B** is the only species that absorbs at 500 nm, and you experimentally measure an absorbance of 0.6. After 1 h has elapsed, you measure the absorbance at 500 nm and now obtain a value of 0.3, which allow you to deduce that the reaction between **A** and **B** has proceeded at a rate of 0.5 moles $h^{-1}$ during this time period.

## Lesson VII.3. Interaction of Infrared Light with Molecules

*Lesson VII.3.1 Infrared Radiation Causes Vibrational Transitions in Molecules*

In Lessons VII.1–2, we saw how UV-vis light interacted with organic molecules to cause electronic transitions. In this lesson, we will learn what happens when lower-energy IR light interacts with organic molecules. Unlike UV-vis light, infrared light does not have enough energy to cause an electron to be promoted to a higher electronic level in the majority of organic molecules. Instead, molecules undergo **vibration** upon interaction with IR light. The wavelength of light that is most commonly used to study the vibrational modes of organic molecule runs from about 2500-20,000 nm. There are several modes of molecular vibration that result when a molecule absorbs infrared energy. Molecular vibrational modes can be broadly divided into **stretching** and **bending** modes. Stretching modes involve changes in the bond lengths. Bending modes involve changes in bond angles. Bond stretching and bond bending in molecules occurs in well-defined combinations at specific energies. Let us begin by looking at the stretching modes. Consider how the C–H bonds in a $CH_2$ unit might undergo stretching in a molecule:

*Symmetric stretch: bonds lengthen and shorten at the same time*

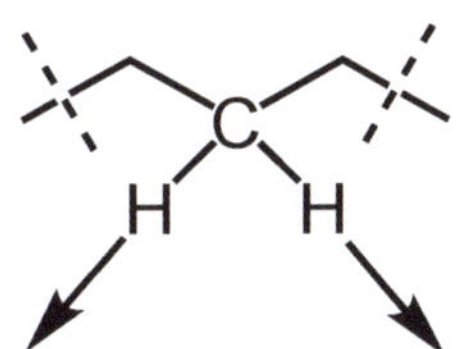


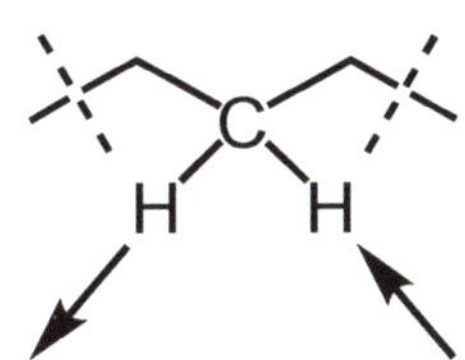


*Asymmetric stretch: One bond lengthens as the other shortens*

There are more possibilities for bending modes. Not only may a given bending mode be symmetric or asymmetric, but the different bending modes may involve the atoms remaining coplanar as they bend, or they may bend in a way that they are no longer coplanar:

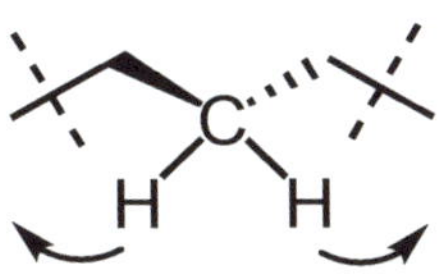


*Symmetric in-plane bending: Both angles increase or decrease at the same time by bending in the plane of the page*

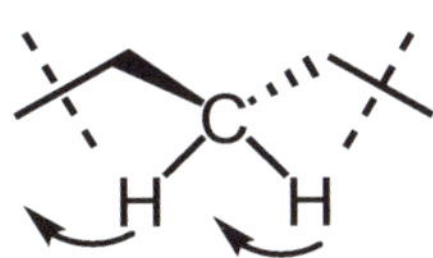


*Asymmetric in-plane bending: One angle increases as the other decreases by bending in the plane of the page*

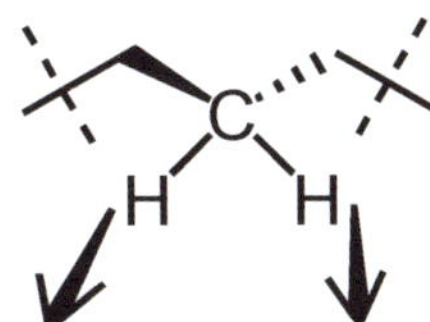


*Symmetric out-of-plane bending: Both angles increase or decrease at the same time by bending out of the plane of page (here, both shown bending towards the viewer)*

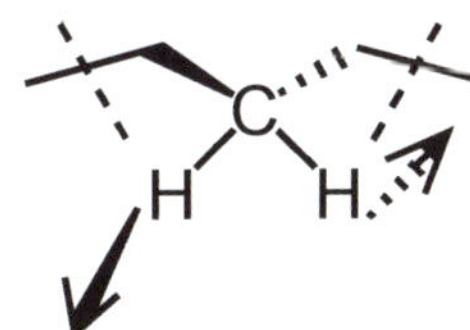


*Asymmetric out-of-plane bending: One angle increases as the other decreases by bending out of the plane of page (here, one bending towards, one away from the viewer)*

### on VII.3.2 Different Types of Bonds Absorb IR Radiation at Different Energies

Bonds can be thought of as springs. The strength of a spring is reflected in how much energy it takes to stretch and compress it. Likewise, the energy at which a bond stretches depends on the strength of the bond. A stronger bond requires more energy to stretch than does a weaker bond. This means that a set of strong bonds absorbs a higher energy of IR light than does a set of weaker bonds. This means that a vibrational mode involving a set of strong bonds will absorb higher energy light than a set of weaker bonds. This phenomenon is invaluable for determining the chemical identity of bonds present in a given sample based on what energies of IR light the sample absorbs.

In addition to the energy of the IR light absorbed, the *amount* of light absorbed per molecule can also provide us with information about the bonds present. Due to some considerations that are outside the scope of this course, one result of how light and charged species interact is that the more polar bonds in a sample absorb more of the IR energy to which the sample is exposed than do less polar bonds. The next lesson will illustrate how the interaction of a molecule with IR light can be used to infer structural information about that molecule.

## Lesson VII.4. Infrared Spectroscopy

*Lesson VII.4.1 The IR Spectrum*

In Lesson VII.3, we learned that a molecule absorbs IR light of an appropriate energy to excite vibrational modes for specific sets of bonds in that molecule. In the current lesson, we will learn how scientists have developed a technique that allows us to use this property of matter to gain a wide range of knowledge about molecules and reactions. The infrared spectrometer instrument setup is very similar to the UV-vis spectrometer we saw in Lesson VII.2, but using IR radiation in place of UV-vis wavelength light:

IR source — *IR radiation* → sample — *any IR radiation not absorbed* → detector

*(some energies of IR radiation are absorbed, some are not)*

A typical IR spectrum provided by the IR spectrophotometer is a plot of transmittance (the % of light that is **not absorbed** by the sample) versus the energy of light (in units of **wavenumbers**, $cm^{-1}$). A higher wavenumber value corresponds to higher energy. If a sample absorbs all of the IR radiation from the IR source, the transmittance would be zero and a downward peak would extend all the way to the bottom of the spectrum:

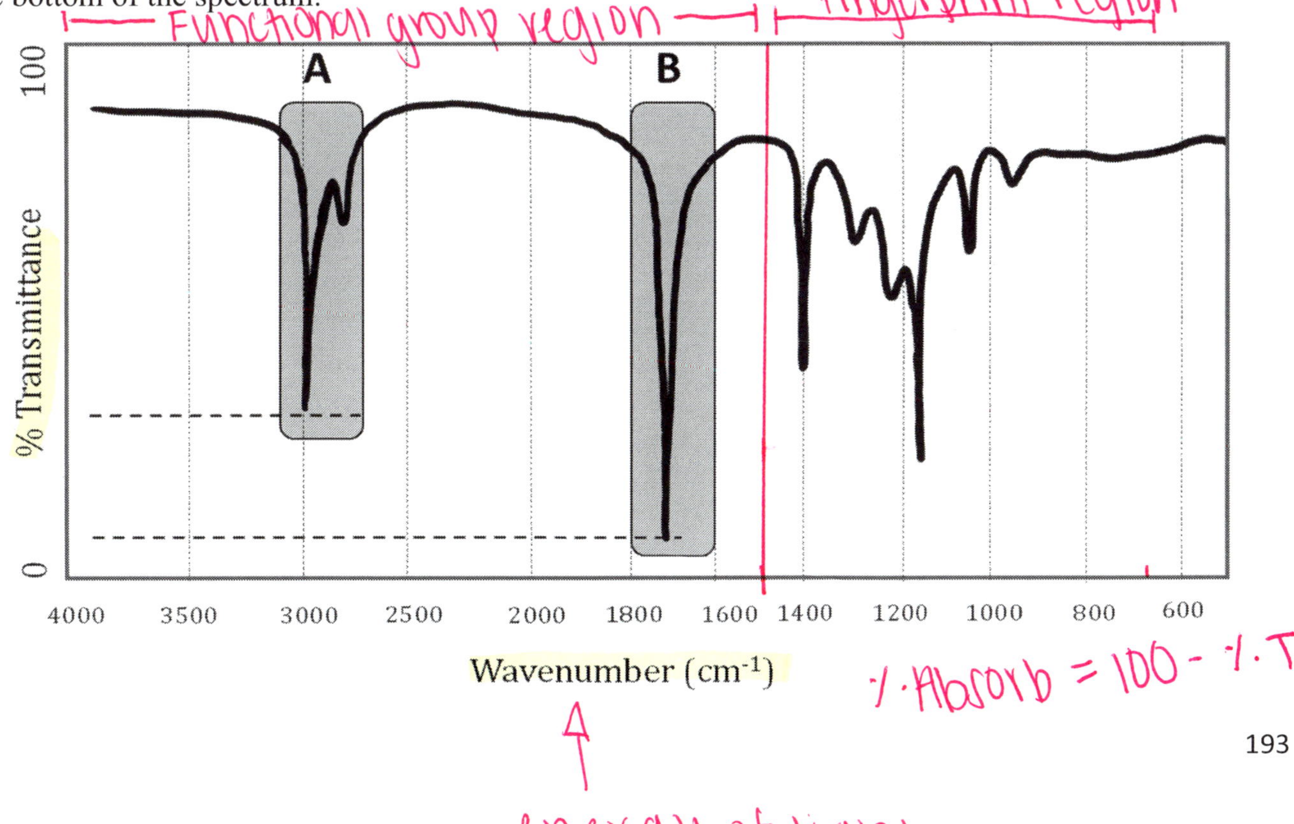

The above spectrum shows the typical *x*-axis range of 4000–500 $cm^{-1}$ (high energy to low energy). Two bands are highlighted. Band **A** corresponds to a set of bonds that absorbs ~75% of the IR light at 3000 $cm^{-1}$ emitted by the source (i.e., 25% transmittance), whereas band **B** corresponds to a set of bonds that absorbs nearly 95% of the IR light at 1750 $cm^{-1}$ (i.e., 5% transmittance).

*Lesson VII.4.2 Information Provided by the IR Spectrum*

The infrared spectrum of a compound can provide information about the types of bonds, and therefore the functional groups, that are present in a given compound. IR spectra are usually less cluttered at the higher-energy region of the IR spectrum (4000–1500 $cm^{-1}$), which is referred to as the **functional group region**. The lower-energy region of the IR spectrum of the spectrum (1500–500 $cm^{-1}$) is referred to as the **fingerprint region**, because the peaks there are characteristic to each given molecule. However, the fingerprint region is significantly more cluttered and is thus more difficult to assign specific transitions. The characteristic peaks for the common functional groups encountered in an undergraduate organic chemistry course are provided below:

| Bond | Energy ($cm^{-1}$) | Intensity |
|---|---|---|
| N≡C | 2255-2220 | m-s |
| C≡C | 2260-2100 | w-m |
| C=C | 1675-1660 | m |
| N=C | 1650-1550 | m |
| benzene ring | 1600 **AND** 1500-1425 | w-s |
| C=O | 1775-1650 | s |
| C—O | 1250-1000 | s |
| C—N | 1230-1000 | m |
| O—H | 3650-3200 | s (br) |
| O—H | 3300-2500 | s (br) |
| N—H | 3500-3300 | m (br) |
| C—H | 3300-2725 | m |

| C-H Bond (Stretch) | Energy ($cm^{-1}$) |
|---|---|
| C≡C—H | 3300-ish |
| C=C—H | 3100-3000 |
| C—C—H | 2950-2850 |
| aldehyde C(=O)—H | 2820-ish and 2720-ish |

**C-H Bond (Bending)**

| C-H Bond (Bending) | Energy ($cm^{-1}$) |
|---|---|
| —$CH_3$, —$CH_2$—, —C(H)— | 1450-1400 |
| trans (H/R, R/H) | 980-960 |
| cis (R/R, H/H) | 730-670 |
| trisubstituted (R/R, R/H) | 840-800 |
| monosubstituted (R/H, H/H) | 990 and 910 |
| disubstituted terminal (R/H, R/H) | 890 |

*Lesson VII.4.3 Interpreting IR Spectra for Simple Molecules*

Knowing the energy at which each of the common organic functional groups absorbs in the IR spectrum, we are now able to look at an IR spectrum and identify what specific functional groups are present or absent. The following examples demonstrate the method by which one deduces the possible structure of a compound from its IR spectrum. Numerous examples to illustrate how IR spectra can be used to gain structural information are provided in Part VIII.

*Lesson VII.4.4 Influence of Resonance and Pi-Conjugation on IR Band Energies*

We know that the resonance hybrid structure is a better representation of the "real" structure of a molecule. The resonance hybrid reflects the fact that the "real" bond order in the hybrid structure is different from that in any of the individual resonance contributors. Consider, for example, the resonance hybrid for a carboxamide that takes into account the two contributors:

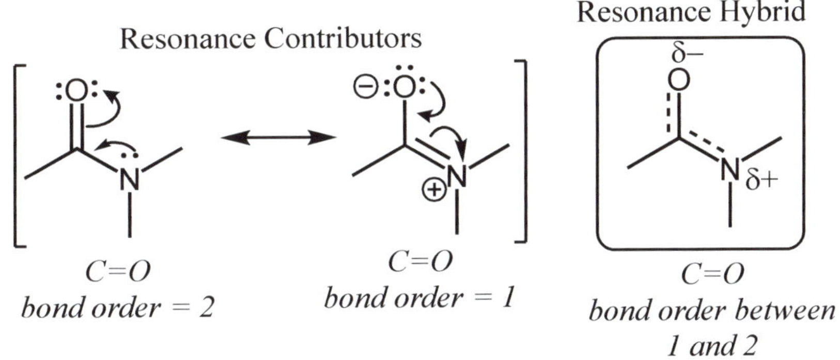


One result of the resonance is that the actual C–O bond order is < 2. For this reason, the C=O bond in a carboxamide is weaker than a C=O bond that does not participate in resonance delocalization, and will thus have an IR band at a lower wavenumber (~1680 $cm^{-1}$). As a rule of thumb, a C=O bond that can engage in resonance delocalization with an adjacent C=C bond will have a peak at ~20 $cm^{-1}$ lower in energy than a C=O bond that does not participate in resonance. Resonance effects can also be observed in the IR absorptions by C=C bonds. Notice in the table of IR absorptions in Lesson VII.4.2 that a C=C bond in a typical alkene absorbs in the 1660–1675 $cm^{-1}$ range, whereas a C=C bond in an arene absorbs at energies below 1600 $cm^{-1}$. This reflects the resonance delocalization of $\pi$-bonds in arene rings, whose highest C–C bond order is 1.5, rather than the 2 for a typical alkene.

## Lesson VII.5. Introduction to Nuclear Magnetic Resonance

*Lesson VII.5.1 Behavior of Nuclei in a Magnetic Field*

Although chemistry primarily focuses on electrons, the behavior of some nuclei can provide useful information about the bonding within a molecule. This information can be obtained due to the fact that some nuclei have *magnetic* dipoles (not to be confused with dipoles in polar molecules). Not all elements and not all isotopes of a single element have a nuclear magnetic dipole, but fortunately many of the elements present in common organic molecules do have isotopes whose nuclei have magnetic dipoles: hydrogen-1 ($^{1}H$), carbon-13 ($^{13}C$), phosphorus-31 ($^{31}P$), and fluorine-19 ($^{19}F$). Nuclei such as these that have magnetic dipoles are called **NMR active** nuclei and can be observed by **Nuclear Magnetic Resonance (NMR) spectrometry**. Under typical conditions, the magnetic dipole vectors for the nuclei in a sample will be oriented in random directions. If an external magnetic field is applied, however, these magnetic vectors will align such that they are pointing in either the same direction (parallel) or opposite direction (antiparallel) of the applied field:

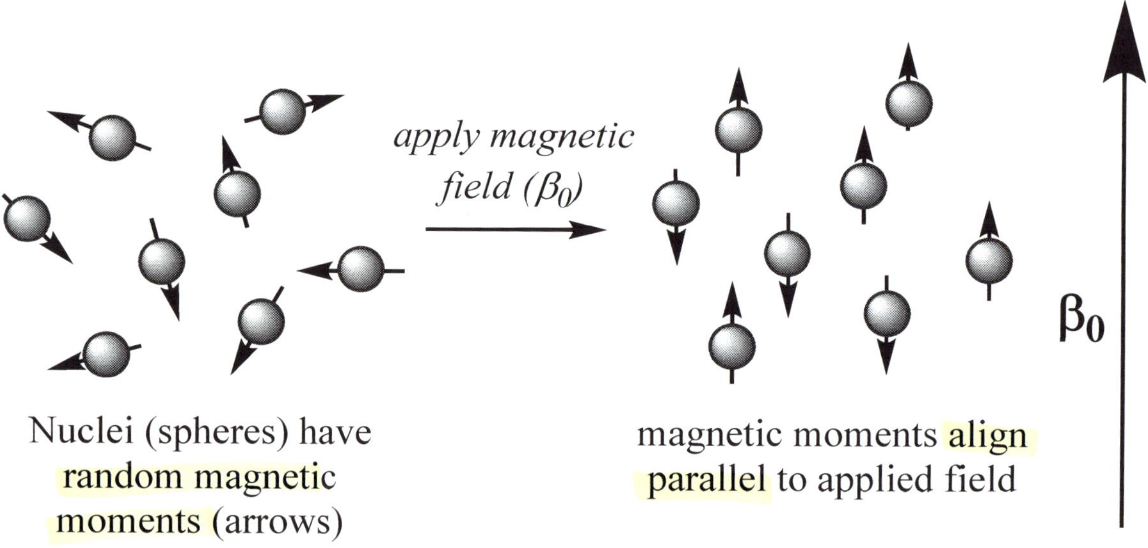


Nuclei that are aligned parallel with a magnetic field will have a lower energy than nuclei aligned antiparallel. The stronger the applied field, the greater the energy difference ($\Delta E$) between these two spin states. In the magnetic field strength used in NMR spectroscopy, the energy difference between these two spin states is comparable to the energy of a photon in the radio frequency range (usually 60–1000 MHz). When **radio frequency (RF) radiation** is transmitted into the sample at an energy that matches $\Delta E$, these photons will be absorbed and will promote nuclei from the lower-energy parallel spin state to the higher-energy antiparallel spin state:

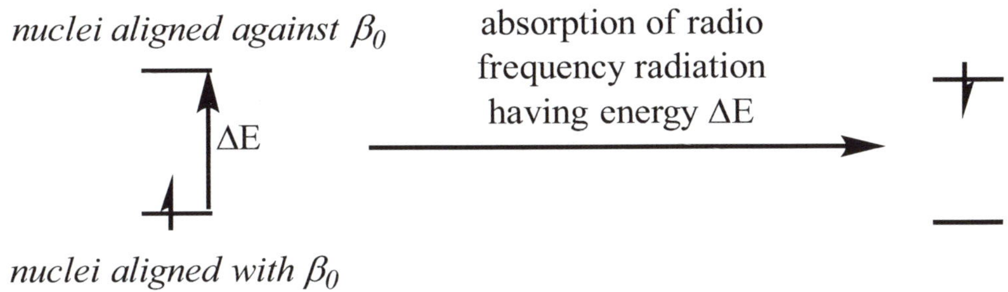


The detector in an NMR spectrometer measures the absorption of the RF energy and can thus report the energy between states for each nucleus in a sample.

*Lesson VII.5.2 The NMR Spectrum*

An NMR spectrometer subjects a sample to a strong magnetic field. A coil around the sample detects absorption of RF photons applied by the instrument. Each instance of the absorption of RF radiation is detected and reported to a computer, whereby this energy absorption is represented as an upwards-pointing peak (also known as a **signal**). The particular magnetic field and radio frequency range required depend on the specific type of nucleus that is being detected. A $^{1}H$ NMR spectrum will have an appearance like this:

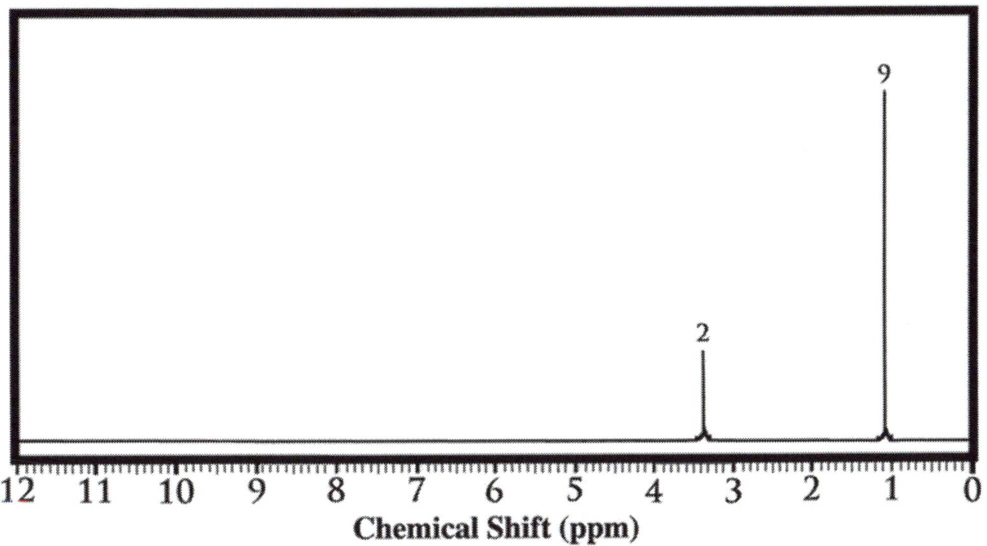


This example features two signals: one at 1.1 ppm and one at 3.3 ppm on the *x*-axis. The units of the *x*-axis are **parts per million (ppm)** and correspond to **chemical shift**, which correlates with energy. In the rest of this Lesson, we will discuss how the chemical shifts of the different nuclei are influenced by (1) the other nuclei and (2) the electrons in the sample. In the few lessons that follow this one, we will study how NMR spectrometry ($^{1}H$ and $^{13}C$ NMR in particular) can be used to elucidate organic molecule structure.

*Lesson VII.5.3 Factors Influencing the Energy Between Spin States*

Not all nuclei in a molecule exhibit the same $\Delta E$ between spin states, because (1) the electrons in a molecule affect how the various nuclei "experience" $\beta_0$ and (2) the electronic environment varies for each nucleus according to its chemical environment. For example, there will be more electron density near a nucleus at the $\delta^-$ end of a polar bond than at the $\delta^+$ end. How does this influence the effective magnetic field strength at a given nucleus? Well, not only does $\beta_0$ applied by the NMR spectrometer interact with the magnetic spin of nuclei, it also interacts with the magnetic spin of electrons. As a result, the electrons generate their own localized magnetic field that opposes $\beta_0$. Thus, as the electron density that surrounds a nucleus increases, the opposing magnetic field those electrons create also increases, which decreases the $\beta_0$ that particular nucleus experiences, and consequently it takes less energy to promote that nucleus to a higher spin state. As a result, **electrons shield the nuclei from the applied magnetic field.**

The NMR spectrometry experiment operates on a timescale slower than the rate of σ-bond rotation and slower than the rate of molecular diffusion or rotation. This means that all nuclei that interconvert by free bond rotation or that can be interconverted by molecular rotation are **magnetically equivalent** and will be promoted at identical energies. On the other hand, nuclei that cannot be interconverted by σ-bond or molecular rotation will have slightly different electron densities around them and will thus require a slightly different energy to be promoted. **Each set of magnetically-inequivalent nuclei will give a signal at a different energy in the NMR spectrum**.

*Lesson VII.5.4 Effect of Neighboring Nuclei on NMR Energies*

As we learned in Lesson VII.5.1, each nucleus has a magnetic dipole that aligns either parallel or antiparallel with the applied magnetic field of an NMR spectrometer. However, just as the magnetic spin of the electrons surrounding a nucleus can affect the $\beta_0$ it "experiences", so too can the magnetic spin a neighboring nucleus. This means that, depending on the direction of nucleus A's magnetic field, nucleus A may shield or reinforce the magnetic field experienced by its neighboring nucleus B:

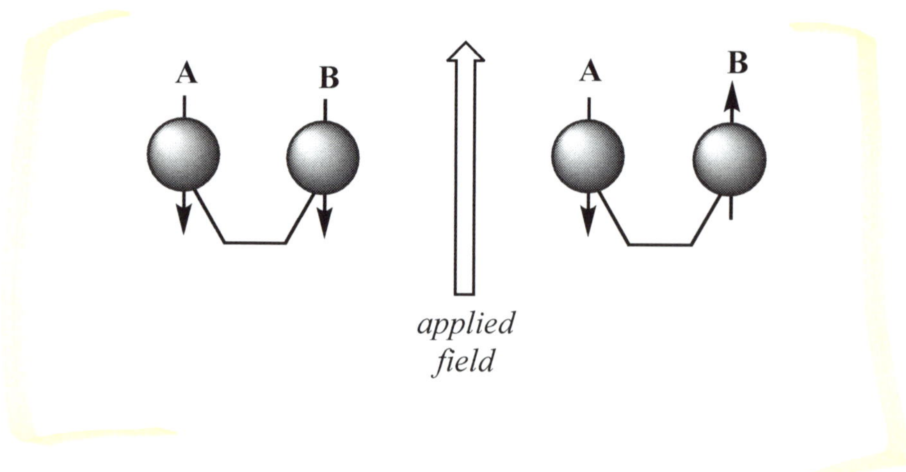

For a nucleus having one nearby nucleus, then, there are two slightly different energies needed to promote the nucleus. The difference in energies is very slight in cases we will encounter in this book, so the result is that the peak for the nucleus is “split” into two peaks. The peak looks like this:

This peak shape is called a doublet. As more and more nuclei are present on neighboring sites, the number of smaller signals into which a signal splits increases, as does the complexity of the peak shape. The number of peaks into which a signal splits is called the **multiplicity** (*m*). For the nuclei studied in organic chemistry, $m = n+1$, where $n$ is the number of nuclei (only counting those capable of causing splitting) on adjacent C atoms. In common organic molecules, a signal that is well-defined will generally not have a multiplicity greater than seven (six splitting atoms on adjacent C atoms). Those signals having a higher multiplicity generally are too broad to readily see the splitting on the strength of spectrometers in common use. The following schematic shows the typical shape of simple **multiplets**:

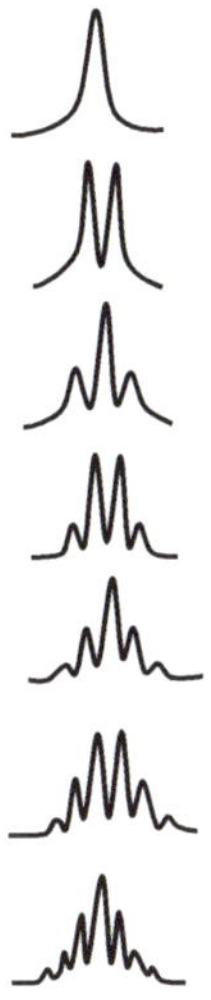

| Type of peak | Ratio of heights |
|---|---|
| singlet | 1 |
| doublet | 1:1 |
| triplet | 1:2:1 |
| quartet | 1:3:3:1 |
| quintet | 1:4:6:4:1 |
| sextet | 1:5:10:10:5:1 |
| septet | 1:6:15:20:15:6:1 |

## Lesson VII.6. Carbon-13 Nuclear Magnetic Resonance Spectrometry

### *Lesson VII.6.1 The NMR Spectrum*

In the current Lesson, we will learn how a $^{13}C$ NMR spectrum can reveal structural information about a molecule. In the previous Lesson, we learned about nuclear magnetic resonance and that the energy at which a nucleus absorbs energy is influenced by its electronic environment. This means that the chemical shifts at which peaks appear in a $^{13}C$ NMR spectrum reveals the types of carbons in the molecule. The diagram below is a representative sample of a $^{13}C$ NMR spectrum:

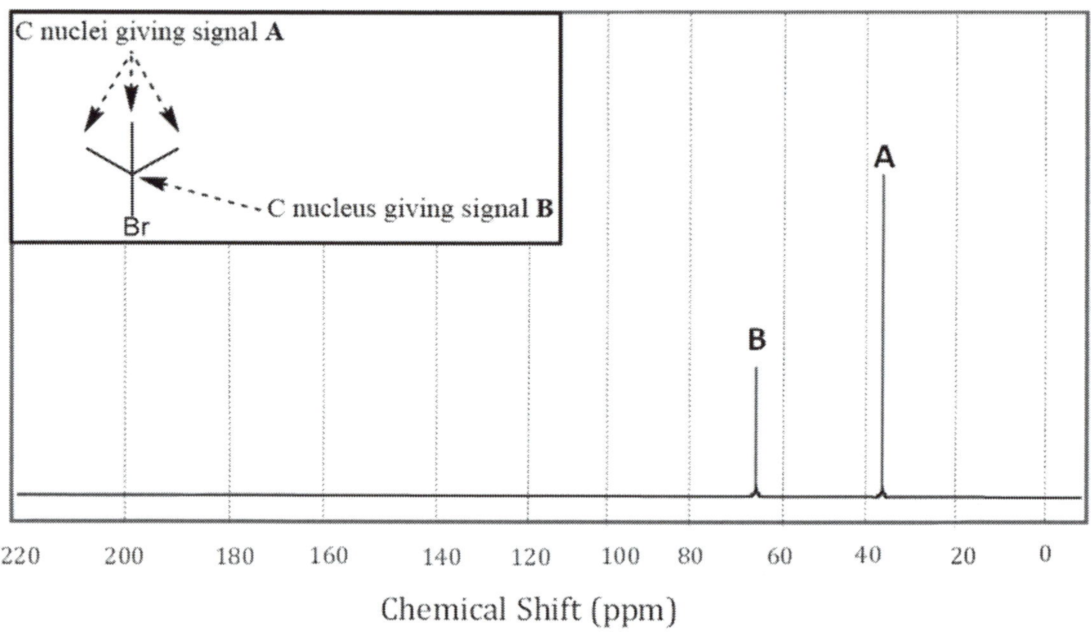


In this spectrum, each $^{13}C$ nucleus (or set of magnetically inequivalent $^{13}C$ nuclei) has a resonance that produces an upward-pointing peak.

### *Lesson VII.6.2 Avoiding Splitting in $^{13}C$ NMR spectra*

We saw in Lesson VII.5 that the interaction of one nuclear magnetic dipole with another can cause an NMR signal to split into a multiplet. This phenomenon can make spectra substantially more difficult to interpret than if each peak was just a singlet. Most $^{13}C$ NMR spectra typically contain only singlets. Why is this? The first reason is that, although a carbon-13 nucleus is NMR-active, 99% of all carbon atoms in the universe are the carbon-12 isotope which is not NMR-active. It is therefore statistically very unlikely for one $^{13}C$ atom to be bonded directly to another $^{13}C$ atom (i.e., 0.01% probability), so C–C splitting is rarely observed. Most organic molecules feature multiple H atoms attached to C. Protons ($^{1}H$) are NMR-active nuclei, so the fact that each $^{13}C$ atom has one or more $^{1}H$ atom would produce a lot of splitting in a $^{13}C$ NMR spectrum and render interpretation nearly impossible. To prevent the splitting of $^{13}C$ NMR signals by $^{1}H$ nuclei, the $^{13}C$ NMR spectrometer is typically run in **proton-**

**decoupled mode**. In this mode, the spectrometer suppresses the [1]H splitting effect so that each [13]C nucleus in a molecule appears as a single, unsplit peak in the [13]C NMR spectrum.

*Lesson VII.6.3 Carbon-13 NMR Spectra Provide Information on Types of C in a Sample*

We know that the electronic environment around an atom differs based on its chemical environment (number and types of bonds). These differences produce predictable shifts in [13]C NMR resonances depending on what substituents the C atoms carry. Chemical shifts for [13]C nuclei in the bonding environments most frequently encountered in organic molecules are listed below:

| Carbon (Shown) | Chemical Shift |
|---|---|
| $Si(CH_3)_4$ | 0 |
| $—CH_3$ | 10-35 |
| $—CH_2—$ | 15-50 |
| —CH(—)— | 20-60 |
| —C(—)(—)— | 30-40 |
| =C< | 100-150 |
| Benzene C—H | 110-175 |
| ≡C— | 60-85 |

| Carbon (Shown) | Chemical Shift |
|---|---|
| O—C | 50-80 |
| N—C | 40-60 |
| C—X | X = I 0-35<br>X = Br 20-65<br>X = Cl 35-80 |
| —C(=O)—Y | Y =<br>H 190-200<br>R 200-220<br>OR 160-180<br>OH 175-185<br>$NR_2$ 165-175 |

Examining the peaks in a [13]C NMR spectrum and comparing the chemical shifts therein to this table can provide significant insight into the structure of the molecule. In the [13]C NMR spectrum of 2-bromo-2-methylpropane on the previous page, the three magnetically-equivalent methyl carbons produce peak **A** at 35 ppm, which falls within in the range (10–35 ppm) suggested in the table above. The quaternary C attached to the bromine produces a peak at 65 ppm, again within the typical range for carbons bearing Br substituents in the table (20–65 ppm). Even if we were not provided with the structure and only had the spectrum, we could easily conclude that the sample molecule contains no alkenes, carbonyls, alkynes, or aromatics because no [13]C resonances characteristic for these functional groups appear in the spectrum.

## Lesson VII.7. Proton Nuclear Magnetic Resonance Spectrometry

### *Lesson VII.7.1 Chemical Shifts in $^1H$ NMR Spectra*

In Lesson VII.6, we saw that $^{13}C$ NMR spectra can provide significant insight into the structures of organic molecules. We can gain even more information from a $^1H$ NMR spectrum, which will look like the following:

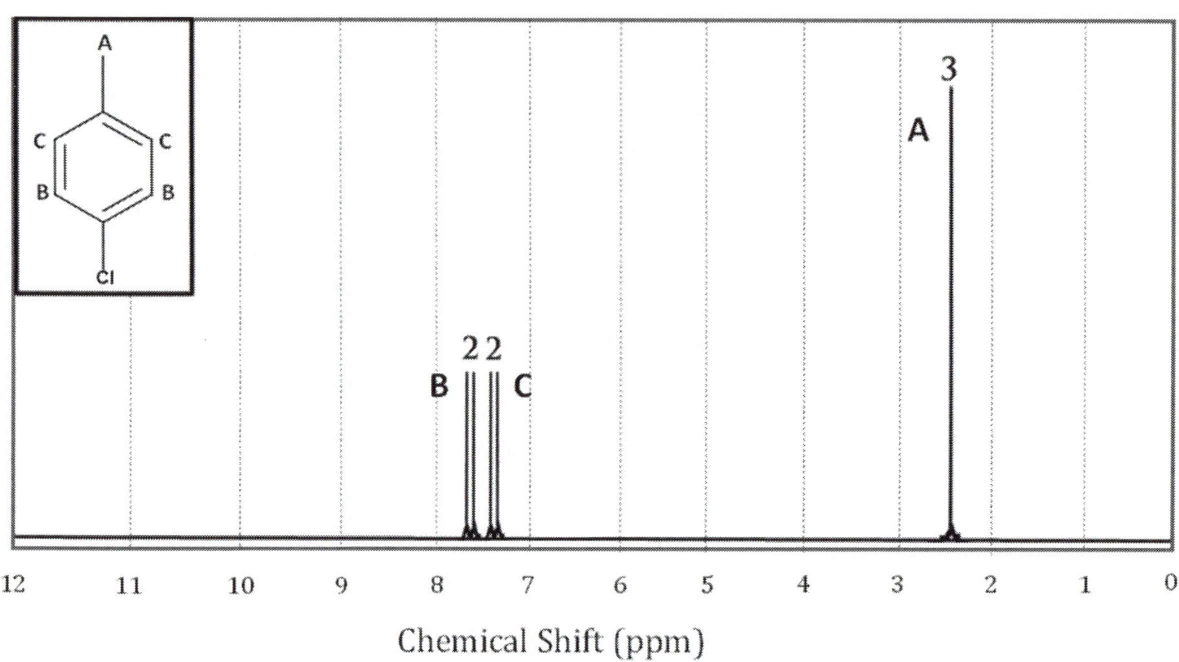


One notable difference between $^{13}C$ NMR spectra and $^1H$ NMR spectra is that the chemical shift scale for $^1H$ signals in typical organic samples fall in the range of about 0–12 ppm (cf. 0–220 ppm for $^{13}C$ NMR). Another difference is that $^1H$ NMR spectra often contain multiplets, whereas $^{13}C$ NMR spectra (by experimental design) only contain singlets.

Although not present in the example above, many $^1H$ NMR samples include tetramethylsilane (TMS) in addition to the molecule of interest, because the TMS speak is set to be 0 ppm as a common reference point for all $^1H$ NMR chemical shifts. Another common reference peak is $CHCl_3$ (7.28 ppm) because chloroform is one of the most commonly used solvents for NMR samples.

Chemical shifts for $^1H$ nuclei in the bonding environments most frequently encountered in organic molecules are listed below:

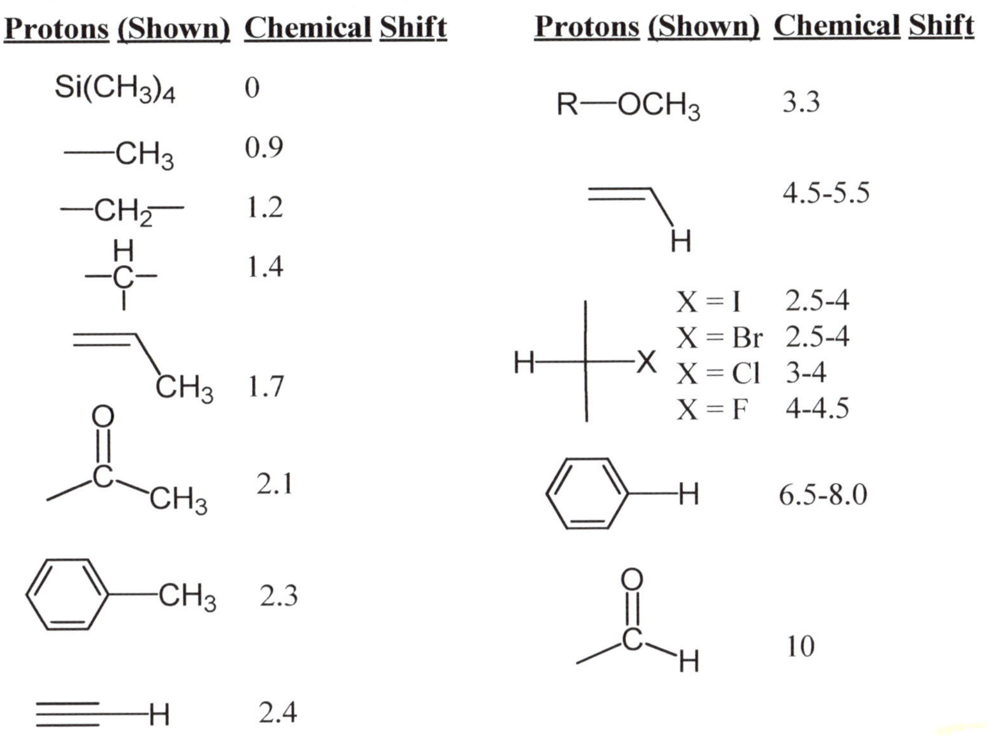

| Protons (Shown) | Chemical Shift |
|---|---|
| $Si(CH_3)_4$ | 0 |
| —$CH_3$ | 0.9 |
| —$CH_2$— | 1.2 |
| —C(H)— | 1.4 |
| $=$—$CH_3$ | 1.7 |
| —C(=O)—$CH_3$ | 2.1 |
| $C_6H_5$—$CH_3$ | 2.3 |
| ≡—H | 2.4 |

| Protons (Shown) | Chemical Shift |
|---|---|
| R—$OCH_3$ | 3.3 |
| $=$—H | 4.5-5.5 |
| H—C—X, X = I | 2.5-4 |
| X = Br | 2.5-4 |
| X = Cl | 3-4 |
| X = F | 4-4.5 |
| $C_6H_5$—H | 6.5-8.0 |
| —C(=O)—H | 10 |

*Lesson VII.7.2 Integration of Proton Nuclear Magnetic Resonance Signals*

In a $^{1}H$ NMR spectrum, the area under each peak (the **integration or integral**) is proportional to the number of protons that absorb at that energy (the number of protons in that set of magnetically equivalent protons in that group in the sample). The same is not true for a proton-decoupled $^{13}C$ NMR spectrum. There are several ways to indicate the integration for a given peak in a $^{1}H$ NMR spectrum. In this book, the integration will be provided as a number printed above the peak. For the spectrum of 4-chlorotoluene (shown in Lesson VII.7.1), the integration of peak **A** is 3, the integration of peak **B** is 2, as is the integration of **C**. The integrations can be very useful in figuring out what type of units are present in a molecule based on NMR data.

In practical settings, the spectrometer does not have a way to know how many protons give a signal, it can only measure the relative integration of one peak with respect to another. As a result, the spectrometer might automatically and arbitrarily provide integrations of 0.60, 0.40 and 0.40 for peaks **A**, **B** and **C** in the spectrum of 4-chlorotoluene. The ratio, 223, must then be calculated by the user (or software used with the spectrometer) to get the values displayed in the spectrum as shown in this book.

*Lesson VII.7.3 Splitting in $^{1}H$ NMR Spectra*

In Lesson VII.5, we saw that the magnetic spin of one nucleus can split the energy of absorption at nearby nuclei. In $^{1}H$ NMR spectra, the signal for a proton on carbon "X" is notably split into $n+1$ subpeaks, where $n$ is the total number of H atoms on the carbons directly adjacent to carbon "X". If the directly adjacent atoms are not carbon (i.e., oxygen or nitrogen), the H atoms on these directly adjacent atoms typically do not produce observable splitting. This property allows us to tell **how many H are on C atoms directly adjacent to the carbon bearing the $^{1}$H nucleus (or set of $^{1}$H nuclei) that produces a particular signal** in a $^{1}$H NMR spectrum. If a particular signal is a triplet ($m = 3$), for example, we know that the protons giving rise to this signal are attached to a carbon which has carbons directly adjacent to it carrying a total of 2 protons among them. Numerous examples to illustrate how the chemical shift, integration and multiplicity of resonances in a $^{1}$H NMR spectrum can be used to gain structural information are provided in Part VIII.

# Lesson VII.8. Introduction to Mass Spectrometry

*Lesson VII.8.1 Formation, Separation, and Detection of Molecular Ions*

Mass spectrometry is a technique that can be used to determine the molecular weight of a molecule as well as the molecular weight of some of the pieces of the molecule that form when it is fragmented by the application of energy. Knowing the molecular weight and the mass of specific fragments of the molecule can be used in conjunction with the other analytical techniques we have seen in Lessons VII.1-7 to assist in characterizing molecules.

Electronic detectors can most easily detect charged, rather than neutral, species. For this reason, the first step in mass spectrometry is creating ions from neutral molecules. The ion that is made by taking one electron away from the neutral molecule has the same molecular weight as the neutral molecule that we are trying to examine. This molecule is called the **molecular ion**. We have seen some ways to make ions by chemical reactions (for example, by heterolysis). A molecular ion can also be formed by "ionizing" a molecule, a process which removes an electron from or adds an electron to a neutral molecule to afford a cationic or anionic molecular ion, respectively. If only a single electron is removed from or added to a neutral molecule, the resulting molecular ion is a "radical cation" or "radical anion".

Mass spectrometric analysis of an analyte is a three-step process involving the formation, separation, and detection of molecular ions. The ionization technique most commonly employed for small organic molecules is "electron ionization" (EI), whereby an analyte molecule in the vapor phase is bombarded by a stream of high-energy electrons, which dislodges an electron from the analyte molecule to afford a cationic molecular ion, which is abbreviated as "$[M]^{+}$". Although there are other mass spectrometry techniques, in this course we will assume all spectra come from the EI process.

After the various ions have been formed, a mass spectrometer employs one or more internal electric and/or magnetic fields to separate the ions according to their mass-to-charge ($m$/z) ratios. An electromagnetic field will deflect the path of a traveling ion to a greater extent if $m$ is smaller (lighter molecules) or $z$ is larger (higher charges). In this manner, a mass spectrometer is able to separate and select ions based on their $m/z$ ratios. The final stage of the mass spectrometry experiment is detection, whereby the separated ions either impact a detector element (which imparts a charge to the detector) or pass by a detector element (which induces an electric current). Because neutral species carry no Coulombic charges, they are undetectable. The typical mass spectrum shows a plot of intensity ($y$-axis) versus $m/z$ ($x$-axis). In this book, $z$ is always one, so the numeric value on the $x$-axis corresponds to the value of the molecular weight of the fragment giving the peak. A typical mass spectrum is shown below:

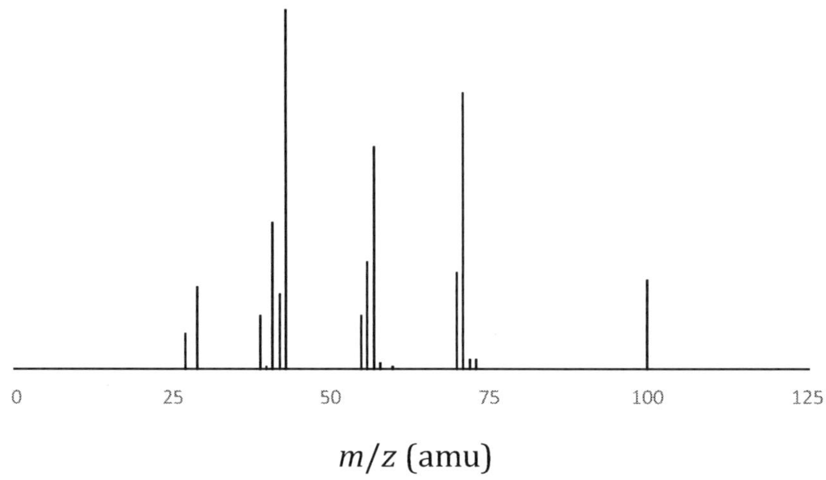


*Lesson VII.8.2 Predicting Molecular Ion Structure*

As a general rule, it is hardest to remove an electron from a σ-bonding orbital than π-bonding orbital, and it is easiest to remove electrons from a non-bonding orbital (i.e., a lone pair). Therefore, ionization will remove an electron from a lone pair before a π-bond, and it will remove an electron from a π-bond before a σ-bond. Removal of an $e^-$ from a lone pair leaves behind a radical cationic charge (•+) on that individual atom. Removal of an $e^-$ from a π-bond leaves behind a cationic charge (+) on one atom of the π-bond and a radical charge (•) on the other atom. Radical stability has a greater impact than cation stability, so the + and • will be preferentially be situated according to maximize radical stability. Keep in mind that ionization is a high-energy process that occurs in the gas phase, so other species with the radical in less stable positions can also be formed by ionization, but usually in lower yields. If the molecule is an alkane, simply draw square brackets around the molecule and place the •+ symbol at the top on the outside of the right bracket (there is no convenient way to represent σ-bonds which contain only 1 electron). In general, drawing the •+ localized to a specific atom or group of atoms is preferred, because it makes understanding fragmentation mechanisms more intuitively accessible.

*Lesson VII.8.3 Isotopic Abundance and Molecular Mass*

When converting mass to moles in organic lab, our calculation of the molecular weight will use 12.011 as the atomic mass of carbon. However, this number is the average mass of a carbon atom. No actual carbon atom weight 12.011 amu. Instead, 98.9% of all the carbon atoms in the Universe have a mass of 12.000 and the remaining 1.1% have a mass of 13.003. Consider this from a different perspective: in a sample of 100 $CH_4$ molecules, 99 will have a carbon with a mass of 12.000 and the 100th will have a carbon with a mass of 13.003. Ionizing this sample of 100 $CH_4$ molecules will generate 99 molecular ions with an *m*/*z* ratio of 16 and 1 molecular ion with an *m*/*z* ratio of 17. Clearly, we will see a huge peak in the mass spectrum at 16 amu and only a small peak at 17 amu.

Although there are some exceptions we will cover in in subsequent lessons, for the purposes of sophomore organic chemistry you only need to consider the atomic mass of the most abundant isotope of each element when calculating the molecular masses of the various molecular ions and descendant fragments. In other words, if you observe a molecular ion peak with an *m*/*z* value of 170.21, and you know from other techniques that the analyte only has hydrogen and carbon atoms, you would only use the atomic mass of hydrogen-1 (1.0) and the atomic mass of carbon-12 (12.0) when calculating the number of hydrogen and carbon atoms present in that ion.

### *Lesson VII.8.4 Isotopic Peaks*

In Lesson VII.8.3, we saw that many of the elements found in organic molecules essentially have only one isotope. Notable absences from the table in that lesson were chlorine and bromine. Whereas fluorine is 100% fluorine-19, chlorine is 76% chlorine-35 and 24% chlorine-37, and bromine is 51% bromine-81 and 49% bromine-81. As a result, a molecular ion or fragment that contains either Cl or Br will exhibit ***two*** prominent peaks, called the "*m*" and "*m*+2" peaks because they differ by 2 units of *m*/*z*. Thus, if you are given a mass spectrum of an unknown compound and you observe the *m* and *m*+2 pattern for some of peaks, you should immediately conclude that the unknown compound contains one or more chlorine or bromine atoms.

## Lesson VII.9. Introduction to Fragmentation Mechanisms

### *Lesson VII.9.1 What is Fragmentation and Why Does It Occur?*

In the context of mass spectrometry (MS), fragmentation is a term that describes when a molecular ion breaks apart into 2 or more lower-molecular weight components. Because fragmentation processes can involve the breakage and formation of σ-bonds, it is useful to think of fragmentation processes as chemical reactions. Bombardment of a molecule with high-energy electrons transfers a significant amount of energy to that molecule, thus a molecular ion formed by EI will often undergo multiple fragment processes.

Why do molecular ions generated by a mass spectrometer undergo fragmentation? Keep in mind that every atom in nearly every organic molecule has an octet, so ionizing a molecule means that some atom(s) will no longer have a complete octet. We know that an atom that lacks a complete octet is much higher in energy than an atom that has a complete octet, so the ionization of a molecule affords a higher energy species, so we have to add a lot of energy to get fragmentation to occur. As we mentioned in the previous lesson, ionization is a very high-energy process that occurs in the gas phase, so a •+ species can fragment to make the more stable + (with less stable •) or the more stable • (with less stable +). In general, the energetically *preferred* fragmentation pathway is determined by cation stability, but other fragmentation pathways are still energetically *accessible* under MS experimental conditions. These fragments formed from the molecular ion can themselves undergo fragmentation reactions, which can result in complicated MS data that provide tremendous insight into molecular structure.

Although there are some exceptions, the molecular ion (which has an odd number of $e^-$) typically fragments into an even-electron cation and an odd-electron radical. The even-electron cations then typically fragment into other even-electron species. Keep in mind that each of the following mechanisms were formulated on the basis of experimental data – chemists made observations on how molecules fragmented and then proposed mechanisms to explain how those fragments formed; it would have been difficult to hypothesize the proper fragmentation of these molecules without the MS data.

### *Lesson VII.9.2 Fragmentation via Bond Heterolysis and/or Bond Homolysis*

Fragmentation of a molecular ion into one or more molecules requires the breakage of bonds (new bonds can also be formed, but some bonds must be broken). Bond breakage can occur by heterolysis or homolysis (1 $e^-$ in a bond is transferred to one atom, and the other $e^-$ is transferred to the other atom).

To illustrate how a molecular ion may fragment via heterolysis, consider a dialkyl ether as an example. Ionization places the •+ on the O, because the $e^-$ in the lone pairs on O are more energetically accessible than the $e^-$ in any of the σ-bonds. One of the C–O bonds then breaks heterolytically to place the 2 $e^-$ on the O, resulting in a carbocation and an O-centered radical. This bond heterolysis is preferred because it affords a C-centered cation instead of an O-centered cation, the former being more stable due to the lower electronegativity of carbon. Do not forget that cations and radicals can rearrange!

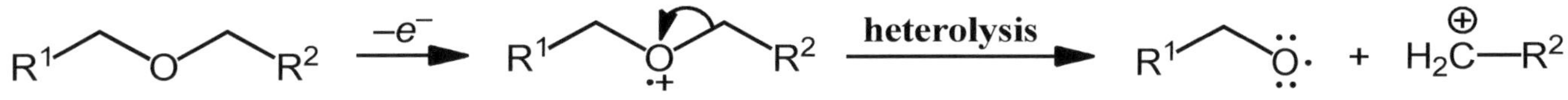

To illustrate how a molecular ion may fragment via homolysis, let us consider an aryl alkyl ether as an example. As before, ionization places the •+ on the oxygen, for the same reasons. Then, the O–alkyl bond breaks homolytically to place 1 $e^-$ on the O, resulting in an O-centered cation and a C-centered radical. This bond homolysis is preferred because the + charge on O is stabilized by resonance delocalization across the arene ring, whereas the carbocation formed by heterolysis would be stabilized by hyperconjugation (which is not as stabilizing as resonance delocalization).

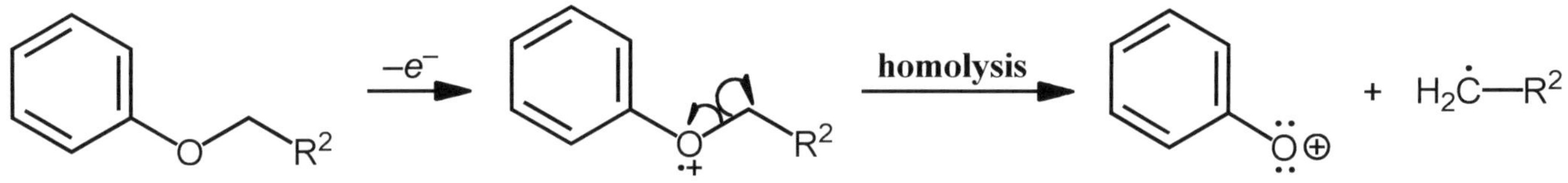

Every fragmentation mechanism involves bond breakage via homolysis or heterolysis, and some mechanisms involve both. Moreover, it is possible to predict products formed by one fragmentation mechanism by invoking only heterolytic bond breakage, and then predict identical products by the same fragmentation mechanism by invoking only hemolytic bond breakage!

*Lesson VII.9.3 Position Labeling Relative to Radical Cations*

When discussing fragmentation in mass spectrometry, the convention is to refer to positions within molecules relative to the atom (or atoms) bearing the •+ character. The atom directly attached to the atom with the •+ character is the α-position, the next atom out is the β-position, the next atom out from that is the γ-position, and so on. If the •+ character is spread across multiple atoms, the α-position is the first atom connected to any atom bearing any • or + character. Most of the time, the α-, β-, γ-, etc., positions are occupied by carbons, which we represent as $C_\alpha$, $C_\beta$, $C_\gamma$, etc. It is important to note that the use of this terminology is both inconsistent and self-contradictory in many organic chemistry textbooks, and this is even the case in the primary literature! In our discussions of fragmentation mechanisms in this text, we will employ consistent position assignments, but it is important for the student to keep in mind that other texts may use other position assignments.

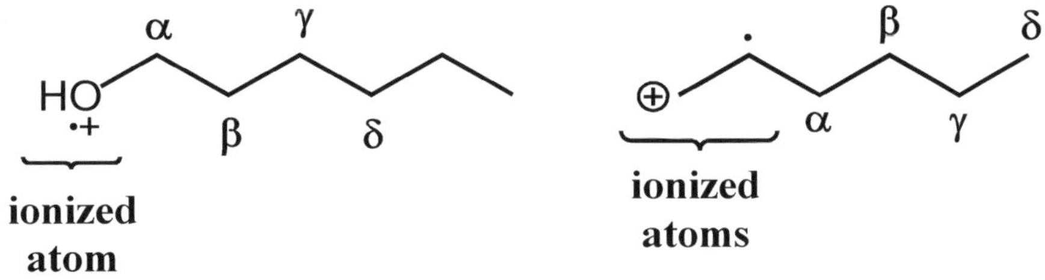

*Lesson VII.9.4 Inductive Cleavage*

The simplest fragmentation mechanism is "inductive cleavage", which refers to breakage of a bond between the atom bearing the •+ character and the α-atom, and it may occur either heterolytically or homolytically. The fragmentation of the dialkyl ether and aryl alkyl ether we showed in Lesson VII.9.2 are examples of inductive cleavage by heterolysis and homolysis, respectively.

*Lesson VII.9. α-Cleavage*

In α-cleavage, the bond between the α- and β-positions is broken, and **the α-atom stays with the original fragment**.

For our first example, let us revisit a dialkyl ether. Ionization again places the •+ on the O atom. Next, the bond between the α-carbon and the β-carbon breaks homolytically:

$$R^1{-}O{-}CH_2{-}CH_2{-}R^2 \xrightarrow{-e^-} R^1{-}\overset{\bullet +}{O}{-}CH_2(\alpha){-}CH_2(\beta){-}R^2 \xrightarrow{\alpha\text{-cleavage}} R^1{-}\overset{\oplus}{O}{=}CH_2 + H_2\dot{C}{-}R^2$$

The driving force for this reaction is that every atom in the cationic fragment has a complete octet.

For our second example, consider a dialkyl ketone. Ionization places the •+ on the O atom, and the bond between the α-carbon and the β-carbon breaks homolytically:

$$R^1{-}C({=}O){-}CH_2{-}R^2 \xrightarrow{-e^-} R^1{-}C_{\alpha}({=}\overset{\bullet +}{O}){-}CH_2(\beta){-}R^2 \xrightarrow{\alpha\text{-cleavage}} R^1{-}C{\equiv}\overset{\oplus}{O} + H_2\dot{C}{-}R^2$$

Again, the driving force for this reaction is that every atom in the cationic fragment has a complete octet.

*Lesson VII.9.6 β-Cleavage*

In β-cleavage, the bond between the β- and γ-positions is broken, whereby **the atoms at the α- and the β-positions stay with the original fragment**.

For an example, let us revisit a dialkyl ketone. Ionization again places the •+ on the O atom, but now the bond between the α-carbon and the β-carbon breaks heterolytically. The 2 $e^-$ from this bond moves in between $C_\alpha$ and $C_\beta$ to form a C=C bond, and the $C_\alpha$–O π-bond breaks heterolytically to place those 2 $e^-$ in an additional lone pair on O, which affords an enol with • character on the O. The other fragment is a carbocation with the + charge on $C_\gamma$. Although the cationic fragment produced by β-cleavage lacks a complete octet, the radical character on the other fragment is stabilized by delocalization. In general, β-cleavage is less common than either α-cleavage or the McLafferty rearrangement that we are about to discuss.

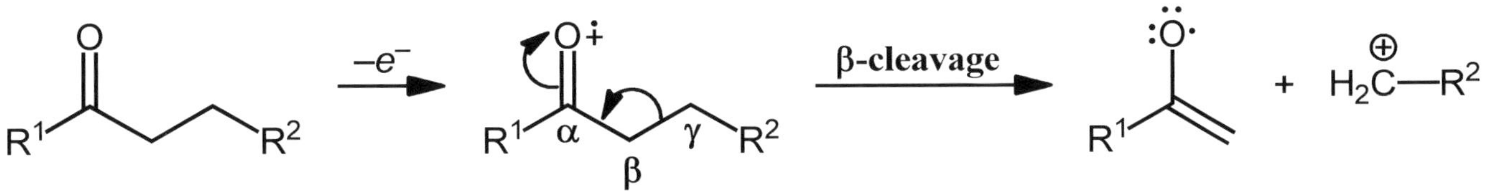

*Lesson VII.9.7 McLafferty Rearrangement*

The McLafferty rearrangement refers to a fragmentation process in which bond cleavage (typically β-cleavage) is accompanied by H transfer (typically from the δ-carbon) and proceeds through a cyclic 6-membered transition state. The McLafferty rearrangement is typically associated with carbonyl-based functional groups, although we will see that other functional groups exhibit similar rearrangements.

We will continue using a dialkyl ketone as our example. Ionization again places the •+ on the O atom, but now one of the alkyl chains is sufficiently long that it places a C–H bond (specifically at $C_\delta$) in close proximity to the $O^{\bullet+}$. A six-membered arrangement of atoms participates in the McLafferty rearrangement to form a stable neutral alkene and a resonance-stabilized cation:

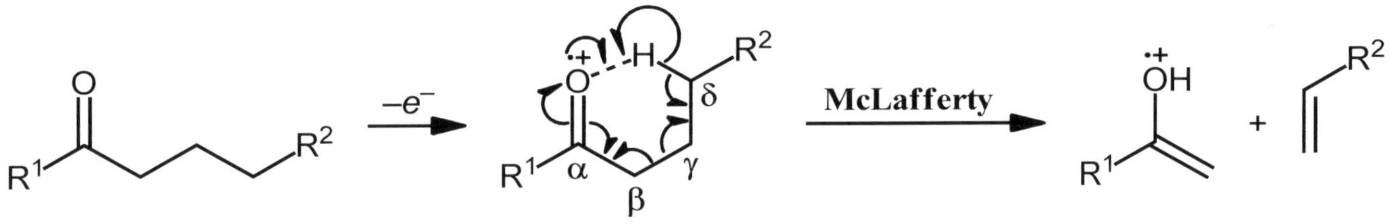

*Lesson VII.9.2.8 Retro Diels–Alder*

Another fragmentation mechanism that proceeds through a cyclic 6-membered transition state is termed "retro Diels–Alder". In retro Diels–Alder, a 6-membered cycle comprising one or more atoms bearing •+ character is converted into one 4-membered chain, which retains the •+ character, and one 2-membered chain, which is neutral.

*Lesson VII.9.2.9 Elimination*

Elimination, in the context of fragmentation during a mass spectrometry experiment, simply refers to a process in which one molecule fragments into two molecules, where one of the molecules is usually (but not always) expelled as a low molecular weight neutral molecule. Sometimes drawing an arrow-pushing mechanism for an elimination process is easy, as is the case with $[R\text{–}CH_2\text{–}C{\equiv}O]^+$, and sometimes it is beyond our abilities in this course, as is the case with $[PhNH_2]^{\bullet+}$, shown below:

$R-CH_2-C\equiv O^{+}$ —elimination→ $R-CH_2^{+}$ + CO

$C_6H_5NH_2^{\bullet +}$ —elimination (complex mechanism)→ $C_5H_6^{\bullet +}$ + HCN

## Lesson VII.10 Fragmentation of Alkanes

### *Lesson VII.10.1 Linear Alkanes*

The $M^+$ peak is typically visible for linear alkanes, although its intensity decreases with increasing alkane length. There is no convenient way to draw the radical cation formed by ionization of an alkane, so we typically represent it with square brackets. The molecular ion can fragment along any of the C–C bonds, usually in a way that places the radical on the smaller fragment and the cation on the larger fragment. A feature that is characteristic for alkanes is a series of peaks separated by 14 amu, the mass of $CH_2$. In general, there will be the $M^+$ peak, then a peak corresponding to $[M-15]^+$, then $[M-29]^+$, and so on. Note that the $[M-15]^+$ peak is derived from the loss of $CH_3^{\bullet}$, the least stable radical, so the $[M-15]^+$ peak is usually very weak and is sometimes absent entirely. Lastly, it is possible for $H_2C{=}CH_2$ to eliminate from one of the cationic fragments. The various fragments for *n*-heptane are shown below.

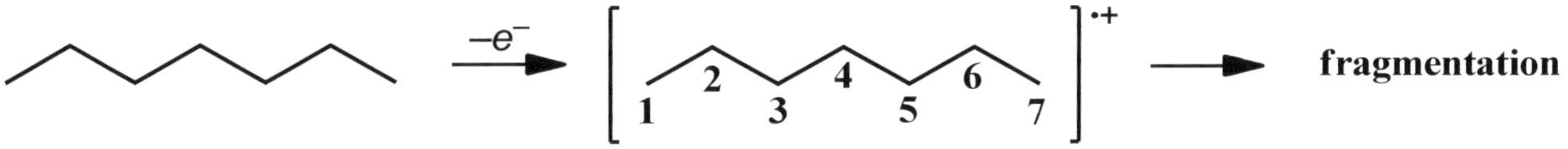

$-e^-$ → $[\ 1\ 2\ 3\ 4\ 5\ 6\ 7\ ]^{\bullet+}$ → **fragmentation**

**$C_1$–$C_2$ homolysis:** $H_3\dot{C}$ + $[M-15]^+$ **and** $H_3C^{\oplus}$ (**15 amu**) +

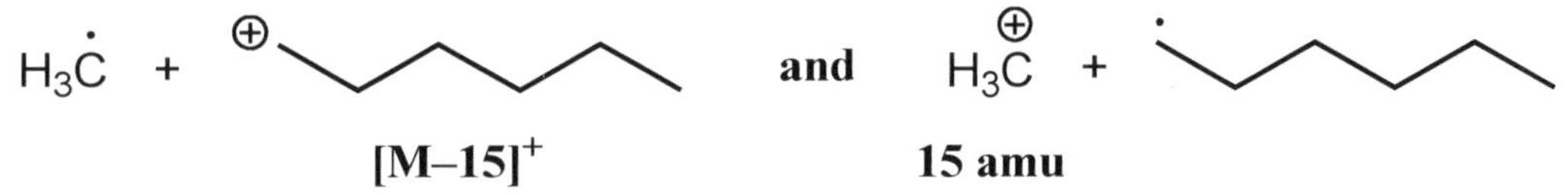

**$C_2$–$C_3$ homolysis:** $CH_3\dot{C}H_2$ + $[M-29]^+$ **and** $CH_3CH_2^{\oplus}$ (**29 amu**) +

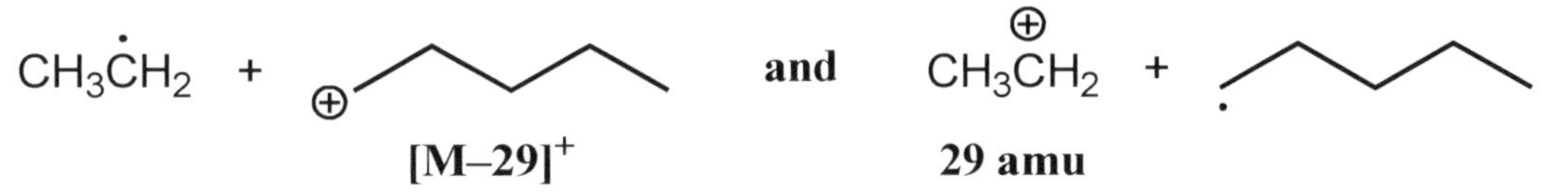

**$C_3$–$C_4$ homolysis:** + $[M-43]^+$ **and** **43 amu** +

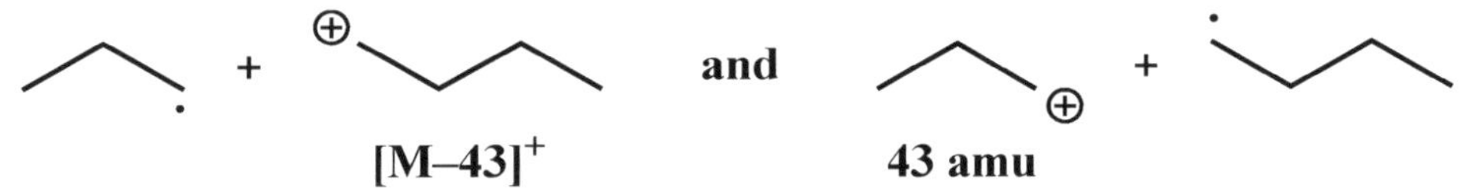

The mass spectrum of *n*-heptane shows prominent peaks for the cationic fragments delineated above:

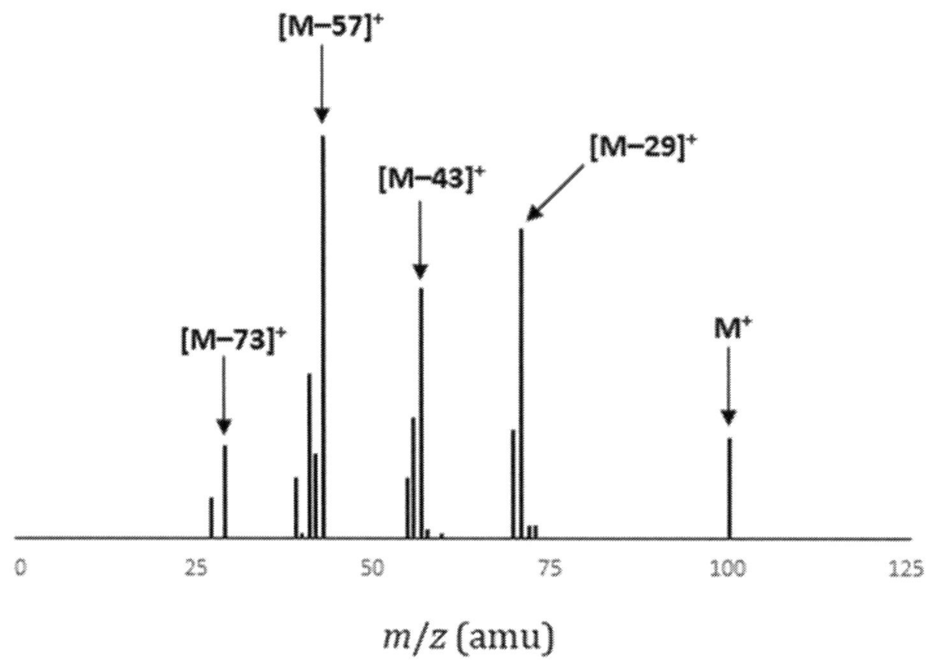


*Lesson VII.10.2 Branched Alkanes*

The $M^+$ peak for branched alkanes is typically smaller than for linear alkanes, and may be absent entirely. Like a linear alkane, the molecular ion for a branched alkane can fragment along any of the C–C bonds, but fragmentation is more favored at the branching sites because it affords more stable carbocations. Similarly, the MS of a branched alkane will exhibit a series of peaks separated by 14 amu, the mass of $CH_2$, but the relative intensities will be different than in a linear alkane. For example, both *n*-heptane and 3-ethylpentane will exhibit $[M–29]^+$ peaks at 71 amu, but this peak in the MS of 3-ethylpentane will have much higher relative intensity because it is derived from a more stable 2° cation. In general, the fragmentation patterns for branched alkanes are highly conserved with those for linear alkanes, with the only differences primarily being different relative peak intensities.

$\xrightarrow{-e^-}$ $[\ ]^{\bullet +}$ $\longrightarrow$ $CH_3\dot{C}H_2$ + $[M–29]^+$

$\xrightarrow{-e^-}$ $[\ ]^{\bullet +}$ $\longrightarrow$ $CH_3\dot{C}H_2$ + $[M–29]^+$

The mass spectrum of 3-ethylpentane shows prominent peaks for the cationic fragments delineated above:

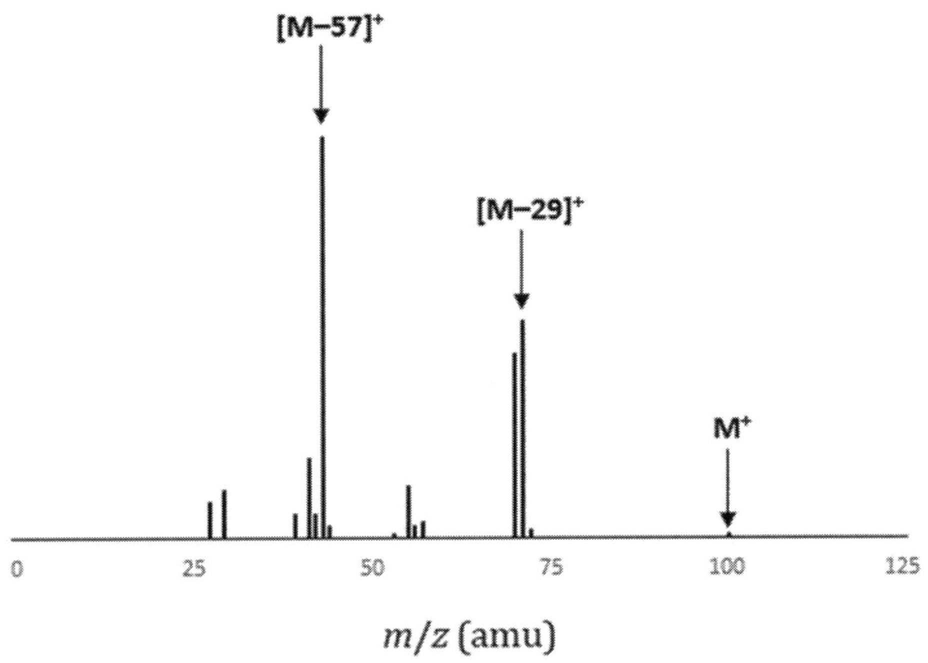


*Lesson VII.10.3 Cycloalkanes*

The M$^+$ peak for cycloalkanes is typically more intense than for either linear or branched alkanes, with intensities higher for unbranched cycloalkanes compared to branched cycloalkanes. With an unbranched cycloalkane, the two major fragments are formed by loss of $H_2C{=}CH_2$ and $CH_3^{\bullet}$, with amu values of [M–28]$^+$ and [M–15]$^+$, respectively. Mechanisms are shown below for the fragmentation of the cyclohexane molecular ion.

[M–28]$^+$

H$_3\dot{C}$

[M–15]$^+$

The mass spectrum of cyclohexane shows prominent peaks for the cationic fragments delineated above:

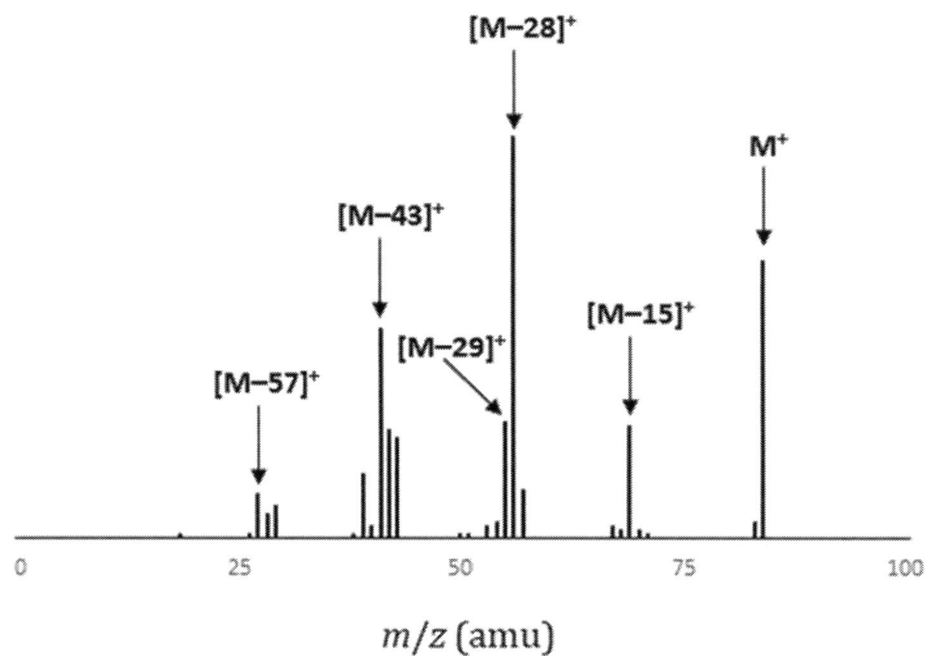


With a branched cycloalkane, loss of the side-chain as an alkyl radical is a major fragmentation pathway, and the resulting carbocation can subsequently eliminate $H_2C{=}CH_2$. However, keep in mind that the parent radical cation can also eliminate $H_2C{=}CH_2$. Mechanisms are shown below for the various fragmentation modes for the ethylcyclopentane molecular ion. In general, if a cycloalkane has an alkyl side chain, it will not exhibit an $[M–15]^+$ peak due to $CH_3^{\bullet}$ release and ring contraction, unless the side chain is a methyl group, in which case an $[M–15]^+$ peak will be present, but instead will arise from side chain heterolysis.

$CH_3\dot{C}H_2$ +

$[M–29]^+$

$[M–57]^+$

$[M–28]^+$

The mass spectrum of ethylcyclohexane shows prominent peaks for the cationic fragments delineated above:

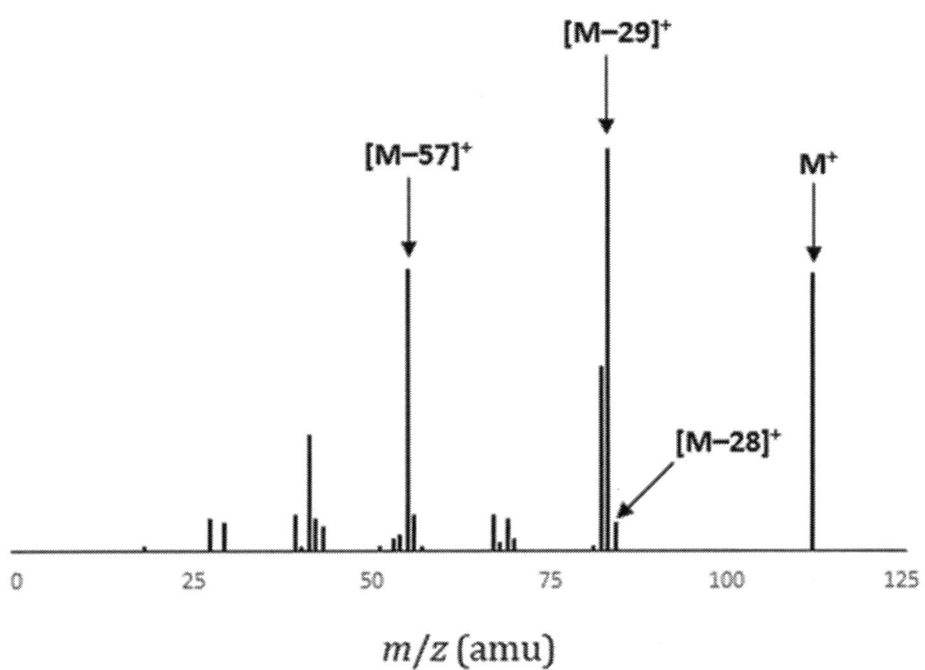

## Lesson VII.11. Fragmentation of Heteroatom-Containing Aliphatics

*Lesson VII.11.1 Halogenated Alkanes*

The molecular ion formed by ionization of a halogenated aliphatic (saturated molecule) has the radical cation character localized on the halogen (it is easier to remove an electron from a lone pair than from a bonding orbital). Remember that the presence and relative intensity of any M+2 peaks will inform you of the identity of the halogen! One fragmentation mechanism involves heterolytic cleavage of the C–X bond, placing both $e^-$ on the departing halogen, affording a carbocation and a halogen radical. This process is most favorable when X = I and is least favorable when X = F. Alternatively, the molecular ion could undergo α-cleavage to afford an alkyl radical and a halonium species. This process will favor formation of the largest possible alkyl radical. When X = I, an additional fragmentation mode is the release of $I^+$, which exhibits a peak at 127 amu.

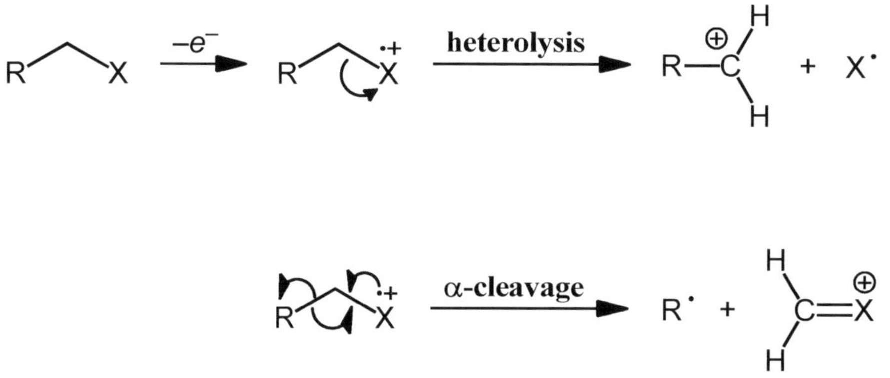

−e⁻   **heterolysis**   + X•

**α-cleavage**   R• +

Elimination of HX from the molecular ion can also occur, which produces another radical cation. This process is most favorable when X = F and is least favorable when X = I. The H-atom can come from either $C_\alpha$ or $C_\delta$, and is referred to as 1,2-elimination or 1,4-elimination, respectively. Note, however, that the mass loss is identical for both elimination mechanisms, so it is impossible to determine which mechanism is occurring using only mass spec.

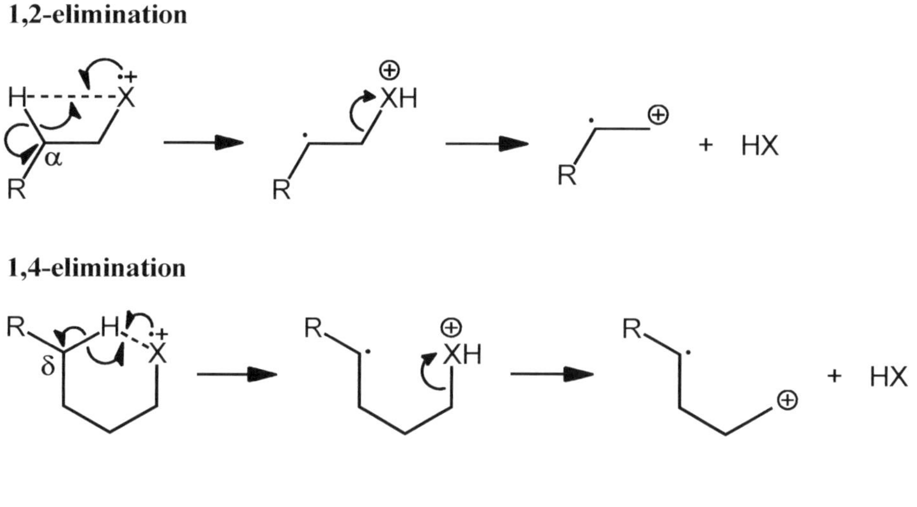

**1,2-elimination**

+ HX

**1,4-elimination**

+ HX

Any alkyl halide that features a hydrogen atom at the δ-position can also undergo a double elimination reaction, which eliminates HX and an alkene, leaving behind a radical cation. Lastly, in alkyl halides that have non-hydrogen substituents at the δ-position, a σ-bond can form between the halogen and $C_\delta$, accompanied by homolytic cleavage of the $C_\delta$–$C_\varepsilon$ bond, to form a cyclic halonium in a process termed δ-cyclization. Remember that the alkyl fragments themselves can also fragment.

**double elimination**

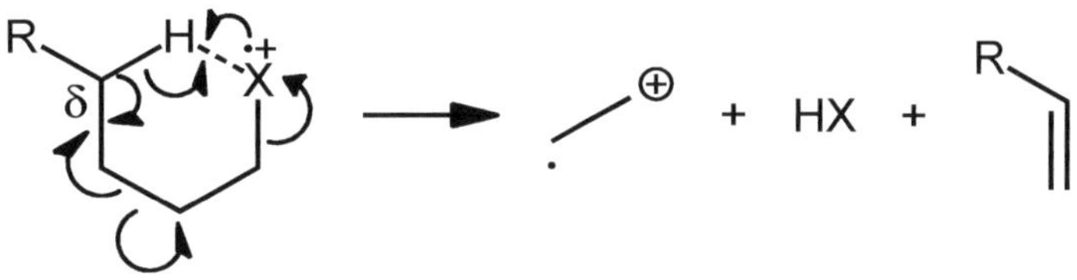

**δ-cyclization**

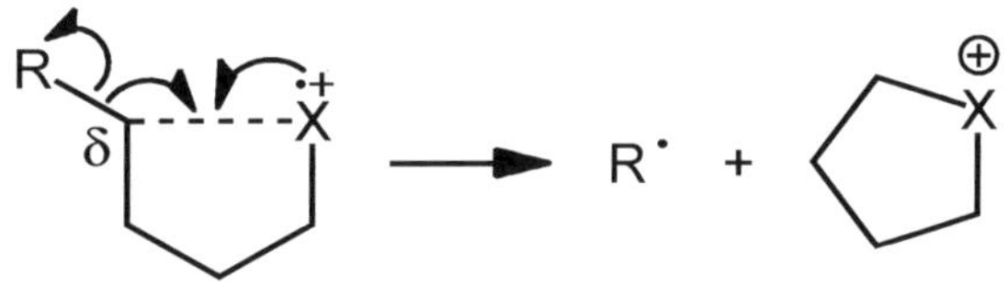

*Lesson VII.11.2 Alcohols*

Ionization of an alcohol places the radical cation character on the oxygen atom. The $M^+$ peak is generally weak for 1° and 2° alcohols, and is often absent for 3° alcohols. Occasionally, a 1° or 2° alcohol may exhibit an $[M–1]^+$ peak due to α-cleavage of an H-atom. Alcohols undergo many of the same fragmentation mechanisms as alkyl halides. Some notable differences are that alcohols rarely undergo heterolysis (to release HO$^\bullet$) or δ-cyclization. Primary alcohols display a prominent peak at 31 amu.

**α-cleavage**

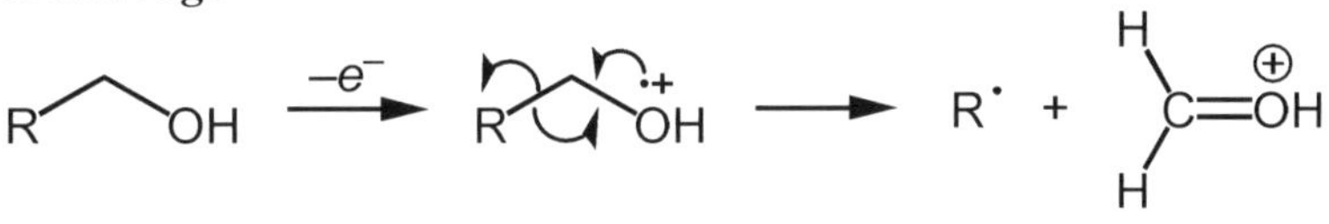

**31 amu**

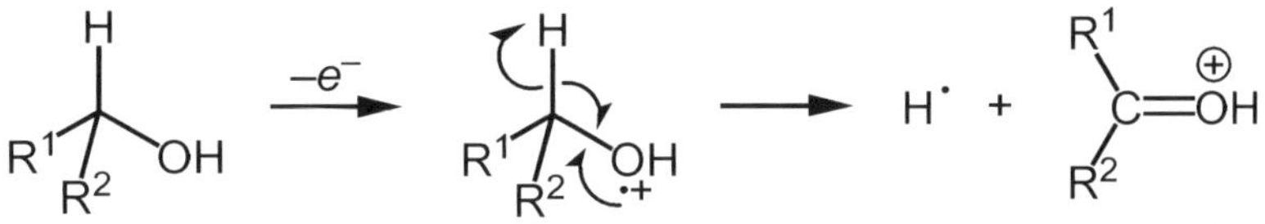

$[M–1]^+$

**1,2-elimination**

+ $H_2O$

**$[M{-}18]^+$**

**1,4-elimination**

+ $H_2O$

**$[M{-}18]^+$**

If an OH is on a cyclopentane or larger ring, an additional fragmentation mechanism is available. After α-cleavage occurs (which has same mass as parent ion), the carbon radical can abstract an H-atom to place the radical character adjacent (and thus in conjugation with) the carbonyl unit. A double bond can then form between this and the adjacent carbon, releasing an alkyl radical fragment. For this reason, cycloalkanols larger than cyclobutanol exhibit a prominent peak at 57 amu.

**57 amu**

The MS of cyclohexanol shows prominent peaks for the cationic fragments delineated above:

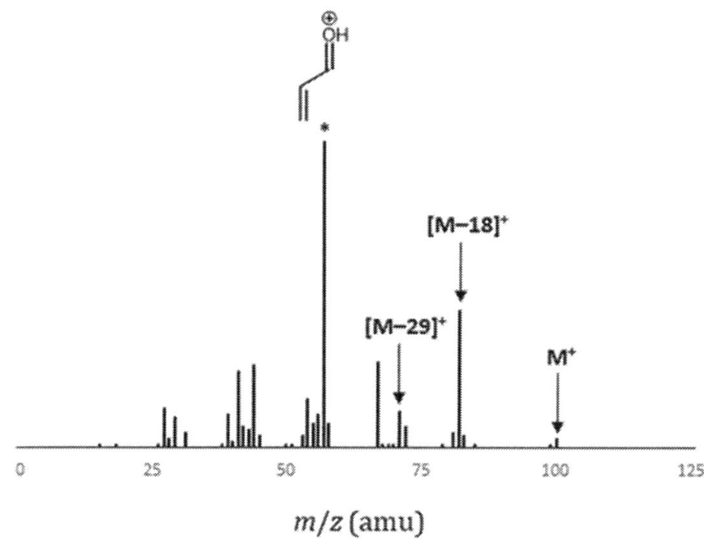

*Lesson VII.11.3 Ethers*

Ionization of an ether places the radical cation character on the oxygen atom. In general, the $M^+$ peak is stronger and there are fewer types of fragmentation mechanisms for an ether compared to an alcohol with the same molecular formula (principal modes are α-cleavage, there is generally no 1,2- or 1,4-elimination). These fragmentation modes can occur on either side of the oxygen, but keep in mind that O–C homolysis will not occur unless there is something that can stabilize an oxygen with only 6 valence electrons (e.g., delocalization across any aryl ring).

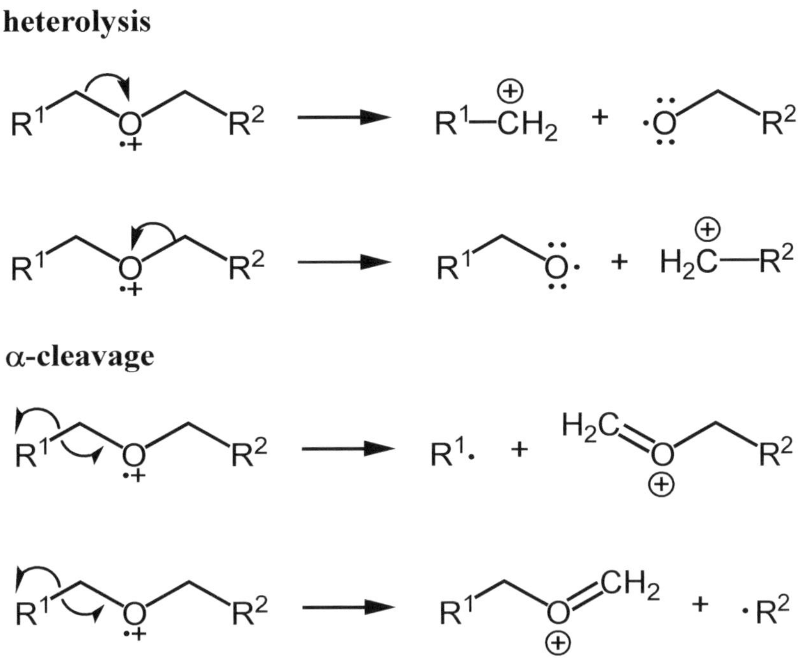

**heterolysis**

**α-cleavage**

Note that ethers of the general formula $R^1CH_2OCH_2R^2$ will give a prominent peak at 31 amu, the same as with 1° alcohols, because the product of α-cleavage can subsequently rearrange. As a result, the presence of a peak at 31 amu is not unambiguous for 1° alcohols.

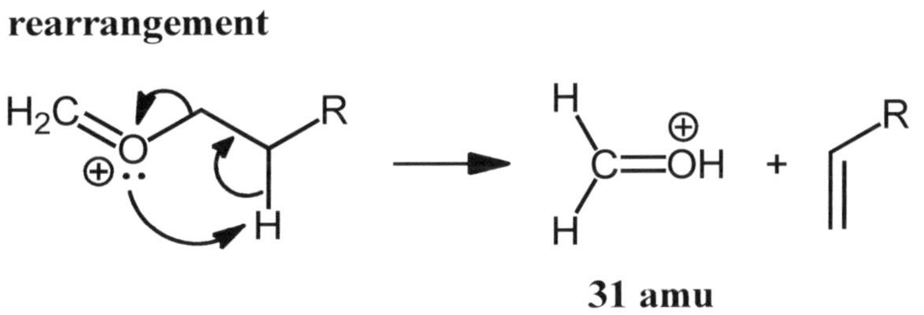

**rearrangement**

**31 amu**

Interestingly, cyclic ethers exhibit peaks at $[M\text{–}30]^+$ due to the elimination of $H_2C\text{=}O$, and those with ring sizes greater than 4 exhibit peaks at 41 amu from the allyl cation! Larger allyl cations can also be observed, depending on the size of the ring. These processes are illustrated for tetrahydropyran here:

**α-cleavage (*i*) then elimination (*ii*)**

*i* *ii*

[M–30]$^+$

**heterolysis (*i*), rearrangement (*ii*), then elimination (*iii*)**

*i* *ii* *iii*

**41 amu**

The mass spectrum of cyclohexanol shows prominent peaks for the cationic fragments delineated above:

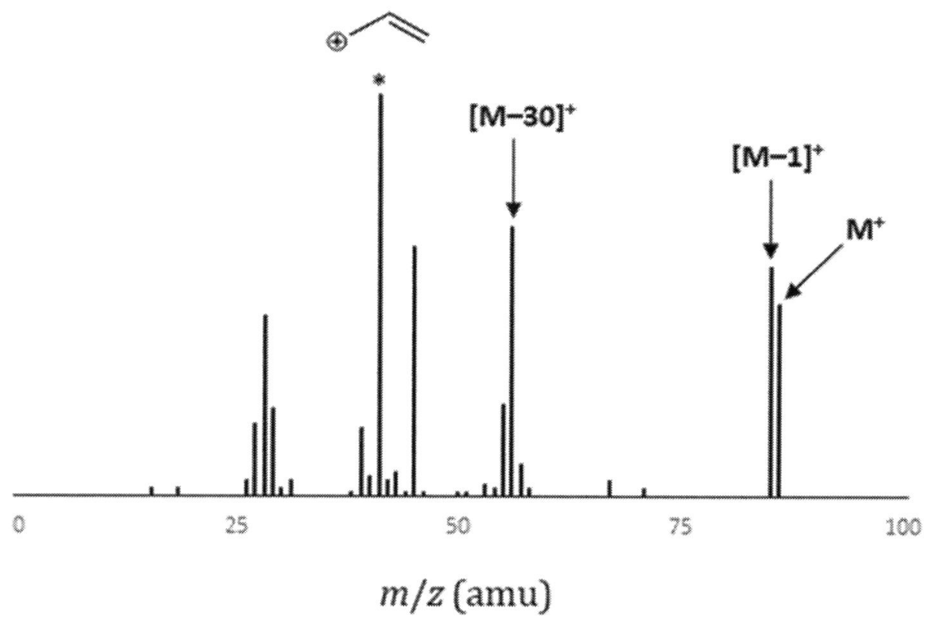


*Lesson VII.11.4 Amines*

The radical cation character in the molecular ion of an amine is localized on the nitrogen atom. The M$^+$ peak is typically weak or absent, but an [M–1]$^+$ peak is often observable. Like ethers, the principal fragmentation modes involve α-cleavage. Primary amines often exhibit peaks at 30 amu.

**α-cleavage**

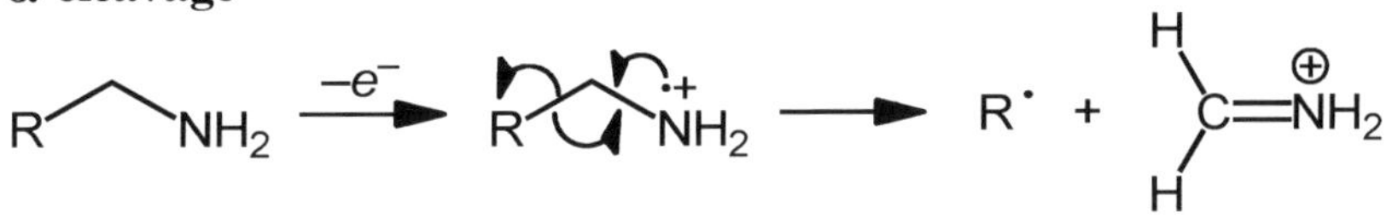

30 amu

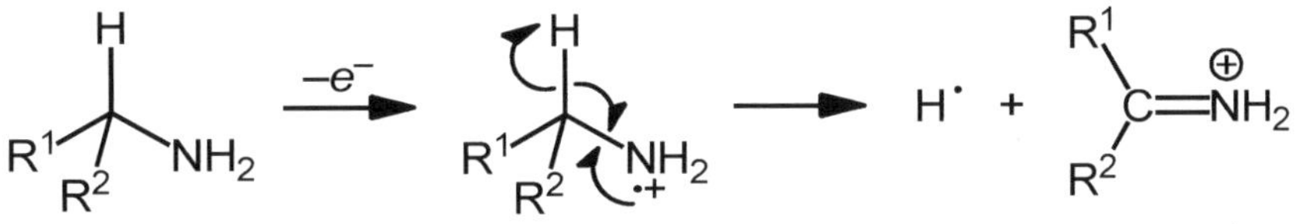

$[M-1]^+$

Amines can undergo δ-cyclization, and will exhibit peaks at 72 amu if there are no substituents at the α-, β-, or γ-positions. However, amines generally do not eliminate $NH_3$ (either by itself or via double elimination).

**δ-cyclization**

72 amu

Cyclic amines display stronger $M^+$ and $[M-1]^+$ peaks, but their fragmentation patterns are much more complicated to interpret. However, a common feature of most cyclic amines is an $[M-28]^+$ peak, corresponding to the elimination of ethylene, as shown for piperidine below:

**elimination**

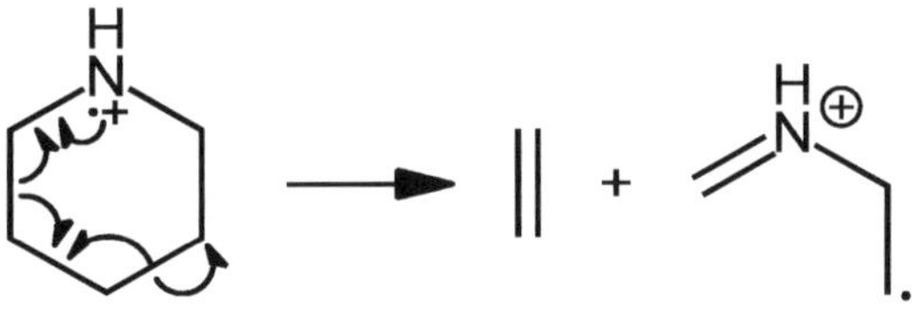

$[M-28]^+$

The mass spectrum of piperidine shows prominent peaks for the cationic fragments delineated above:

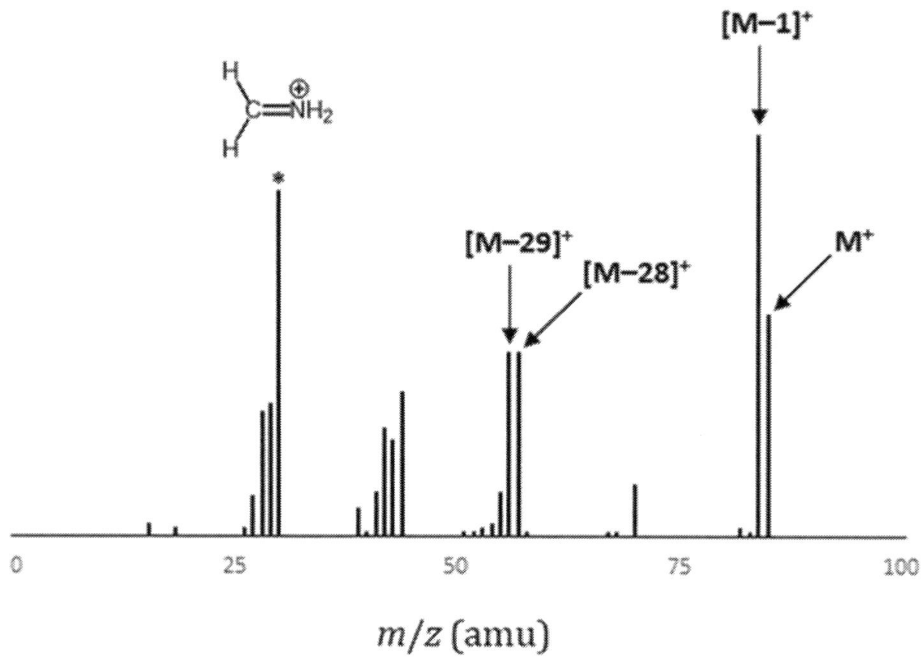
[M−1]+
[M−29]+
[M−28]+
M+
*
0
25
50
75
100
m/z (amu)

## Lesson VII.12. Fragmentation of Carbonyl-Containing Molecules

### *Lesson VII.12.1 Aldehydes*

The molecular ion of any carbonyl-containing functional group carries the radical cation character on the oxygen atom belonging to the carbonyl unit. In aldehydes, the $M^+$ and $[M–1]^+$ peaks are usually very weak or absent. The fragmentation modes are generally limited to α-cleavage to form an alkyl radical and the $[O≡C–H]^+$ cation (29 amu). However, 29 amu also corresponds to the molecular weight for the $CH_3CH_2^+$ cation, so the presence of a peak at 29 amu is not diagnostic for either $[O≡C–H]^+$ or $CH_3CH_2^+$ cations. Alternatively, α-cleavage can form an $[M–29]^+$ alkyl cation and $[O≡C–H]^•$ radical. Do not forget that any alkyl cation fragments formed can themselves fragment or rearrange.

**α-cleavage**

R—ĊH₂ + O≡C–H (⊕)

**29 amu**

**$[M–29]^+$**

β-cleavage can occur in aldehydes to give $[M–43]^+$ peaks, but this is not as favorable as α-cleavage. If α-substituents are present, the mass loss will be greater than 43 amu.

**β-cleavage**

**$[M–43]^+$**

Aldehydes longer than propanal can also undergo a McLafferty rearrangement to afford a neutral alkene and 44 amu radical cation (or higher if α-substituents are present). Interestingly, aldehydes with long alkyl chains also exhibit $[M–18]^+$ peaks due to the elimination of $H_2O$, but the mechanism by which this occurs is unclear.

**McLafferty**

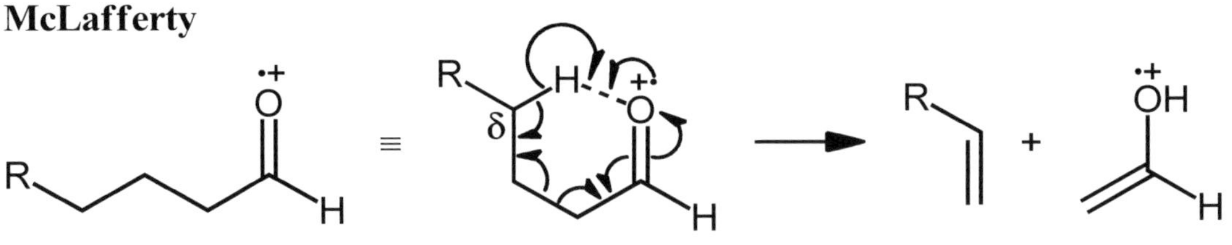

**44 amu**

*Lesson VII.12.2 Ketones*

In contrast to aldehydes, $M^+$ peaks are visible for most ketones. Ketones fragment in ways similar to aldehydes, namely via α-cleavage, but no $H_2O$ elimination peaks are observed even with very long alkyl chains. Note that α-cleavage can occur on both sides of the carbonyl unit. Methyl alkyl ketones show prominent peaks at 43 amu corresponding to the acylium cation $[O{\equiv}C{-}CH_3]^+$, but this mass is identical to that of the $CH_3CH_3CH_2^+$ cation, so this peak is not diagnostic for either species. For some $[O{\equiv}C{-}R]^+$ fragments, CO can eliminate to afford an alkyl cation that is 28 amu lighter, but this is a relatively minor pathway, and 28 amu is also identical to the mass of $H_2C{=}CH_2$, so it is not diagnostic for either elimination pathway.

**α-cleavage**

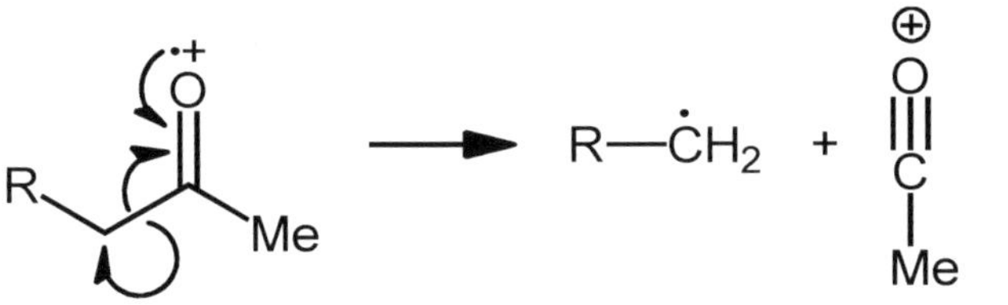

**43 amu**

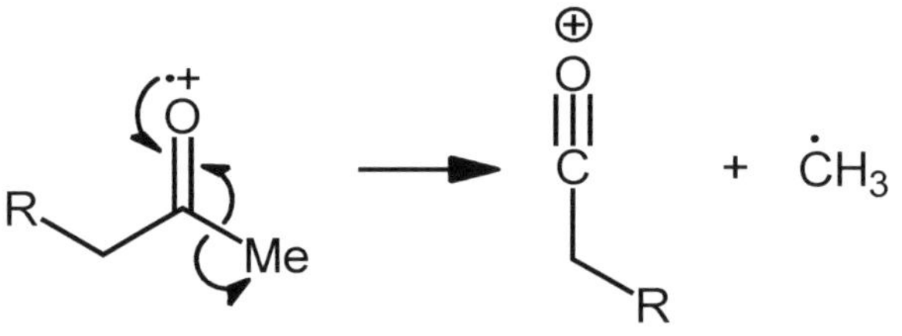

**$[M{-}15]^+$**

**elimination**

Ketones can undergo β-cleavage, but it is not as common as α-cleavage. If one alkyl substituent has 3 or more carbons, a McLafferty rearrangement is also possible.

**β-cleavage**

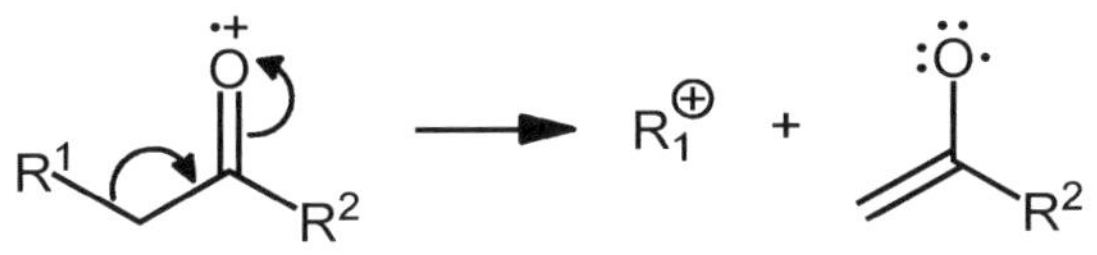

**McLafferty**

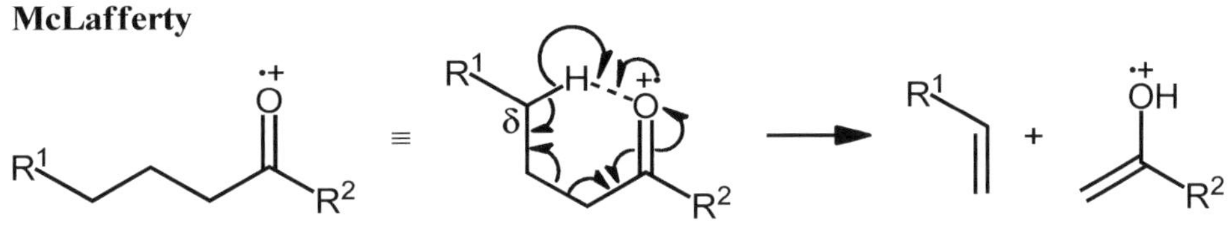

Cyclic ketones exhibit much stronger $M^+$ peaks than acyclic ketones. Although the fragmentation patterns are more complex, the fragmentation mechanisms are similar to those with cyclic alcohols. We show cyclohexanone below as an example. One $C_\alpha$–$C_\beta$ bond breaks homolytically to afford a cationic O≡C moiety and a carbon radical. This acyclic species can rearrange by 3 different mechanisms. The radical can abstract $H^\bullet$ from adjacent to the O≡C moiety, which then fragments to eliminate an alkyl radical and yield a 55 amu cation. This 55 amu cation can subsequently eliminate CO.

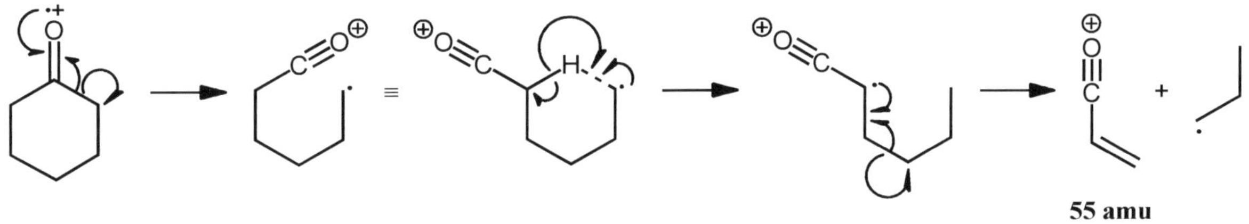

**55 amu**

Alternatively, this $H^\bullet$ abstraction can be followed by a cyclization and elimination of an alkyl radical to afford an 83 amu cation. This cation can also subsequently eliminate CO.

$\dot{C}H_3$ +

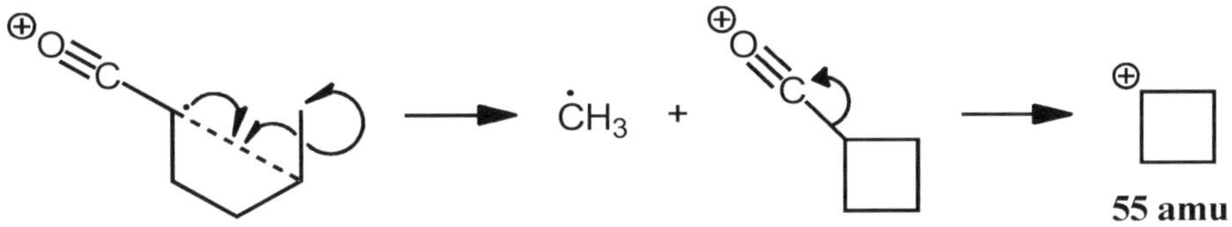

**83 amu**

**55 amu**

Lastly, the initial radical cation can eliminate $H_2C{=}CH_2$ (derived from $C_\beta$ and $C_\gamma$) to afford another radical cation, which can further eliminate CO to yield an alkyl radical cation.

42 amu

70 amu

The mass spectrum of cyclohexanone does shows prominent peaks for the cationic fragments delineated above:

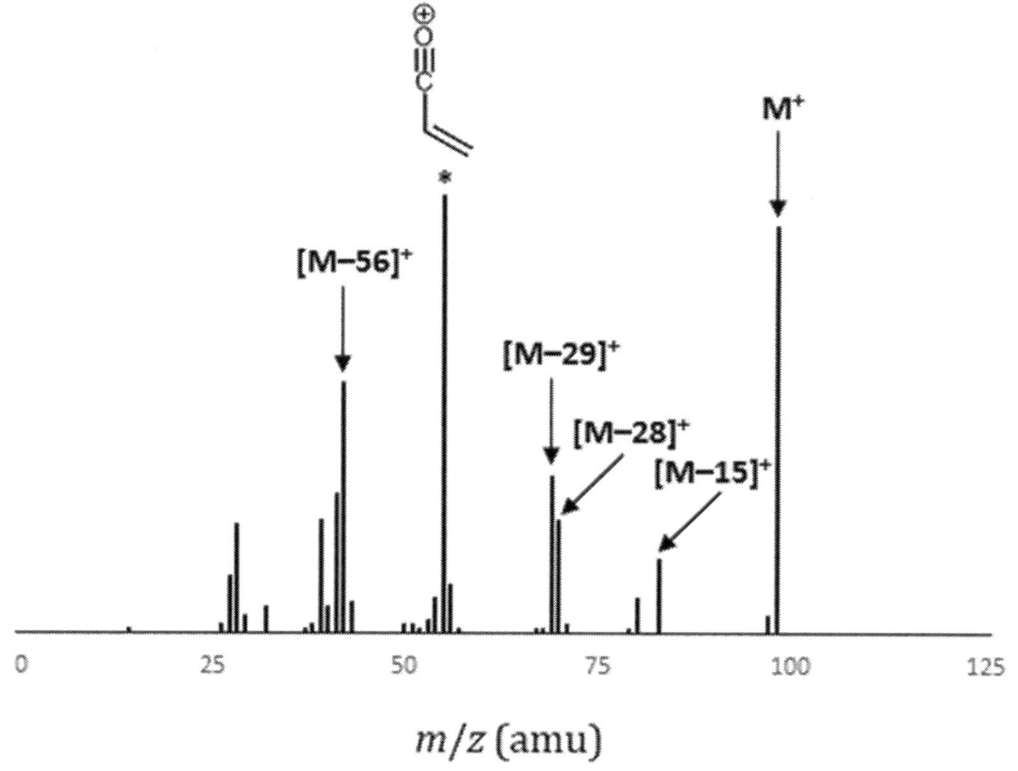


*Lesson VII.12.3 Carboxylic Acids*

Carboxylic acids generally do not exhibit observable $M^+$ peaks, and drawing mechanisms for some of the more prominent fragments is non-trivial. An $[M–17]^+$ peak from α-cleavage to release $HO^{\bullet}$ is often present, albeit with low intensity. Alternatively, α-cleavage on the other side of the carbonyl would release a $[CO_2H]^{\bullet}$ radical and afford an $[M–45]^+$ alkyl cation. However, this is the same mass loss as if α-cleavage to release $HO^{\bullet}$ were followed by elimination of CO.

**α-cleavage**

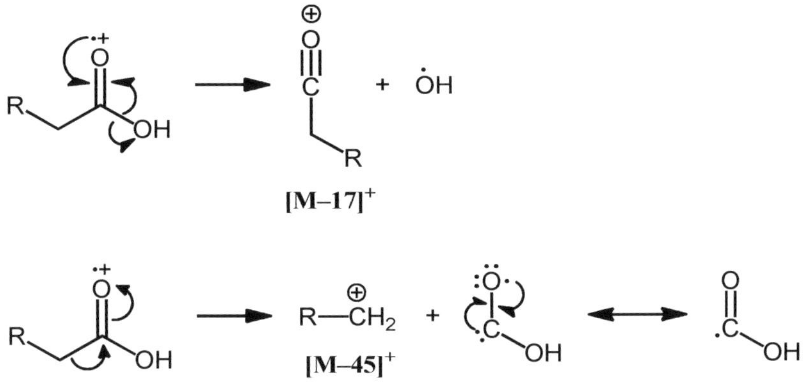

[M–17]$^+$

[M–45]$^+$

β-cleavage can occur in carboxylic acids to give [M–59]$^+$ peaks, and this is usually more favorable than α-cleavage. In general, the most intense peak observed for carboxylic acids is at 60 amu, which is produced by a McLafferty rearrangement.

**β-cleavage**

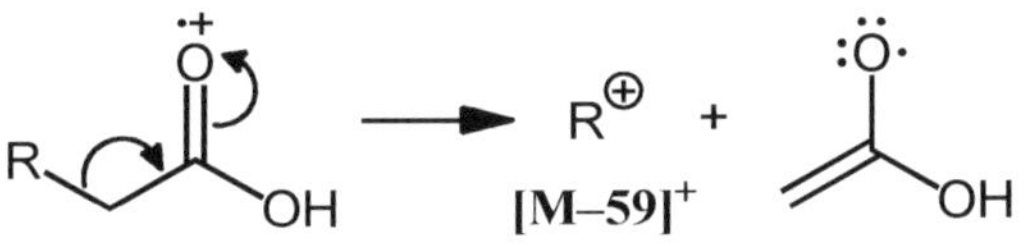

[M–59]$^+$

**McLafferty**

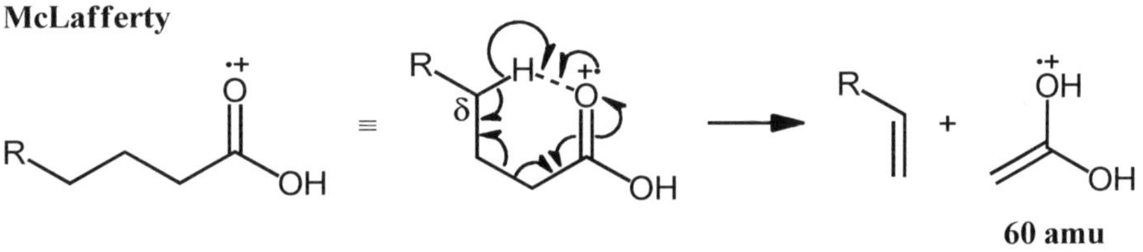

60 amu

Interestingly, carboxylic acids show peaks consistent with γ-cyclization, despite the fact that neither aldehydes nor ketones do. Butyric, pentanoic, and hexanoic acids, for example, exhibit prominent [M–15]$^+$, [M–29]$^+$, and [M–43]$^+$ peaks, respectively.

**γ-cyclization**

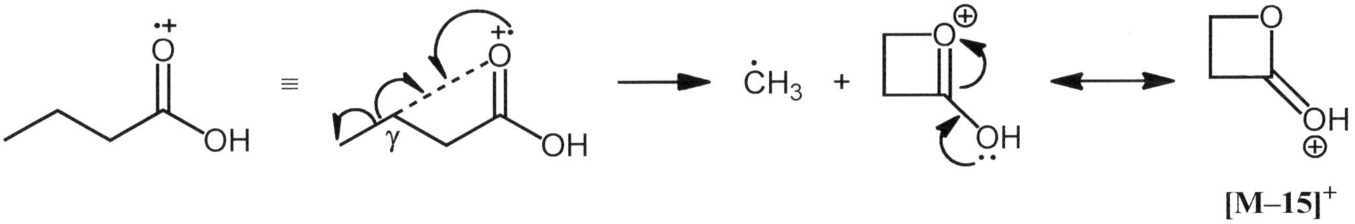

[M–15]$^+$

The mass spectrum of butanoic acid shows prominent peaks for the cationic fragments delineated above:

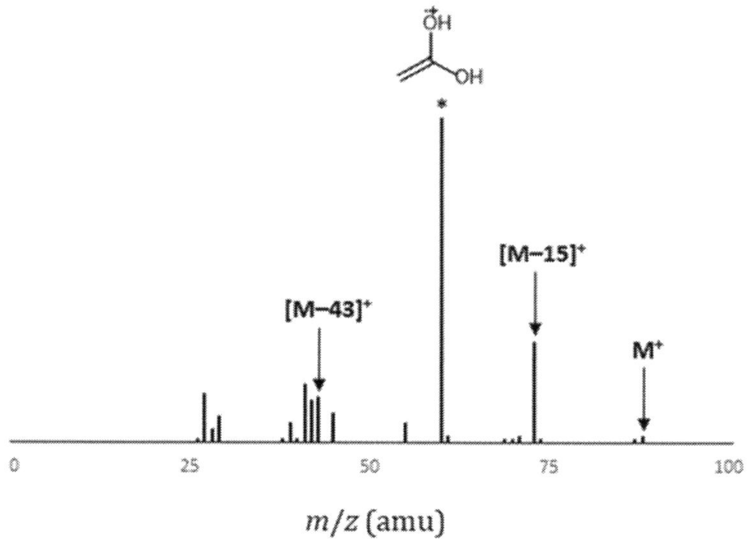


*Lesson VII.12.4 Esters*

Esters may exhibit weak $M^+$ peaks and they undergo a greater variety of fragmentation pathways than carboxylic acids. For example, α-cleavage can occur on either side of the carbonyl unit of an ester, and the alkyl chain can depart as either a radical or a cation, so there are 3 permutations of α-cleavage. In contrast to aldehydes, ketones, and acids, β-cleavage is rarely observed.

**α-cleavage**

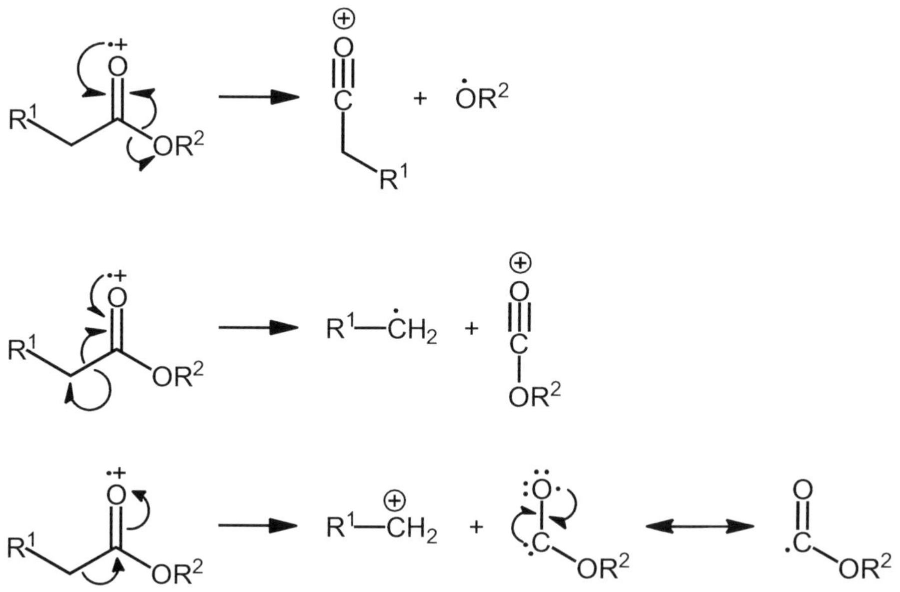

The most prominent peak observed with an ester is derived from a McLafferty rearrangement, but keep in mind that this can occur on either side of the carbonyl unit if the alkyl chains are sufficiently long.

**McLafferty (carbonyl side)**

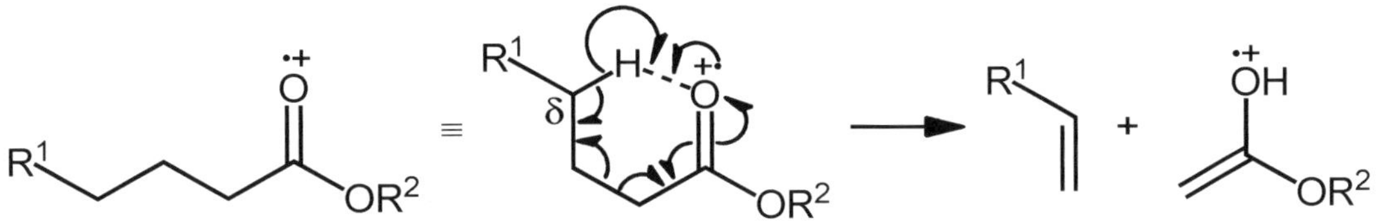

**McLafferty (oxygen side)**

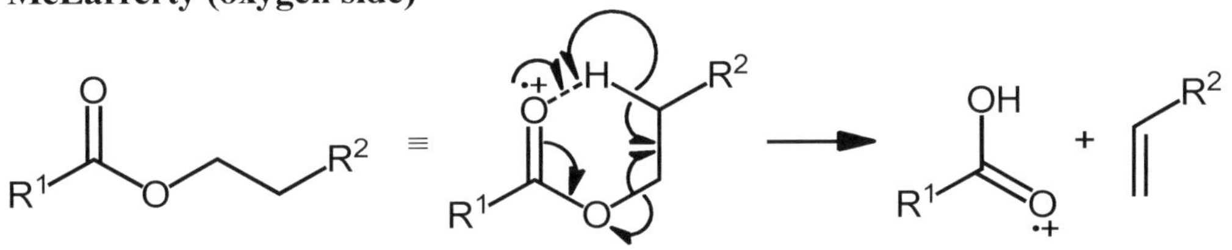

If the alkyl chains on both the "carbonyl side" and the "oxygen side" of sufficient length, then McLafferty rearrangement will occur exclusively on the "oxygen side". If McLafferty rearrangement does occur on the "oxygen side", it will be accompanied by a so-called "McLafferty + 1" peak, which will have significantly greater intensity than the regular McLafferty peak.

**McLafferty + 1**

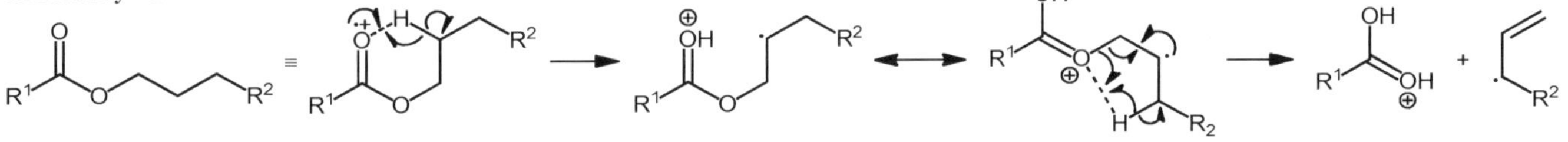

Cyclic esters are more commonly known as lactones, and these exhibit much stronger $M^+$ peaks than their acyclic counterparts. Lactone fragmentation is determined by whether α-cleavage occurs on the "carbonyl side" or the "oxygen side". If α-cleavage occurs on the "carbonyl side", it will be followed by elimination of $CO_2$ to give an alkyl radical cation at $[M–44]^+$. Consider the fragmentation of δ-valerolactone:

**α-cleavage (*i*) then elimination (*ii*)**

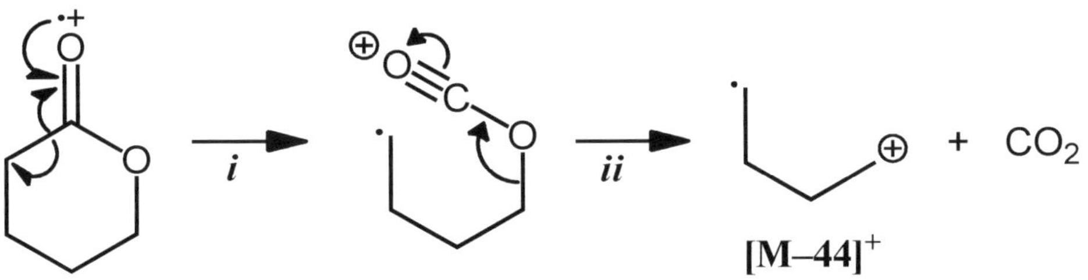

$[M–44]^+$

If α-cleavage instead occurs on the "oxygen side", it will be followed by elimination of $H_2C{=}O$, like we observed with cyclic ethers, to give an alkyl radical and cationic carbonyl at $[M{-}30]^+$. This species can fragment further by eliminating CO to produce an alkyl radical cation at $[M{-}58]^+$.

**α-cleavage (*i*) then elimination (*ii,iii*)**

*i* → *ii* → $H_2C{=}O$ + $[M{-}30]^+$ *iii* → $[M{-}58]^+$ + CO

The mass spectrum of δ-valerolactone shows prominent peaks for the cationic fragments delineated above:

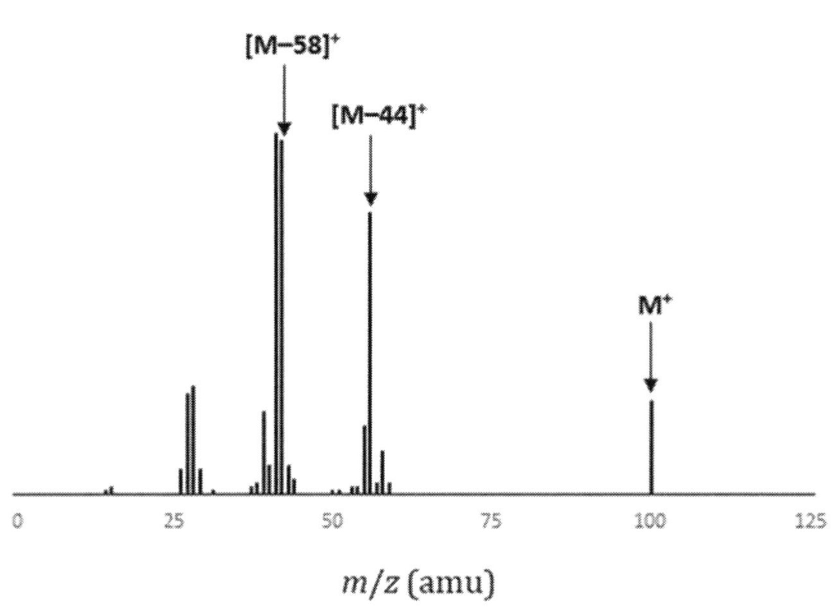


*Lesson VII.12.5 Amides*

Primary amides (those without N-alkyl substituents) typically do not exhibit $M^+$ peaks, and the major fragmentation pathways are α-cleavage to form the $[O{=}C{=}NH_2]^+$ ion at 44 am and McLafferty rearrangement to yield a 59 amu radical cation. β-cleavage is generally not observed. Like carboxylic acids, 1° amides exhibit peaks consistent with γ-cyclization: butyramide, valeramide, and hexanamide, for example, exhibit prominent $[M{-}15]^+$, $[M{-}29]^+$, and $[M{-}43]^+$ peaks, respectively.

**α-cleavage**

$R\text{—}\dot{C}H_2$ + 

**44 amu**

**McLafferty**

**59 amu**

Similar to carboxylic acids, 1° amides also exhibit peaks consistent with γ-cyclization: butyramide, valeramide, and hexanamide, for example, exhibit prominent $[M-15]^+$, $[M-29]^+$, and $[M-43]^+$ peaks, respectively. The fragmentation of amides is illustrated by butyramide:

**γ-cyclization**

$\dot{C}H_3$ +

$[M-15]^+$

The mass spectrum of cyclohexanol shows prominent peaks for the cationic fragments delineated above:

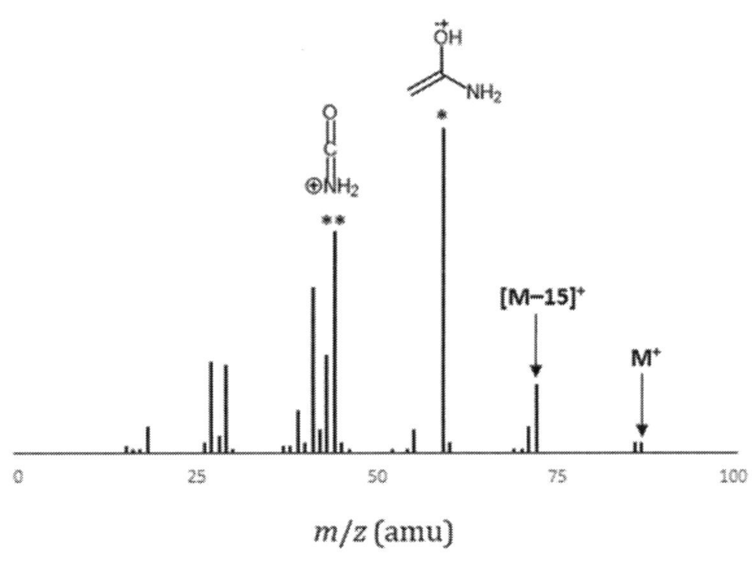

If *N*-alkyl substituents are present, then the $M^+$ peak is typically strong, and is often accompanied by an $[M-1]^+$ peak. The fragmentation of 2° and 3° amides is significantly more complex than their 1° counterparts. To understand their fragmentation patterns, it is useful to consider that there two distinct isomers can be formed by ionization, one with the radical cation character localized on the oxygen and the other with the •+ on the nitrogen.

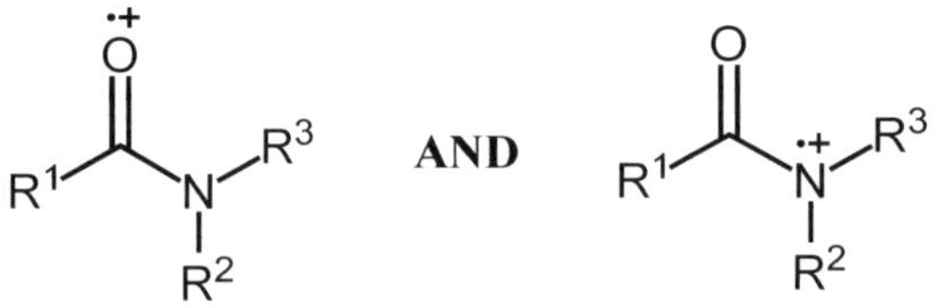

As expected, fragmentation of the isomer with the •+ on the oxygen is similar to what is observed with 1° amides (i.e., McLafferty, and γ-cyclization), however it generally does not undergo α-cleavage on the carbonyl side of the amide functional group.

**McLafferty**

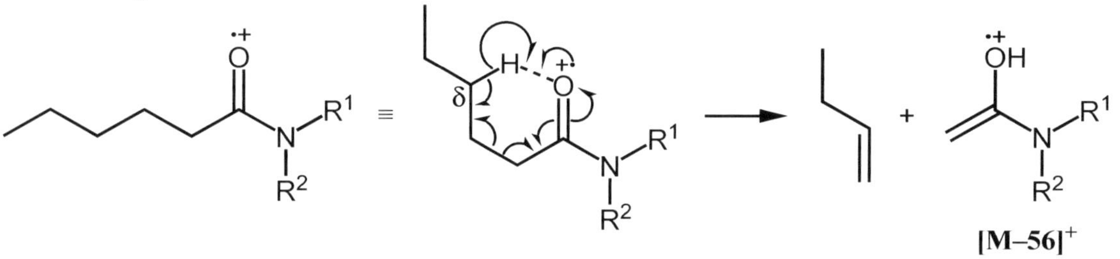

**γ-cyclization**

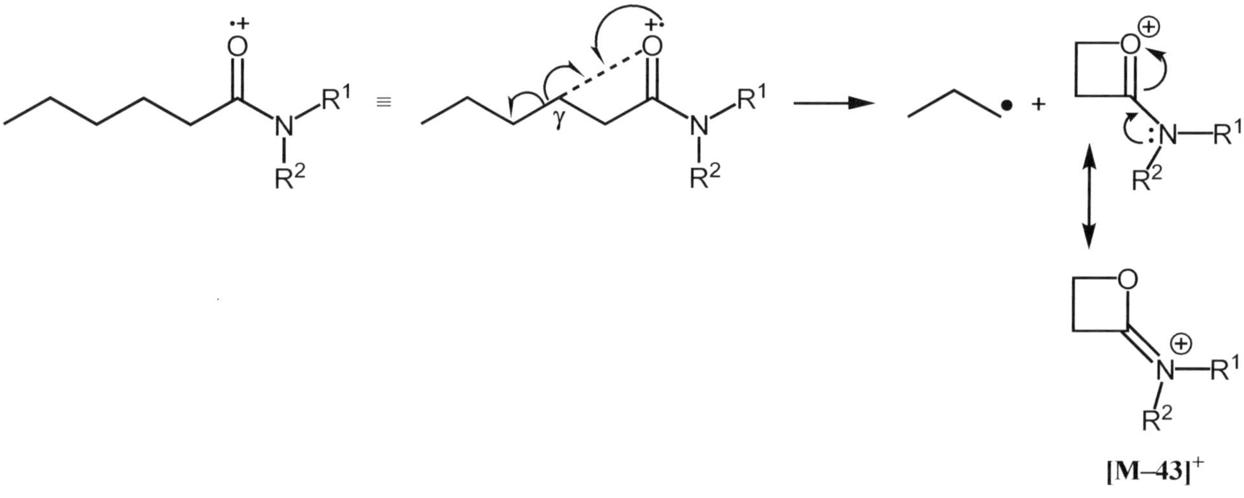

The isomer with the •+ on the nitrogen can undergo homolysis and α-cleavage.

**homolysis**

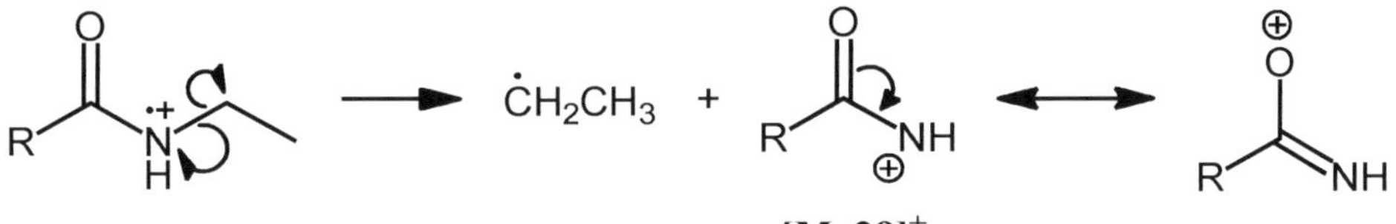

$\dot{C}H_2CH_3$ +

$[M-29]^+$

**α-cleavage**

$\dot{C}H_3$ +

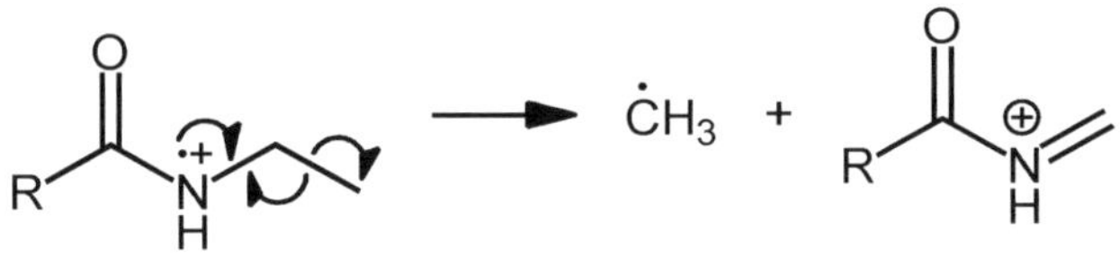

$[M-15]^+$

Depending on the N-alkyl substituent(s), some 2° and 3° amides exhibit peaks at 30 and 44 amu, corresponding to the $[H_2C{=}NH_2]^+$ and $[O{=}C{=}NH_2]^+$ ions, respectively. However, attempting to draw an arrow pushing mechanism for the formation of these fragments can be extremely challenging.

## Lesson VII.13. Fragmentation of Arenes

### *Lesson VII.13.1 Alkyl Substituents Only*

If only an alkyl substituent is present on an arene, the radical cation character of the molecular ion resides in the π-system of the arene. The $M^+$ peak is typically strong, and the principal fragmentation mode is α-cleavage to form, after rearrangement, the tropylium cation (91 amu). Interestingly, $[C_7H_7]^+$ can release acetylene (H–C≡C–H, ΔM = –26) to afford the $[C_5H_5]^+$ cation, which can also eliminate acetylene to afford the cyclopropenyl cation.

**α-cleavage (*i*) then rearrangement (*ii*)**

R –e⁻ R *i* R˙ + ⊕ *ii* + 91 amu

**Elimination**

+ 91 amu H–C≡C–H + (+) 65 amu H–C≡C–H + (+) 39 amu

If there are hydrogen atoms at the γ-position of the alkyl chain, the molecular ion can undergo a McLafferty rearrangement. Note this affords a 92 amu radical cation, which we know will readily lose H˙ to yield the 91 amu benzyl cation, which can subsequently rearrange to tropylium.

**McLafferty (*i*) then elimination (*ii*)**

H γ R *i* R + H H 92 amu *ii* H˙ + 91 amu

An arene with two alkyl substituents can additionally undergo homolysis of one of the arene–alkyl bonds, the driving force being the formation of a substituted tropylium (which can then undergo multiple successive eliminations of acetylene).

**homolysis (*i*), rearrangement (*ii*), then elimination (*iii*)**

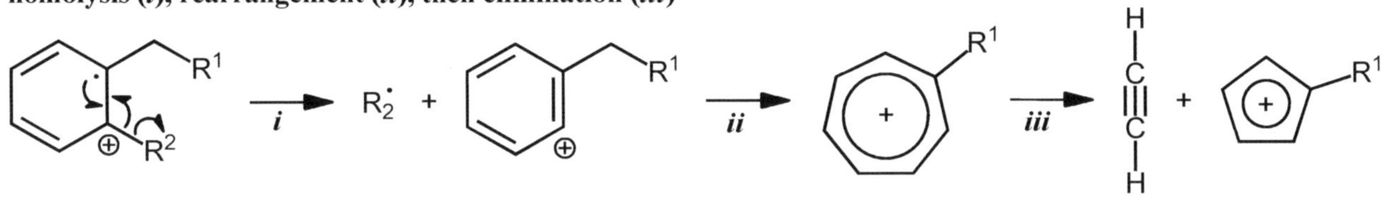

*Lesson VII.13.2 Heteroatom Attached to Arene*

In an aryl alkyl ether, the radical cation character is primarily localized on the oxygen atom. The $M^+$ peak for these compounds is typically very strong. The parent ion can undergo homolysis of the oxygen–alkyl bond to generate the phenoxonium cation $[PhO]^+$, and this + charge is stabilized by delocalization across the ring. The phenoxonium cation can then eliminate C≡O to afford $[C_5H_5]^+$, and we have already seen that this ion can fragment further.

**homolysis (*i*) then elimination (*ii*)**

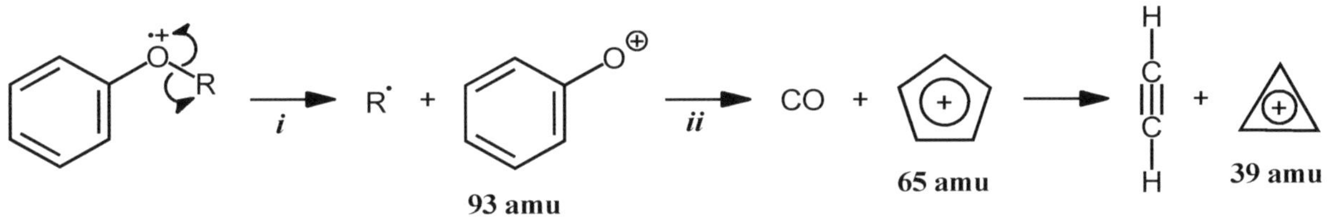

93 amu

65 amu

39 amu

When the alkyl substituent is ethyl or larger, heterolysis of the aryl–oxygen bond can also occur, affording the phenyl cation (77 amu). Similar to $[C_7H_7]^+$, the phenyl cation can eliminate acetylene to generate $[C_4H_3]^+$ (51 amu).

**heterolysis (*i*) then elimination (*ii*)**

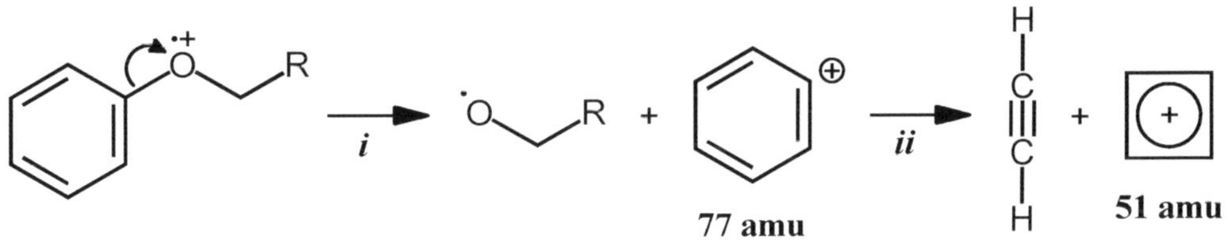

77 amu

51 amu

When the alkyl substituent is methyl, a rearrangement to eliminate $H_2CO$ occurs instead, and the resulting phenyl radical cation (78 amu) will readily lose $H^•$ to afford the phenyl cation.

**elimination**

H H

78 amu

77 amu

The mass spectrum of anisole reflects the fragmentation pathways described above:

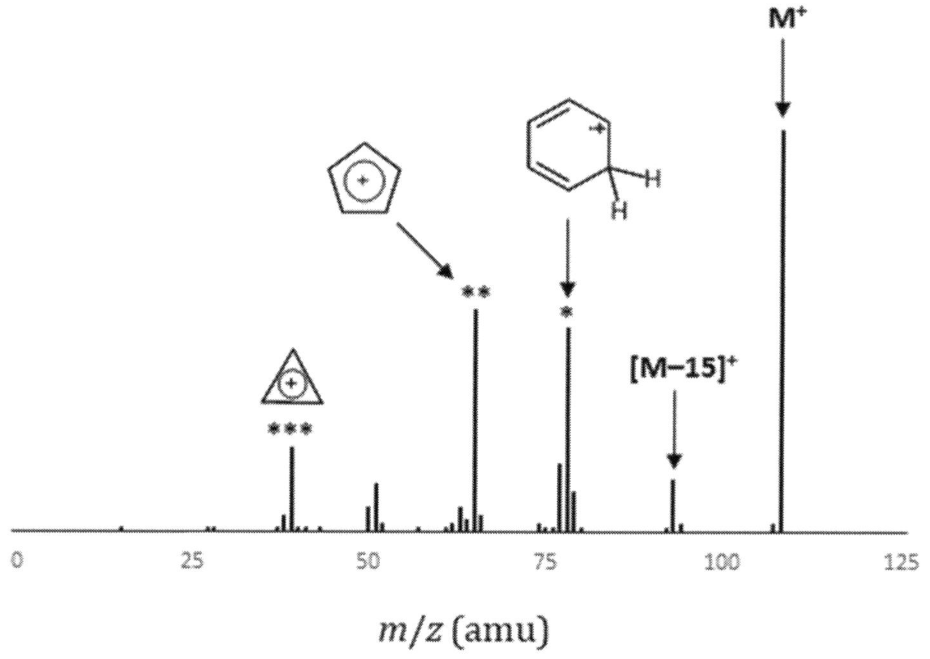


Aryl alcohols exhibit more complicated fragmentation patterns that cannot be represented as easily with arrow pushing mechanisms. Phenol, for example, displays a prominent $M^+$ peak, but it also displays an $[M–28]^+$ peak, which corresponds to the elimination of CO. This affords a radical cation with a mass of 66 amu, which can lose $H^{\bullet}$ to yield the $[C_5H_5]^+$ cation (65 amu) and its descendant ions.

**elimination**

CO +

65 amu

66 amu

Aryl alcohols with alkyl substituents also undergo α-cleavage at the alkyl substituent, followed by a rearrangement to give hydroxytropylium (107 amu). This ion can eliminate CO to afford the $[C_6H_7]^+$ ion (79 amu), which can then eliminate $H_2$ to afford the phenyl cation (77 amu).

**α-cleavage (*i*), rearrangement (*ii*), then elimination (*iii*)**

$R^{\bullet}$ + … → **107 amu** → CO + **79 amu** → $H_2$ + **77 amu**

The mass spectrum of phenol reflects the fragmentation pathways described above:

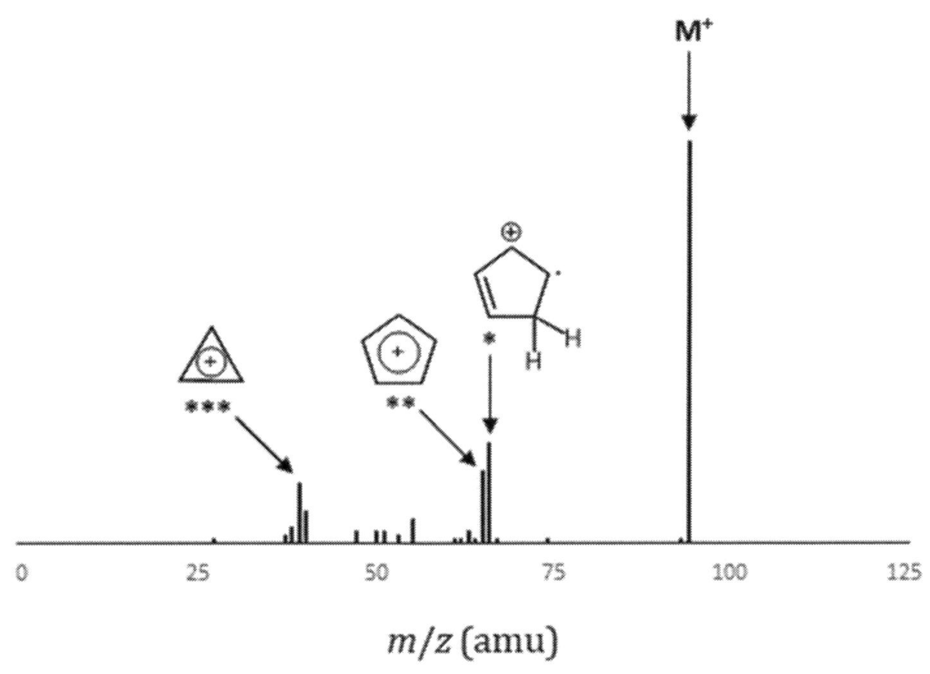


Anilines present strong $[M–1]^+$ peaks due to loss of $H^{\bullet}$ in addition to prominent $M^+$ peaks. Note that the structure of the $[M–1]^+$ fragment is similar to that of the phenoxonium cation, which we previously learned could eliminate CO. Analogously, the $[M–1]^+$ fragment of an aniline can eliminate H–C≡N (27 amu) to afford a peak at $[M–28]^+$. The parent ion can also eliminate HCN to afford a peak at $[M–27]^+$, which can subsequently lose $H^{\bullet}$ and be converted into the $[M–28]^+$ fragment. In the case of aniline itself, we note these fragments are identical to the 65 and 66 amu fragments observed with phenol.

**elimination**

H· + [M–1]⁺ → HCN + [M–28]⁺

$[M-1]^+$ $[M-28]^+$

HCN + $[M-27]^+$ → H· + $[M-28]^+$

$[M-27]^+$ $[M-28]^+$

The mass spectrum of aniline reflects the fragmentation pathways described above:

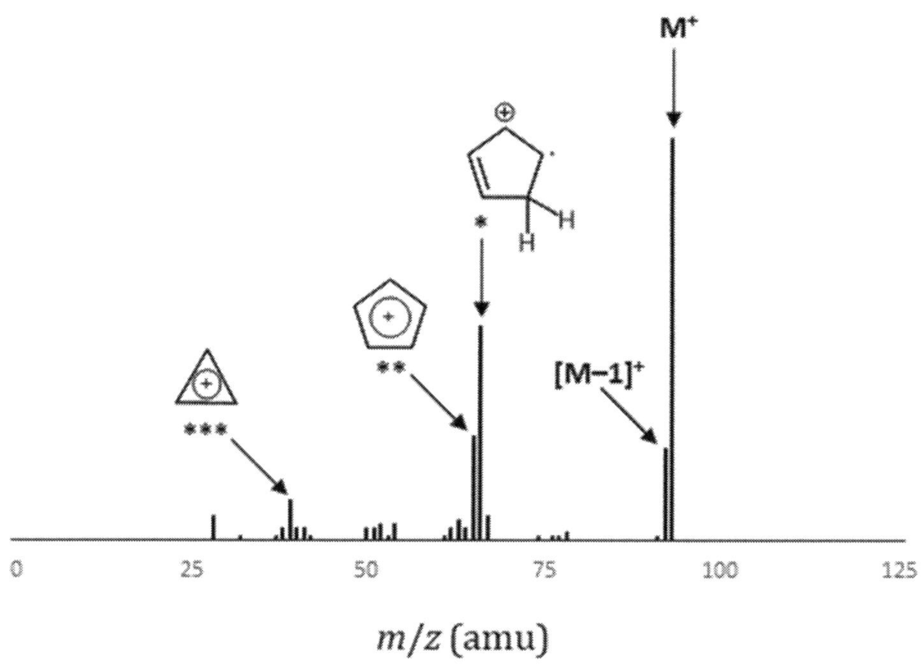


Anilines with alkyl substituents on the arene can undergo α-cleavage followed by rearrangement to give the aminotropylium ion (106 amu). Similar to hydroxytropylium, aminotropylium can eliminate HCN to yield the familiar $[C_6H_7]^+$ ion, which can further fragment by losing $H_2$ to give the phenyl cation at 77 amu.

**α-cleavage (*i*), rearrangement (*ii*), then elimination (*iii*)**

106 amu

77 amu

79 amu

Anilines with *N*-alkyl substituents likewise undergo α-cleavage, rearrangement, and elimination to give the same peaks at 106, 79, and 77 amu, but generating an arrow-pushing mechanism for these processes is more challenging.

*Lesson VII.13.3 Heteroatom at a Benzylic Position*

The molecular ion for a compound with a heteroatom at a benzylic position features radical cation character at that heteroatom, which typically results in a strong $M^+$ peak. Benzyl alcohols can lose $HO^•$ via heterolysis and $H^•$ via α-cleavage. Both heterolysis and α-cleavage can subsequently be followed by rearrangement to afford tropylium and hydroxytropylium ions, respectively, which can undergo further elimination reactions (as we saw in the last lesson).

**heterolysis (*i*) then rearrangement (*ii*)**

$[M–17]^+$

91 amu

**α-cleavage (*i*) then rearrangement (*ii*)**

$[M–1]^+$

107 amu

The mass spectrum of benzyl alcohol reflects the fragmentation pathways described above:

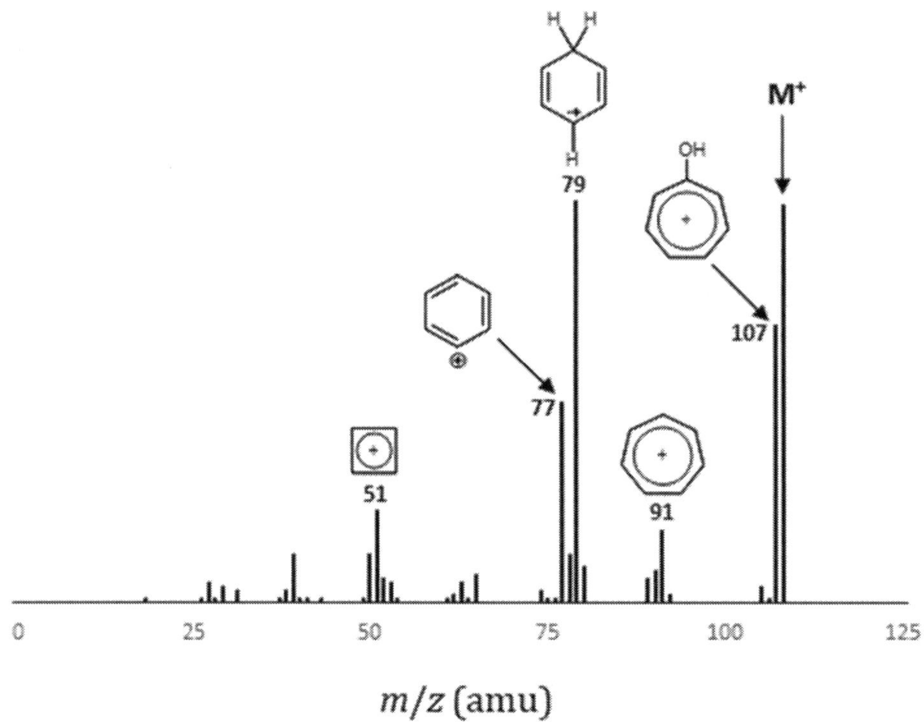


Benzylamines likewise lose $H_2N^{\bullet}$ via heterolysis and $H^{\bullet}$ via α-cleavage, which can be followed by rearrangement to afford tropylium and aminotropylium ions, respectively. Do not forget that these ions can also undergo subsequent elimination reactions.

**heterolysis (*i*) then rearrangement (*ii*)**

$NH_2$ $-e^-$ → $NH_2$ → *i* $^{\bullet}NH_2$ + [M–16]$^+$ → *ii* 91 amu

**α-cleavage (*i*) then rearrangement (*ii*)**

H $NH_2$ → *i* $H^{\bullet}$ + H $NH_2$ [M–1]$^+$ → *ii* $NH_2$ 106 amu

The mass spectrum of benzyl amine reflects the fragmentation pathways described above:

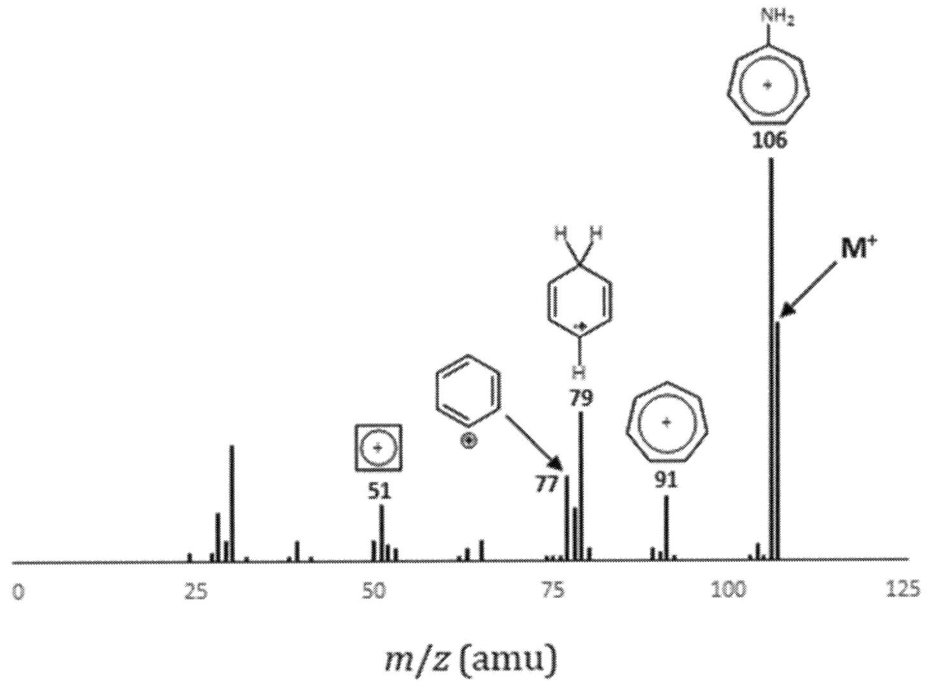


If an O-alkyl or N-alkyl substituent is present, the tropylium peak will be significantly more intense because RO$^\bullet$ and RHN$^\bullet$ are significantly more stable that HO$^\bullet$ and $H_2N^\bullet$, which makes heterolysis of the C–O and C–N bonds more favorable. A prominent peak at 92 amu may also be observed, which can be assigned to a fragment formed by a McLafferty-like process, i.e., intramolecular H$^\bullet$ transfer via a cyclic 6-membered transition state followed by elimination (shown below for E = O or NR'). As might be expected, this 92 amu radical cation will readily lose H$^\bullet$ to afford the familiar tropylium ion.

**rearrangement (*i*) then elimination (*ii*)**

92 amu

91 amu

If an alkyl substituent is instead present on the arene of a benzylic alcohol or benzylamine, additional peaks appear at $[M–18]^+$ due to loss of $H_2O$ or $[M–17]^+$ due to loss of $NH_3$, respectively, via elimination. With 2-methylbenzyl alcohol, for example, the most intense peak is at 104 amu, corresponding to $[M–18]^+$. This fragment is formed from the molecular ion via intramolecular H$^\bullet$ transfer via a 6-membered cyclic transition state followed by dissociation of $H_2O$ to afford a radical cation.

**rearrangement (*i*) then elimination (*ii*)**

*i* *ii*

**104 amu**

Unsurprisingly, the relative intensity of the peak at 104 amu decreases when the methyl substituent goes from the 2- to the 3- to the 4-position. The mass spectrum of 2-methylbenzyl alcohol reflects the fragmentation pathways described above:

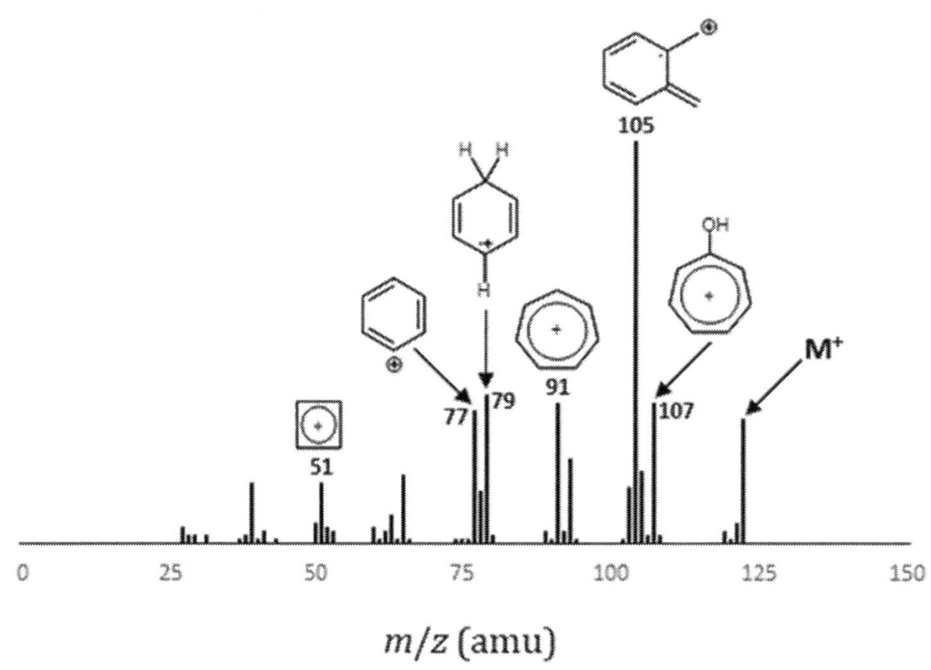


A similar mechanism occurs in the elimination of $NH_3$ from appropriately substituted benzylamines.

*Lesson VII.13.4 Carbonyl at a Benzylic Position*

To discuss the fragmentation of molecules with carbonyl groups at the benzylic position, it is first convenient to introduce some shorthand terms to simplify our discussion in this Lesson. We will divide these molecules along the location of the carbonyl subunit. The side with the aryl ring will be referred to as the "aryl side" and the other side will be cleverly and creatively termed the "non-aryl side". Note that the identity of Y will vary based on the identity of the functional group: aldehydes (H), ketones (R), carboxylic acids (OH), esters (OR), and amides ($NH_2$, NHR, $NR_2$).

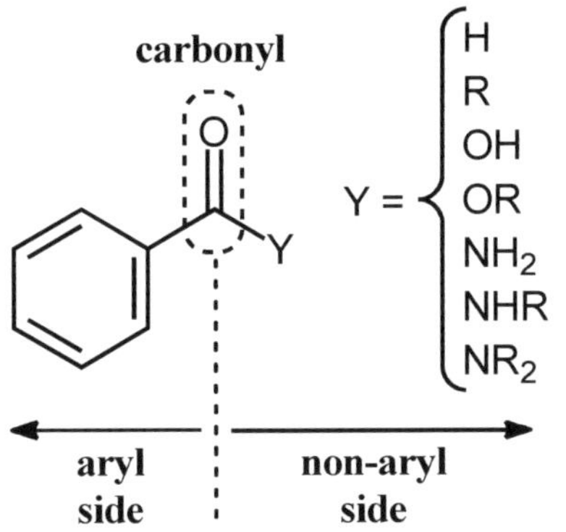

carbonyl

Y = H, R, OH, OR, $NH_2$, NHR, $NR_2$

aryl side | non-aryl side

Each of these classes of molecules will undergo some fragmentation processes that are identical to what we have previously covered and some that are new. Essentially, the fragmentation pathways of the non-aryl side are identical to what we have previously covered with the non-aryl molecules. As a result, in this Lesson, we only need to discuss the fragmentation pathways involving the aryl side, and these are conserved for all identities of Y.

The primary fragmentation pathway for these types of molecules is α-cleavage along the C–Y bond, which releases $Y^{\bullet}$ and the $[Ph–C{\equiv}O]^+$ cation. This cation can eliminate CO to afford the familiar phenyl cation at 77 amu, which itself can undergo subsequent elimination reactions.

**α-cleavage (*i*) then elimination (*ii*)**

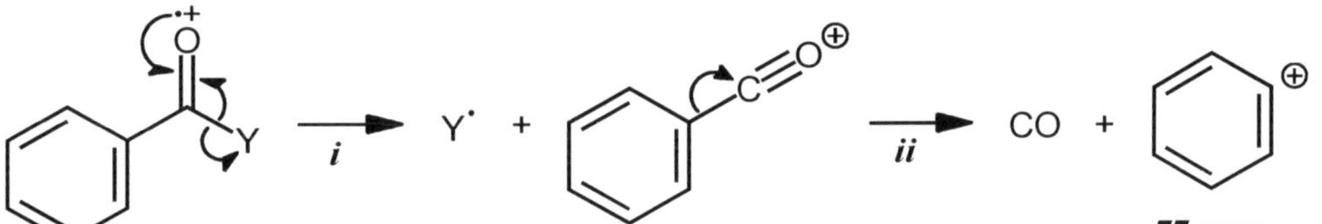

$Y^{\bullet}$ + ... → CO + ...

**77 amu**

If a substituent –EH is located *ortho* to the carbonyl-based functional group, where E can be $CH_2$, CHR, $CR_2$, O, NH, or NHR, then an additional fragmentation pathway becomes available. First, H–Y can eliminate, leaving behind a ketene (i.e., a molecule containing a C=C=O moiety) radical cation. We can draw another resonance structure for this radical cation, which illustrates the next fragmentation reaction, specifically the elimination of CO.

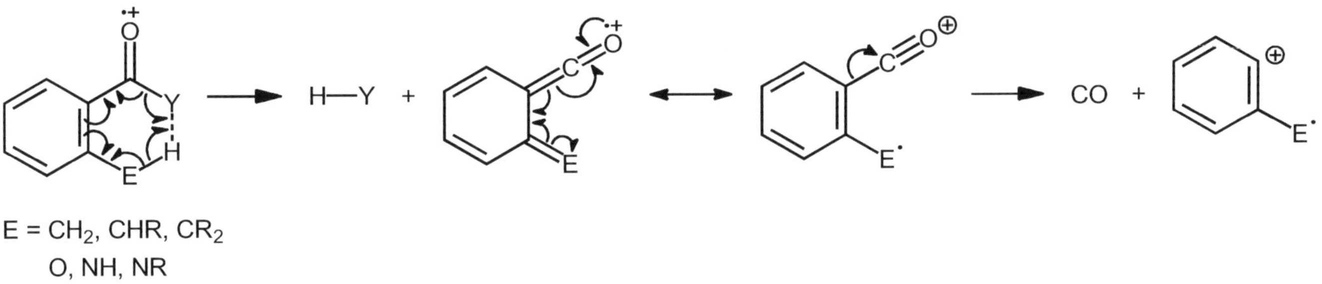

H—Y + ... ⟷ ... → CO + ...

E = $CH_2$, CHR, $CR_2$
O, NH, NR

## Lesson VII.14. Fragmentation of Alkenes

*Lesson VII.14.1 Linear Alkenes*

Ionization of an alkene removes an electron from the C–C π-bond, which places the radical on one of the alkene carbons and the + on the other alkene carbon. Because these electrons are so much easier to remove than those in C–C and C–H σ-bonds, the $M^+$ peak for an alkene is generally much stronger than for an alkane. To understand the fragmentation modes, let us consider 1-hexene specifically. We can draw the molecular ion with the + on the less substituted carbon and the • on the more substituted carbon. We can then see how α-cleavage would release an allyl cation (41 amu).

**α-cleavage**

$-e^-$

41 amu

Due to the high energies involved in the mass spectrometry experiment, the position of the C=C bond will freely migrate over all the carbons. As a result, ***every*** alkene will show a peak at 41 amu due to the allyl cation. Aside from that, the fragmentation pattern for alkenes is similar to alkanes: there will be a peak at $[M-15]^+$ due to loss of $^{\bullet}CH_3$, at $[M-29]^+$ due to loss of $^{\bullet}CH_3CH_2$, etc. Drawing a mechanism for these fragments can only be done after the position of the C=C bond has moved.

**rearrangement (*i*) then α-cleavage (*ii*)**

*i* *ii* $\dot{C}H_3$ +

$[M-15]^+$

All alkenes exhibit an $[M-28]^+$ peak due to the elimination of $H_2C=CH_2$, which can occur via a McLafferty-like rearrangement (shown below for 1-pentene).

**elimination**

H

$[M-28]^+$

Alkenes longer than pentene can, in addition to eliminating $H_2C=CH_2$, also eliminate longer-chain alkenes, such as propene, butene, etc., depending on the initial chain length:

**elimination**

[M–42]$^{+}$

It is important to emphasize that **all** alkenes exhibit an [M–28]$^{+}$ peak, but it is not always possible to draw a McLafferty-like rearrangement or other mechanism that would make sense given the initial molecule (e.g., 2-methyl-2-butene). The mass spectrum of 1-hexene and 1-pentene illustrate the above-mentioned fragmentation pathways.

MS for 1-hexene:

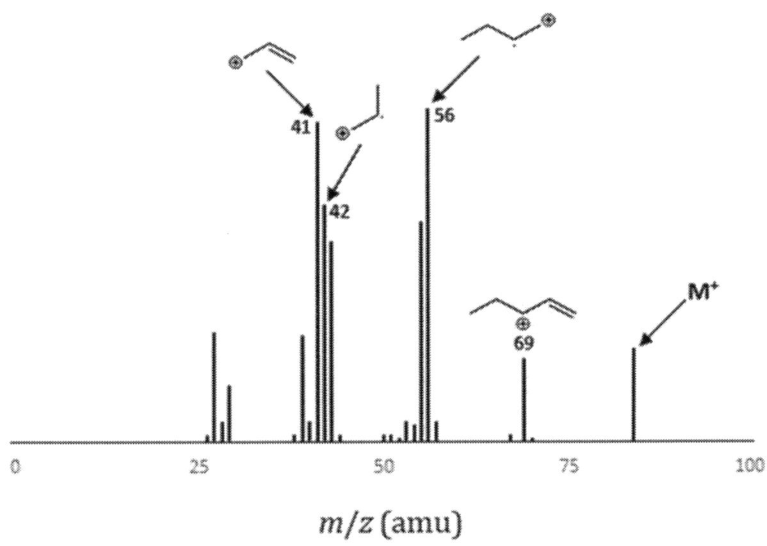


MS for 1-pentene:

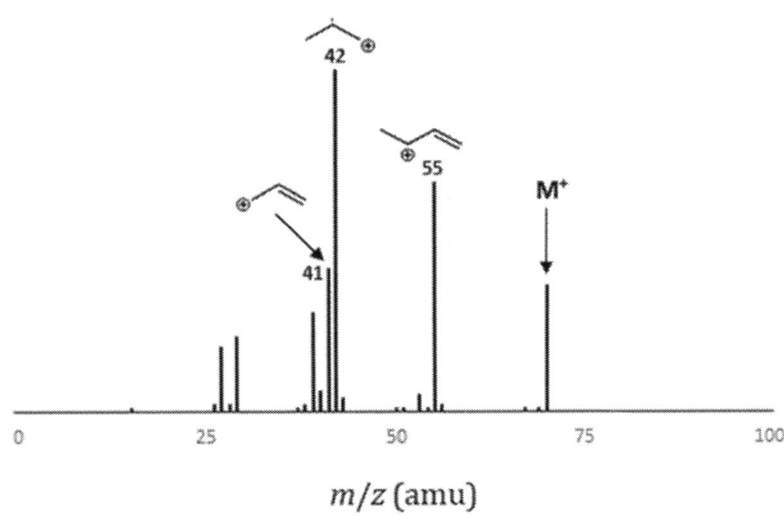

*Lesson VII.14.2 Branched Alkenes*

Branched alkenes also exhibit $M^+$ peaks, as well as many of the same fragmentation peaks as linear alkenes, e.g., 41 amu (allyl cation), $[M{-}15]^+$, $[M{-}28]^+$, $[M{-}29]^+$, etc. However, the relative intensity of the α-cleavage peaks will differ between branched and linear alkenes. One reason is that α-cleavage will occur more readily at any branched sites (similar to alkane behavior), thus the corresponding peaks will be more intense.

**rearrangement (*i*) then α-cleavage (*ii*)**

*i* → *ii* → $\dot{C}H_3$ + $[M{-}15]^+$

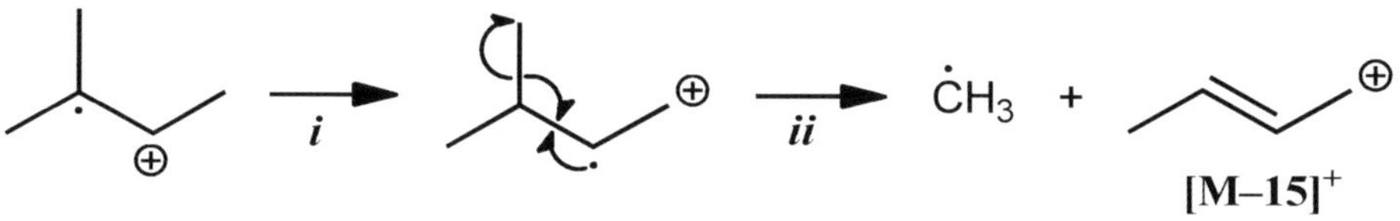

Analogously, a carbocation will be stabilized to a greater extent at a more substituted carbon, so any α-cleavage process that affords an allylic cation at a more substituted carbon will be more favorable.

**rearrangement (*i*) then α-cleavage (*ii*)**

*i* → *ii* → $\dot{C}H_2CH_3$ + $[M{-}29]^+$

Molecules with a branch attached to an alkene carbon can occasionally exhibit $[M{-}1]^+$ peaks with weak intensity due to α-cleavage of $H^{\bullet}$.

**α-cleavage**

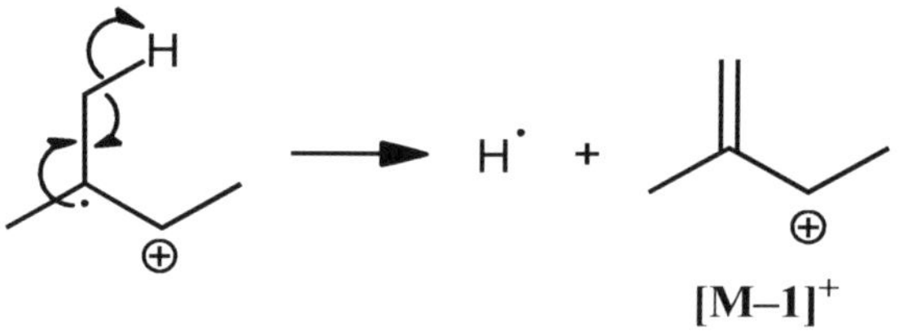

H → $H^{\bullet}$ + $[M{-}1]^+$

The MS for 2-methyl-2-butene demonstrates fragmentation in branched alkenes, and is shown on the following page.

MS for 2-methyl-2-butene

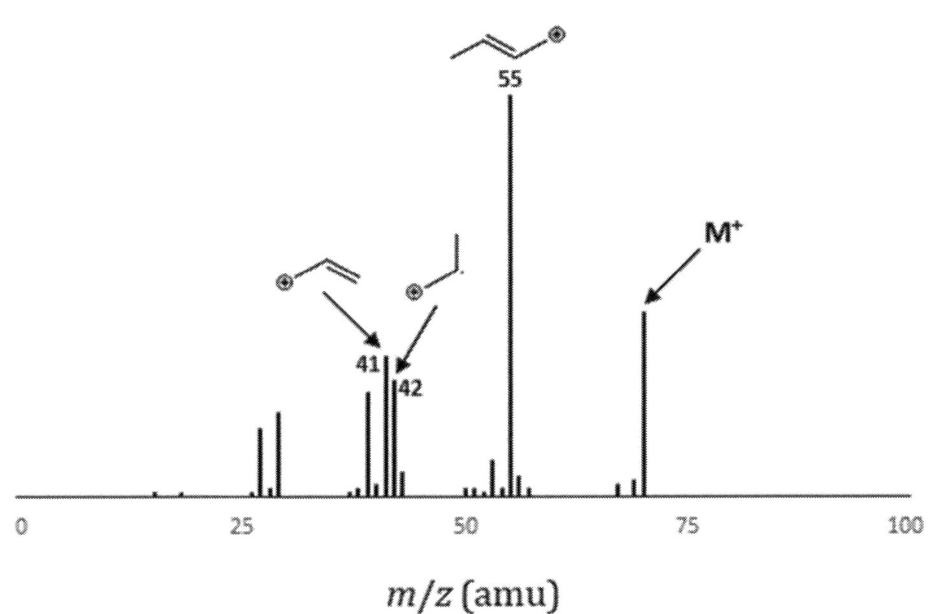

*Lesson VII.14.3 Cycloalkenes*

Cycloalkenes exhibit more intense $M^+$ peaks than acyclic alkenes, for the same reasons that $M^+$ peaks are more intense for cycloalkanes than acyclic alkanes. Cycloalkenes also exhibit significantly $[M-1]^+$ peaks than acyclic alkenes.

Interestingly, cycloalkenes still exhibit peaks at $[M-15]^+$, $[M-29]^+$, etc., consistent with elimination of alkyl radicals, despite the fact that there are no $CH_3$ groups in any of these molecules. Shown below are mechanisms for the elimination of $^{\bullet}CH_3$ and $^{\bullet}CH_2CH_3$ from cycloheptene.

**α-cleavage (*i*), rearrangement (*ii*), then elimination (*iii*)**

$\dot{C}H_3$ + $[M-15]^+$

**α-cleavage (*i*), rearrangement (*ii*), then elimination (*iii*)**

$\dot{C}H_2CH_3$ + $[M-29]^+$

The MS for cycloheptene further demonstrates these fragmentation pathways:

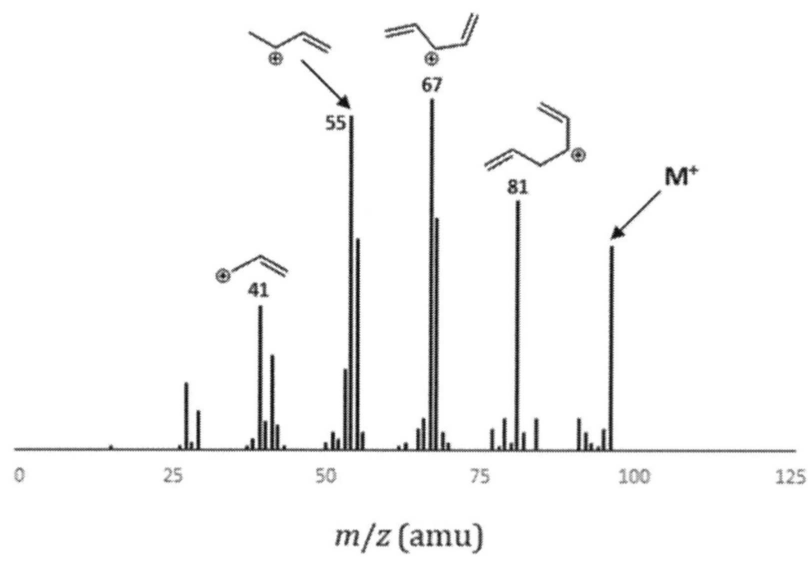

However, cyclopentene also exhibits $[M–15]^+$ and $[M–29]^+$ peaks, and cyclohexene also exhibits an $[M–29]^+$ peak, even though it is not possible to draw analogous mechanisms to form these fragments from these molecules. Likewise, **all** cycloalkenes exhibit peaks at 41 amu from the allyl cation.

Cyclohexene can undergo a retro Diels-Alder reaction to eliminate $H_2C{=}CH_2$ and yield an $[M–28]^+$ peak. It should be noted that this type of reaction can only occur with 6-membered rings. It is sometimes convenient to draw the radical cation using the square bracket convention, as it more readily depicts the diene fragment generated upon retro Diels-Alder release of the dienophile. Substituted cyclohexenes may eliminate alkenes larger than ethylene.

**retro Diels-Alder**

$[M–28]^+$

Here is the MS for cyclohexene:

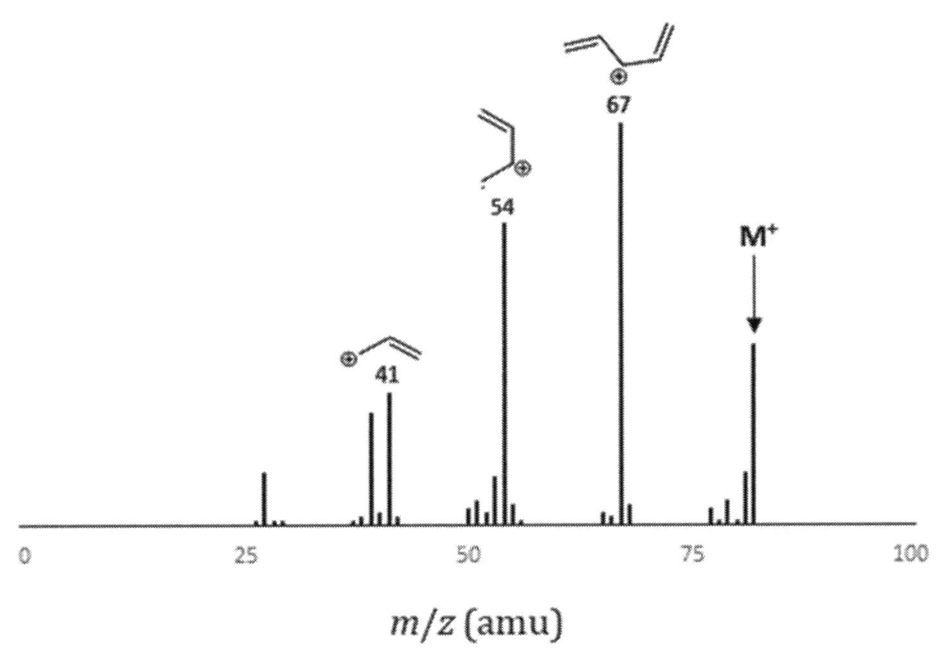


Interestingly, cycloalkenes larger than cyclohexene still exhibit $[M–28]^+$ peaks, despite the fact that they lack the necessary geometry to undergo retro Diels-Alder. One possible way to draw a mechanism is to consider rearrangement of the initial radical cation (feasible at the high energies of typical mass spectrometry experiments). For example, we could draw the rearrangement of an 8-membered ring into an ethyl-substituted 6-membered ring, which would be capable of retro Diels-Alder. It is then possible to draw retro Diels-Alder reactions that eliminate either ethylene (28 amu) or 1-butene (56 amu). Indeed, cyclooctene exhibits intense $[M–28]^+$ and $[M–56]^+$ peaks.

**rearrangement (*i*) then retro Diels-Alder (*ii*)**

*i* → *ii* → + $[M–28]^+$

**rearrangement (*i*) then retro Diels-Alder (*ii*)**

*i* → *ii* → + $[M–56]^+$

The MS for cyclooctene reflects these fragmentation pathways:

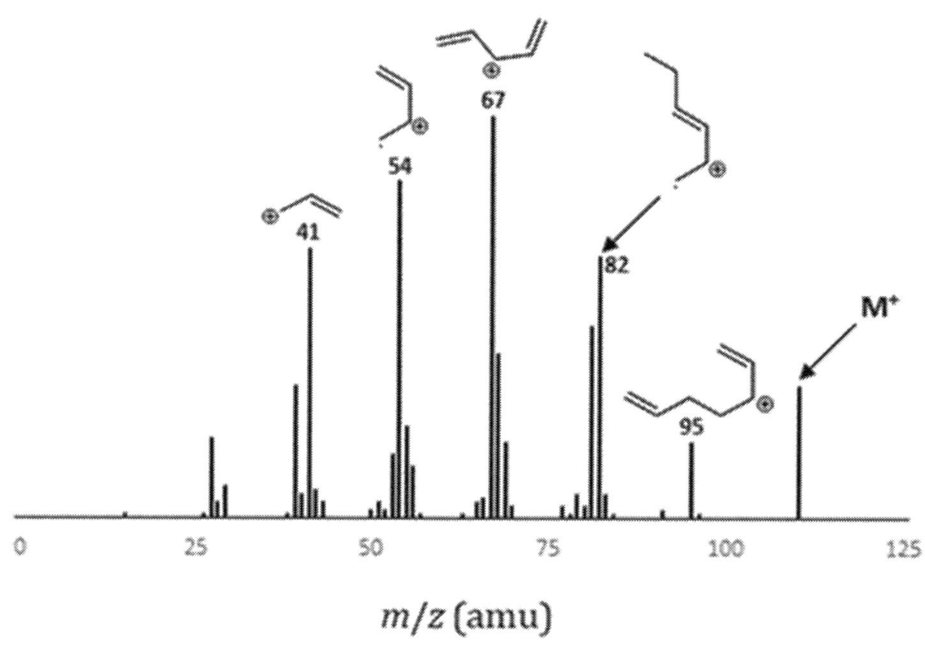

*Lesson VII.14.4 Branched Cycloalkenes*

Cycloalkenes with alkyl substituents on the ring undergo all of the same fragmentation reactions as those without alkyl substituents. However, a branched cycloalkene also undergoes α-cleavage to release the alkyl substituent as R•, and it is this fragmentation process that usually produces the most intense peak. Keep in mind that the position of the C=C bond freely migrates, so the location of the • and + may shift. As an example, 3-methyl-1-cyclohexene is shown below.

**rearrangement (*i*) then α-cleavage (*ii*)**

$-e^-$ → *i* → *ii* → $\dot{C}H_3$ + [M–15]+

The MS for 3-methyl-1-cyclohexene further demonstrates these fragmentation pathways:

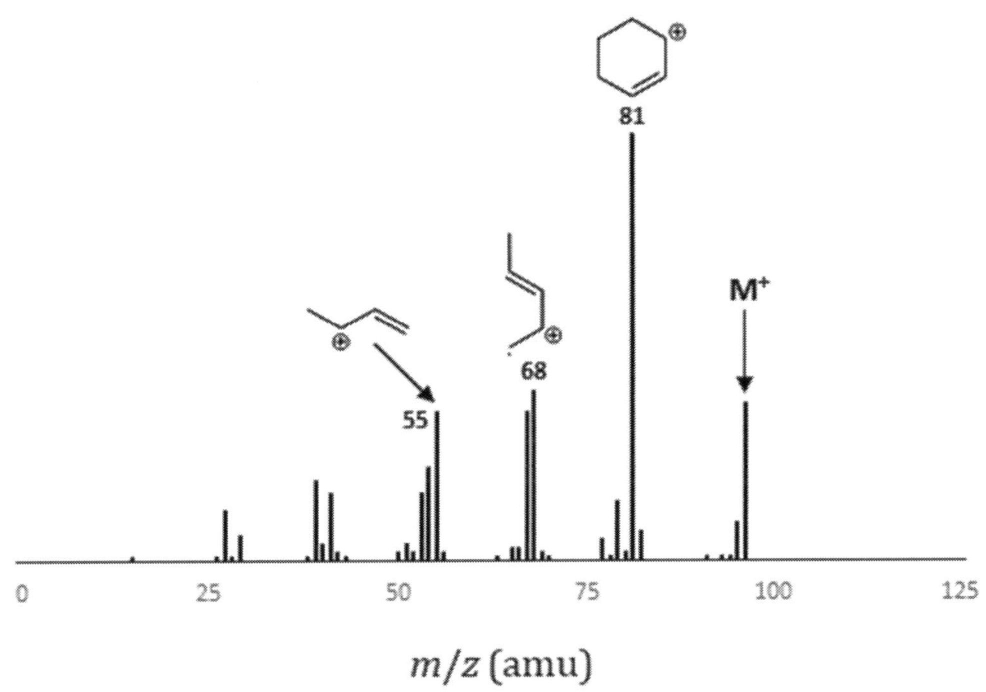

## Lesson VII.15. Fragmentation of Other Unsaturated Molecules

### *Lesson VII.15.1 Alkynes*

Similar to an alkene, ionization of an alkyne removes an electron from a C–C $\pi$-bond, which places the radical on one alkyne carbon and the + on the other. The $M^+$ peak for an alkyne is more intense than for an alkane, but the $M^+$ peak is much weaker for terminal alkynes than for internal alkynes. Whereas alkenes typically do not exhibit $[M-1]^+$ peaks, alkynes frequently exhibit $[M-1]^+$ peaks, which can be attributed to $\alpha$-cleavage of a C–H bond to afford a propargylic carbocation.

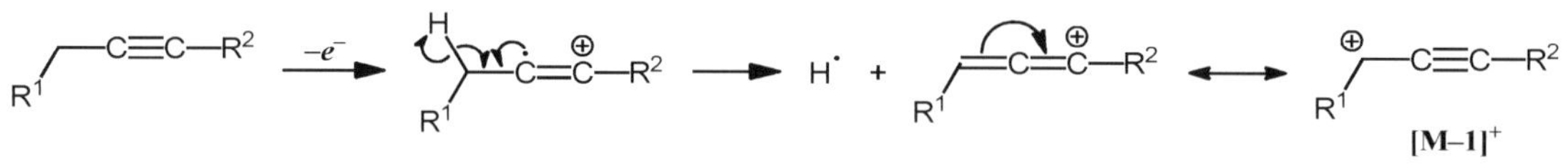

**$\alpha$-cleavage**

**$[M-1]^+$**

Alternatively, $\alpha$-cleavage could instead release an alkyl radical. If $R^2 = H$, a characteristic peak can be observed at 39 amu. However, the position of the C≡C bond can migrate, so a peak at 39 amu is even observed with internal alkynes, therefore its presence cannot be interpreted as evidence that the analyte is a terminal alkyne.

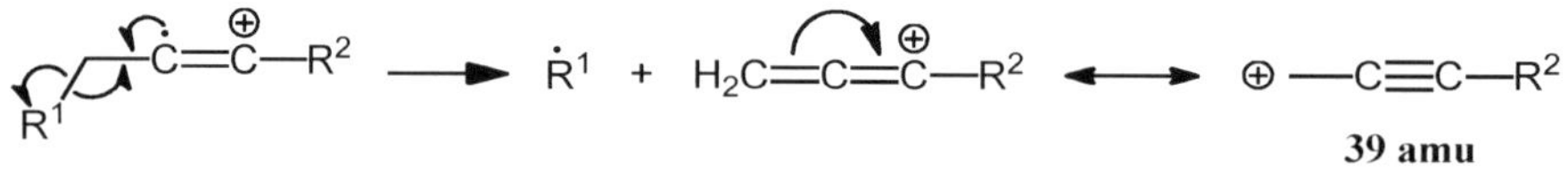

**$\alpha$-cleavage**

**39 amu**

Keep in mind that the alkyl components of alkynes can still undergo alkane fragmentation reactions, so much of the mass spectrum of an alkyne will comprise $[M-15]^+$, $[M-29]^+$, etc., peaks.

### *Lesson VII.15.2 Nitriles*

The •+ on a nitrile is localized on the N, rather than spread over a C–N $\pi$-bond, because $e^-$ in lone pairs are higher in energy (and thus easier to remove) than $e^-$ in $\pi$-bonds. However, mass spectra and fragmentation modes of nitriles are highly conserved with those of alkynes, although $M^+$ peak intensities are generally lower for nitriles than alkynes. Nitriles readily undergo $\alpha$-cleavage to release $H^\bullet$ or alkyl radicals, although peaks from the latter are typically of lower intensity than in alkynes.

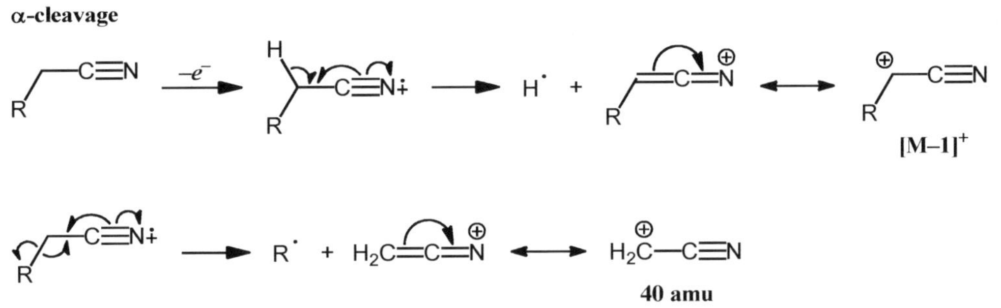

**α-cleavage**

**[M–1]$^+$**

**40 amu**

In addition, a nitrile can undergo a McLafferty-like rearrangement to eliminate an alkene and afford a radical cation at 41 amu, which is often one of the most intense peaks.

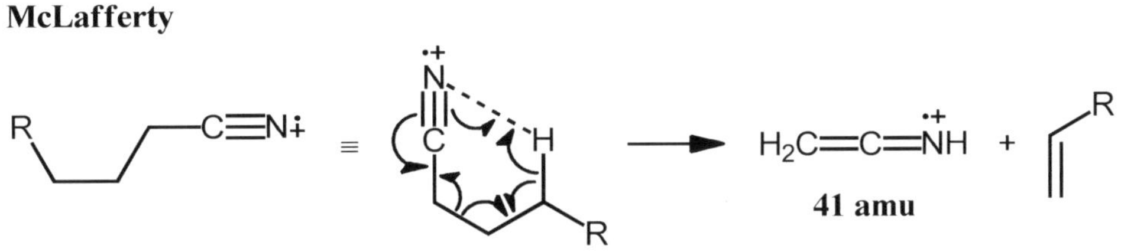

**McLafferty**

**41 amu**

Lastly, a nitrile can also eliminate HCN (giving an M–27 peak) to leave behind an alkyl radical cation, but this peak is generally weak.

## Part VIII. Spectroscopy Practice with Solutions

## Lesson VIII.1. Charts for IR and NMR Spectral Analysis

### *Lesson VIII.1.1 Typical Wavenumbers in IR Spectra*

This table provides some useful general ranges for where certain types of bonds are likely to show up in IR spectra.

| Bond | Energy ($cm^{-1}$) | Intensity |
|---|---|---|
| N≡C | 2255-2220 | m-s |
| C≡C | 2260-2100 | w-m |
| C=C | 1675-1660 | m |
| N=C | 1650-1550 | m |
| benzene ring | 1600 **AND** 1500-1425 | w-s |
| C=O | 1775-1650 | s |
| C—O | 1250-1000 | s |
| C—N | 1230-1000 | m |
| O—H | 3650-3200 | s (br) |
| O—H | 3300-2500 | s (br) |
| N—H | 3500-3300 | m (br) |
| C—H | 3300-2725 | m |

| C-H Bond (Stretch) | Energy ($cm^{-1}$) |
|---|---|
| C≡C—H | 3300-ish |
| C=C—H | 3100-3000 |
| C—C—H | 2950-2850 |
| aldehyde C(=O)—H | 2820-ish and 2720-ish |

**C-H Bond (Bending)**

| Group | Energy ($cm^{-1}$) |
|---|---|
| —$CH_3$, —$CH_2$—, methine C—H | 1450-1400 |
| trans (HRC=CRH) | 980-960 |
| cis (RHC=CHR) | 730-670 |
| trisubstituted ($R_2C$=CRH) | 840-800 |
| monosubstituted (RHC=$CH_2$) | 990 and 910 |
| disubstituted terminal ($R_2C$=$CH_2$) | 890 |

*Lesson VIII.1.2 Flowchart for Determining Functional Groups Present in Monofunctional Compounds*

If only one functional group is present in a particular compound, the following simplified flowchart is a good starting point for determining which functional group it is:

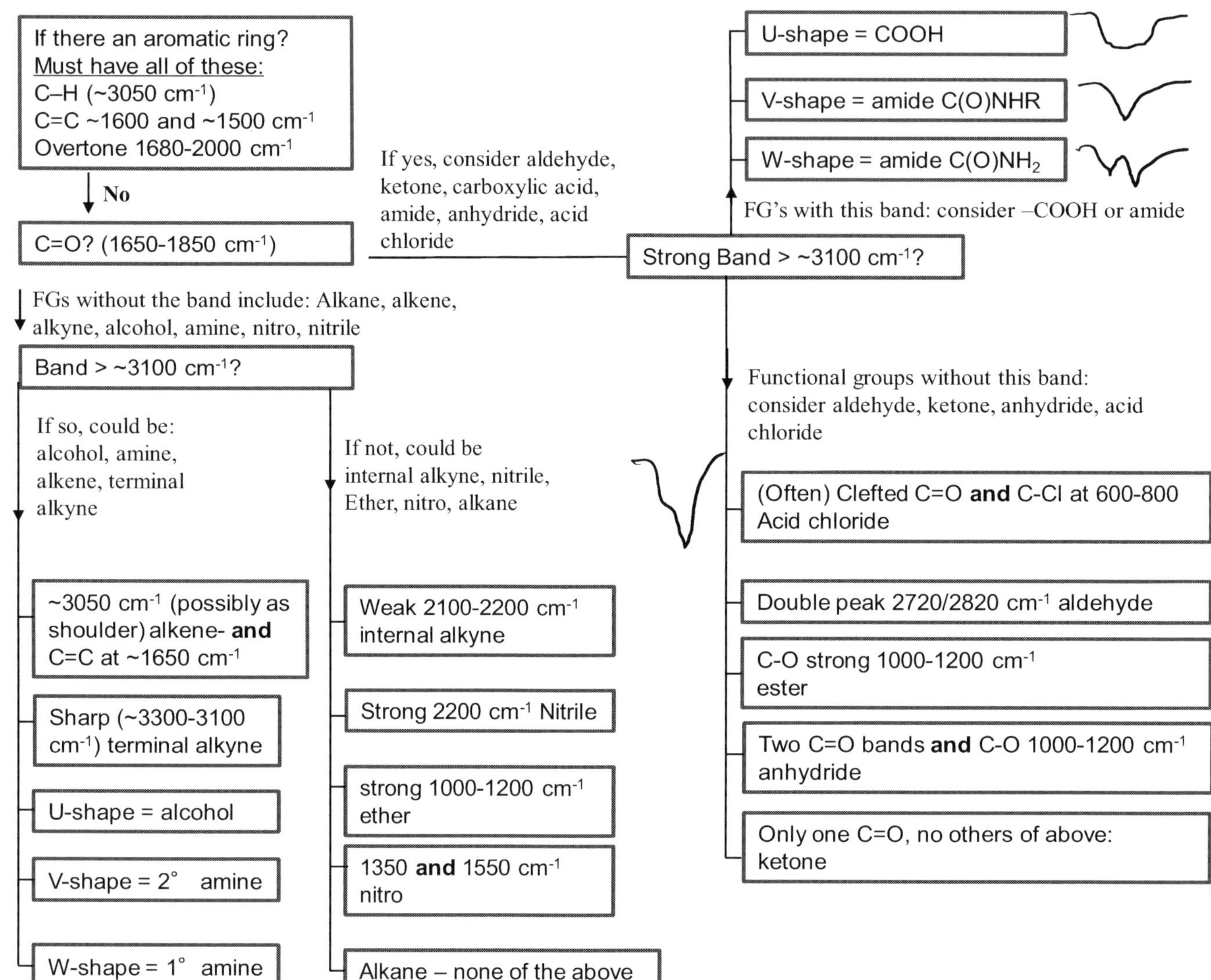

*Lesson VIII.1.3 Typical $^{13}C$ NMR Chemical Shifts*

This diagram gives you a good idea for determining the general functiona sample from the observed $^{13}C$ NMR chemical shifts. These are very rough esti spectrum into four areas:

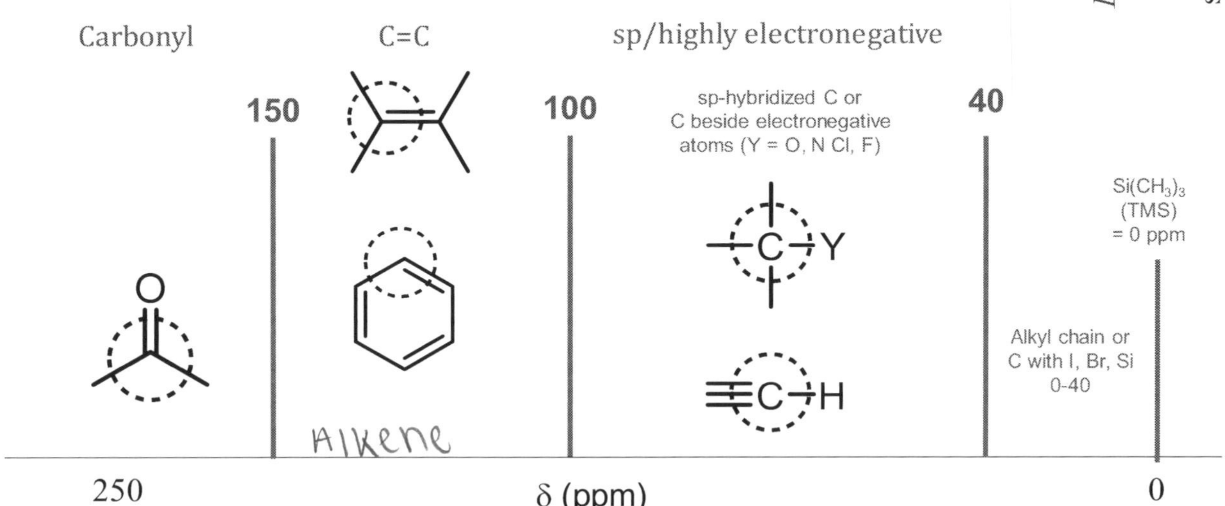


For more fine-tuned determinations than are possible with the above diagram, this table provides useful information on the commonly-observed $^{13}C$ NMR chemical shifts for different types of carbons:

| Carbon (Shown) | Chemical Shift | Carbon (Shown) | Chemical Shift |
|---|---|---|---|
| $Si(CH_3)_4$ | 0 | O—C | 50-80 |
| —$CH_3$ | 10-35 | N—C | 40-60 |
| —$CH_2$— | 15-50 | C—X | X = I 0-35<br>X = Br 20-65<br>X = Cl 35-80 |
| H–C (tertiary) | 20-60 | | |
| C (quaternary) | 30-40 | | |
| =C< | 100-150 | O=C–Y | Y =<br>H 190-200<br>R 200-220<br>OR 160-180<br>OH 175-185<br>$NR_2$ 165-175 |
| (benzene ring) C—H | 110-175 | | |
| ≡C— | 60-85 | | |

**Additional note:**

Many common $^{13}C$ NMR spectra also have a peak at 0.00 ppm due to the presence of tetramethylsilane, $Si(CH_3)_4$, a standard, as well as a usually intense triplet at 77 ppm due to the presence of deuterated chloroform, $CDCl_3$, a solvent often used for NMR spectra.

### .4 Typical $^1H$ NMR Chemical Shifts

his diagram gives you a good idea for determining the general functional groups present in a le from the observed $^1H$ NMR chemical shifts:

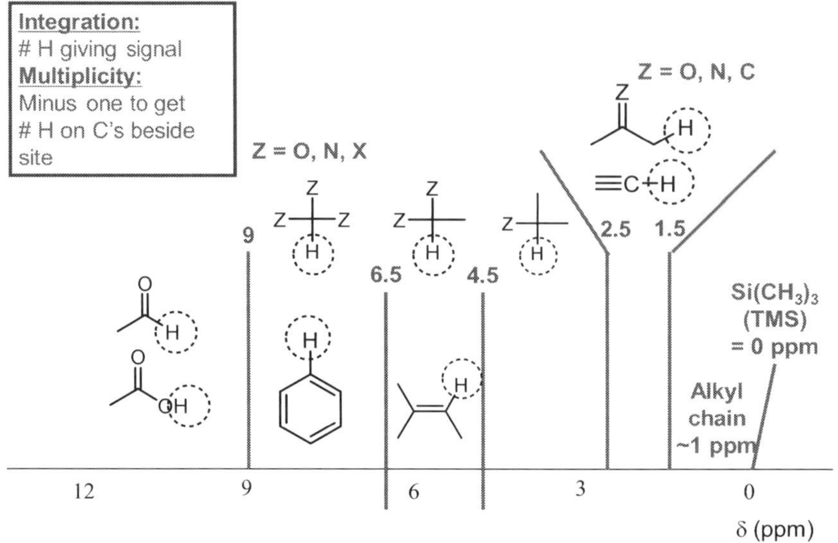


For more fine-tuned determinations than are possible with the above diagram, this table provides useful information on the commonly-observed $^1H$ NMR chemical shifts for different types of protons:

| Protons (Shown) | Chemical Shift | Protons (Shown) | Chemical Shift |
|---|---|---|---|
| $Si(CH_3)_4$ | 0 | $R—OCH_3$ | 3.3 |
| $—CH_3$ | 0.9 | =CH–H (vinyl H) | 4.5-5.5 |
| $—CH_2—$ | 1.2 | H–C–X: X = I | 2.5-4 |
| H–C (methine) | 1.4 | X = Br | 2.5-4 |
| =C–$CH_3$ (allylic) | 1.7 | X = Cl | 3-4 |
| C(=O)–$CH_3$ | 2.1 | X = F | 4-4.5 |
| Ph–$CH_3$ | 2.3 | Ph–H | 6.5-8.0 |
| ≡–H | 2.4 | C(=O)–H | 10 |

O-H ~2

**Additional note:**

Many common $^1H$ NMR spectra also have a peak at 0.00 ppm due to the presence of tetramethylsilane, $Si(CH_3)_4$, a standard, as well as at 7.28 ppm due to the presence of chloroform, $CHCl_3$, present in the deuterated chloroform that is a solvent often used for NMR spectra.

## Lesson VIII.2. Infrared Spectroscopy Practice Problems

For each of the following, determine which of the potential structures is most likely to have produced the provided IR spectrum.

IR Spectroscopy Problem 1

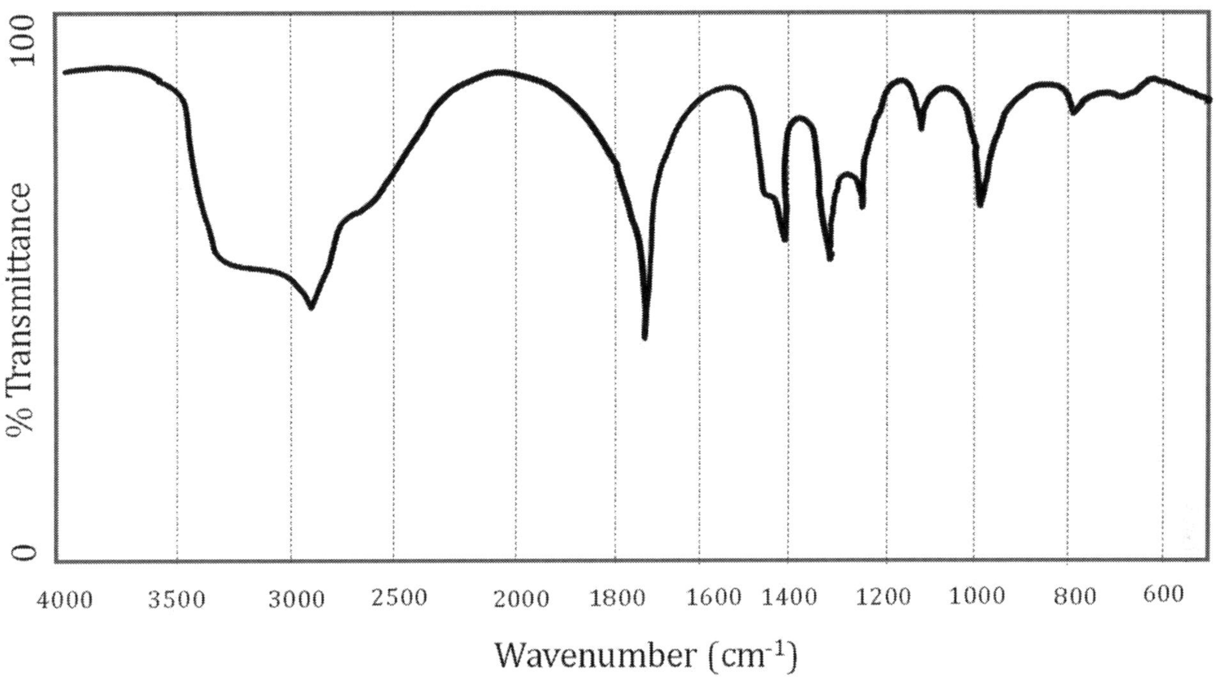


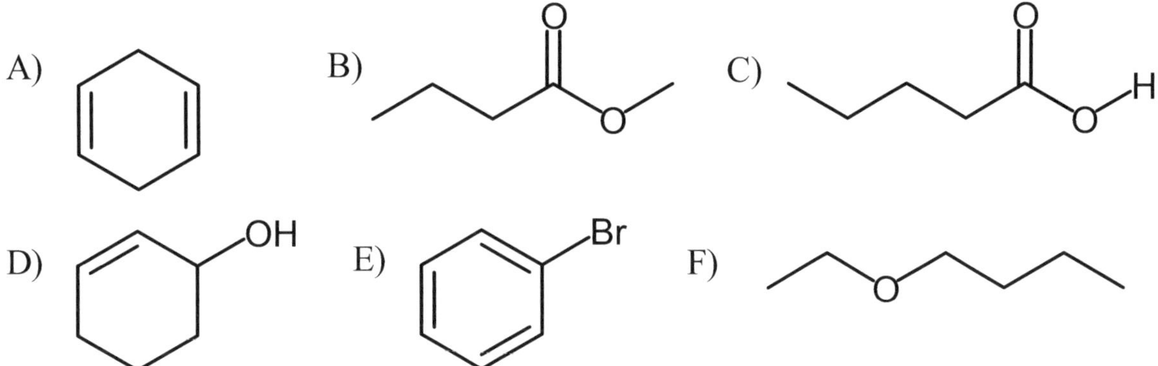

IR Spectroscopy Problem 2

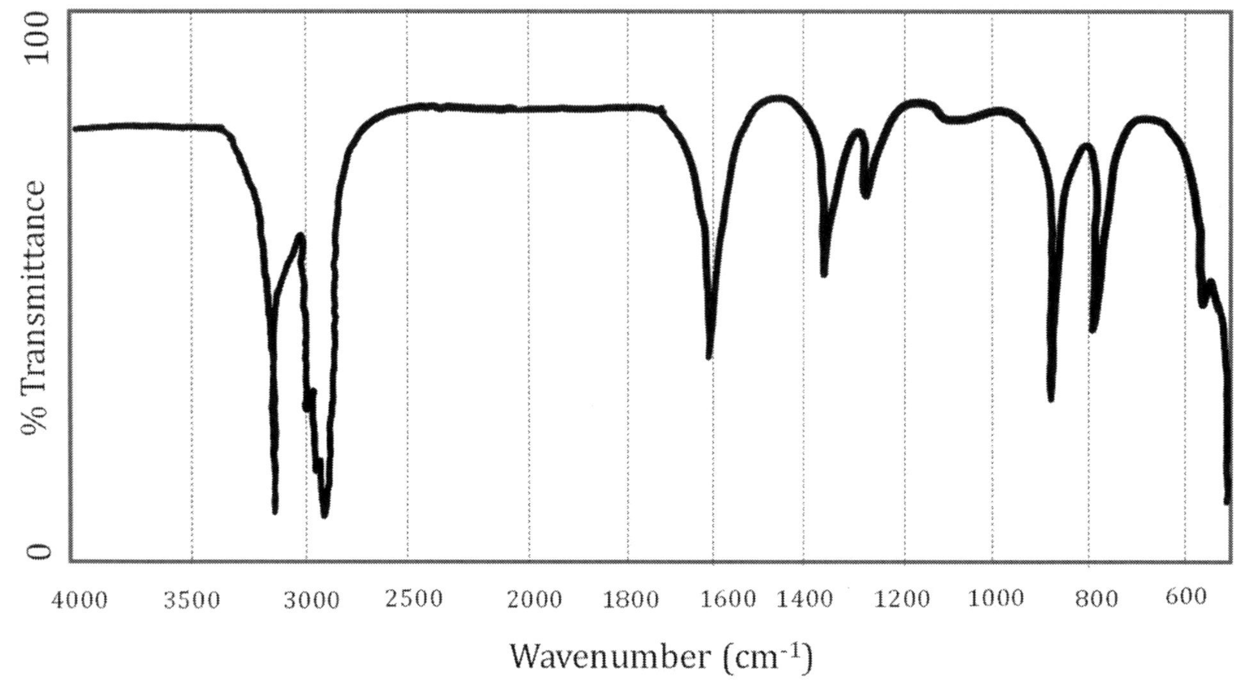


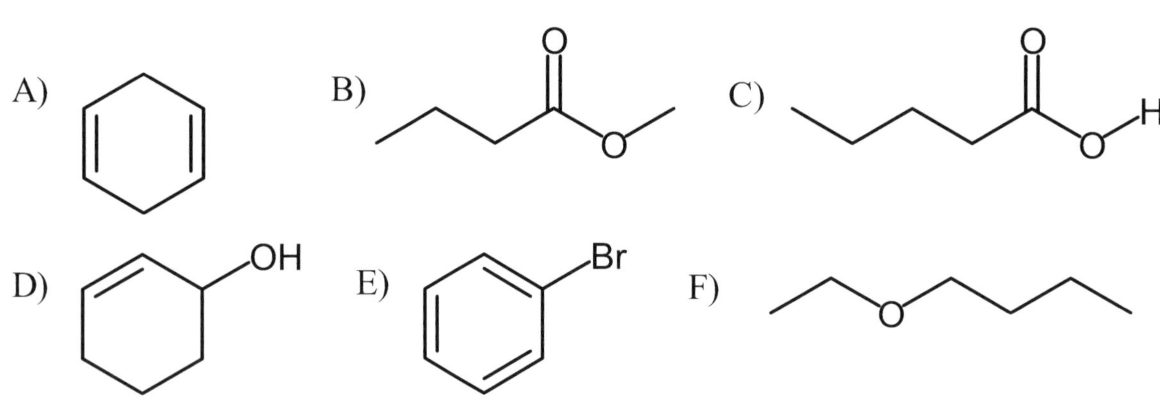

IR Spectroscopy Problem 3

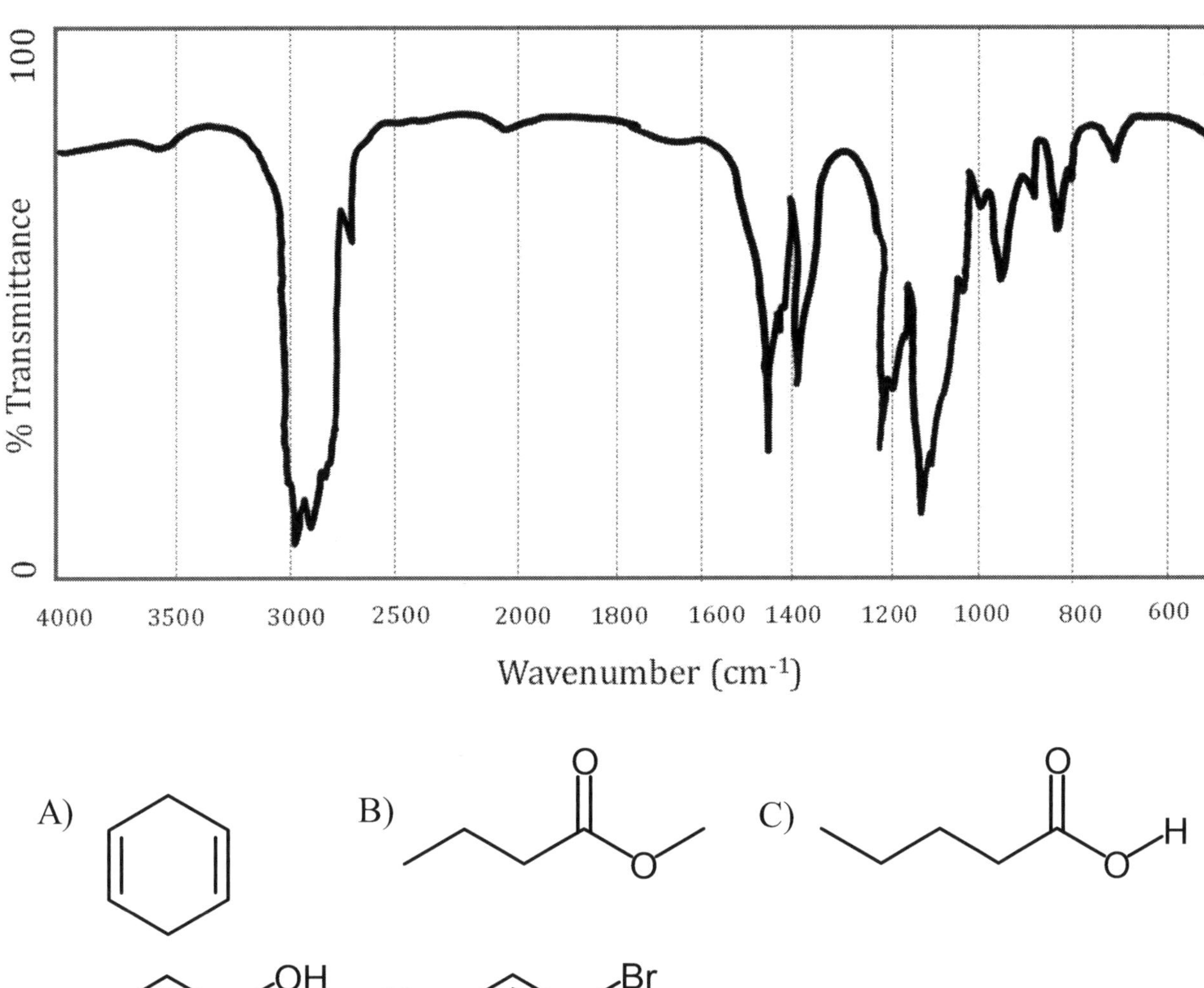

IR Spectroscopy Problem 4

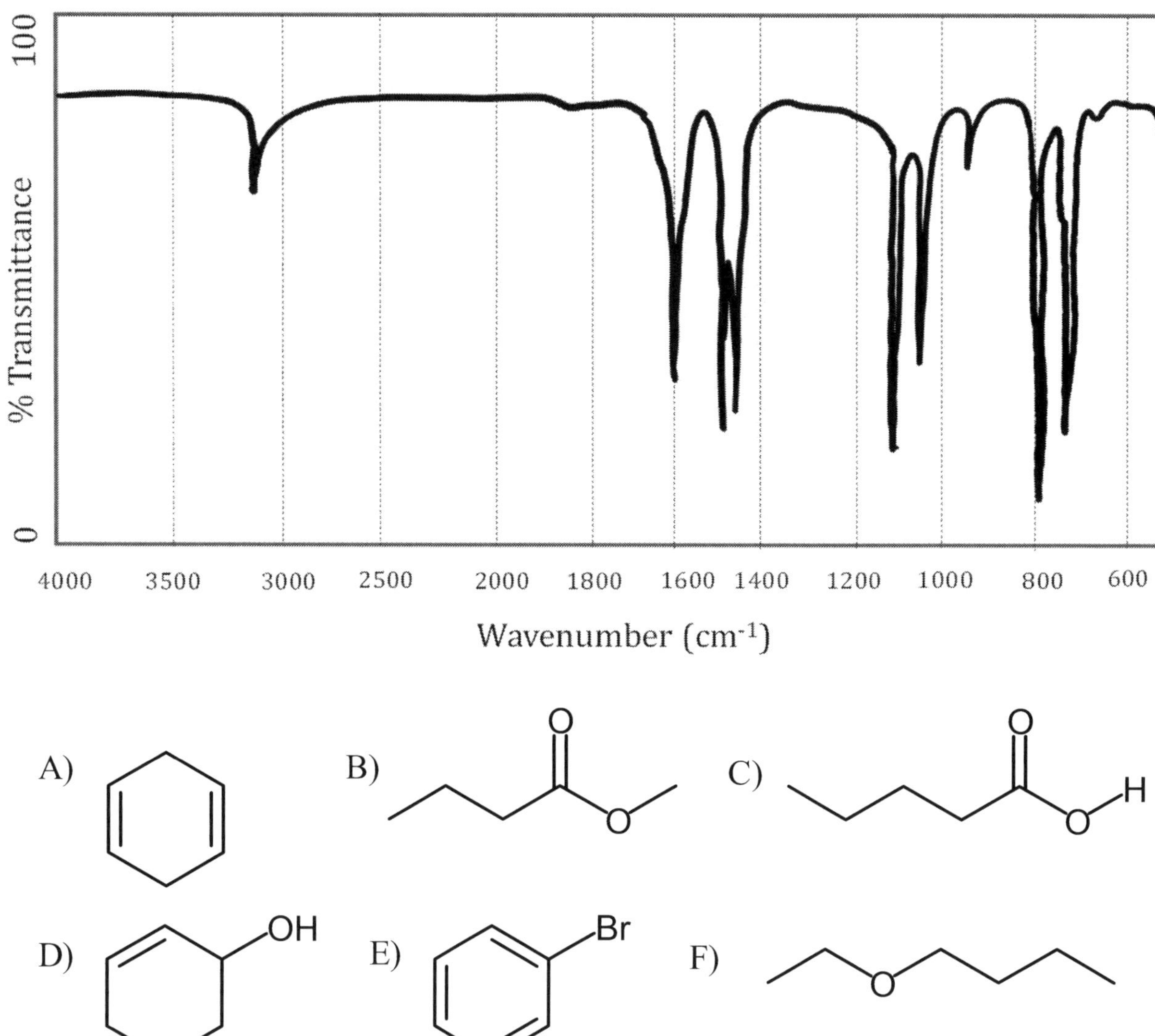

IR Spectroscopy Problem 5

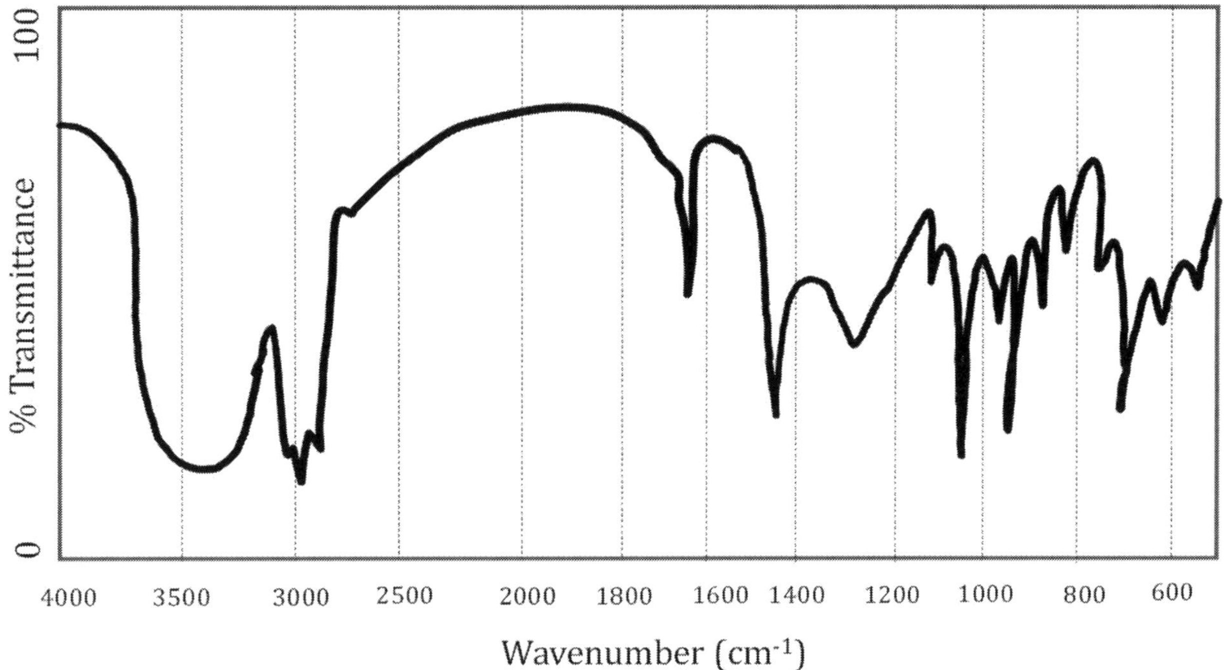


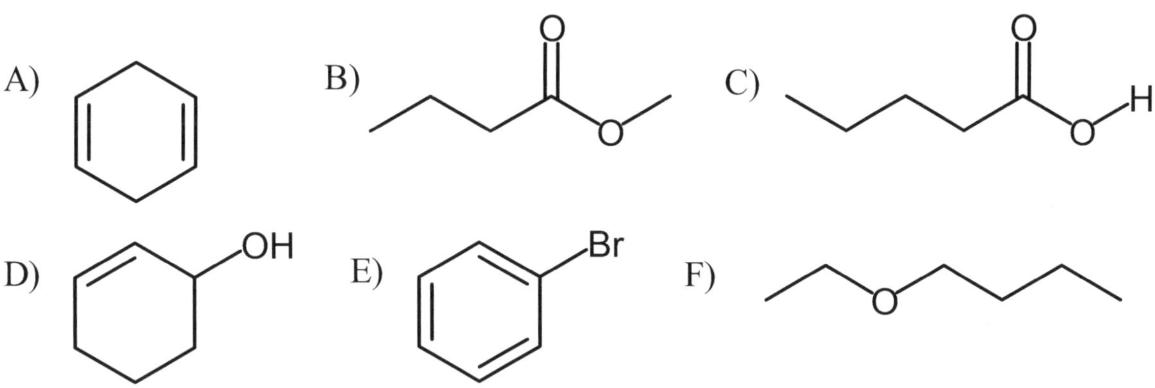

IR Spectroscopy Problem 6

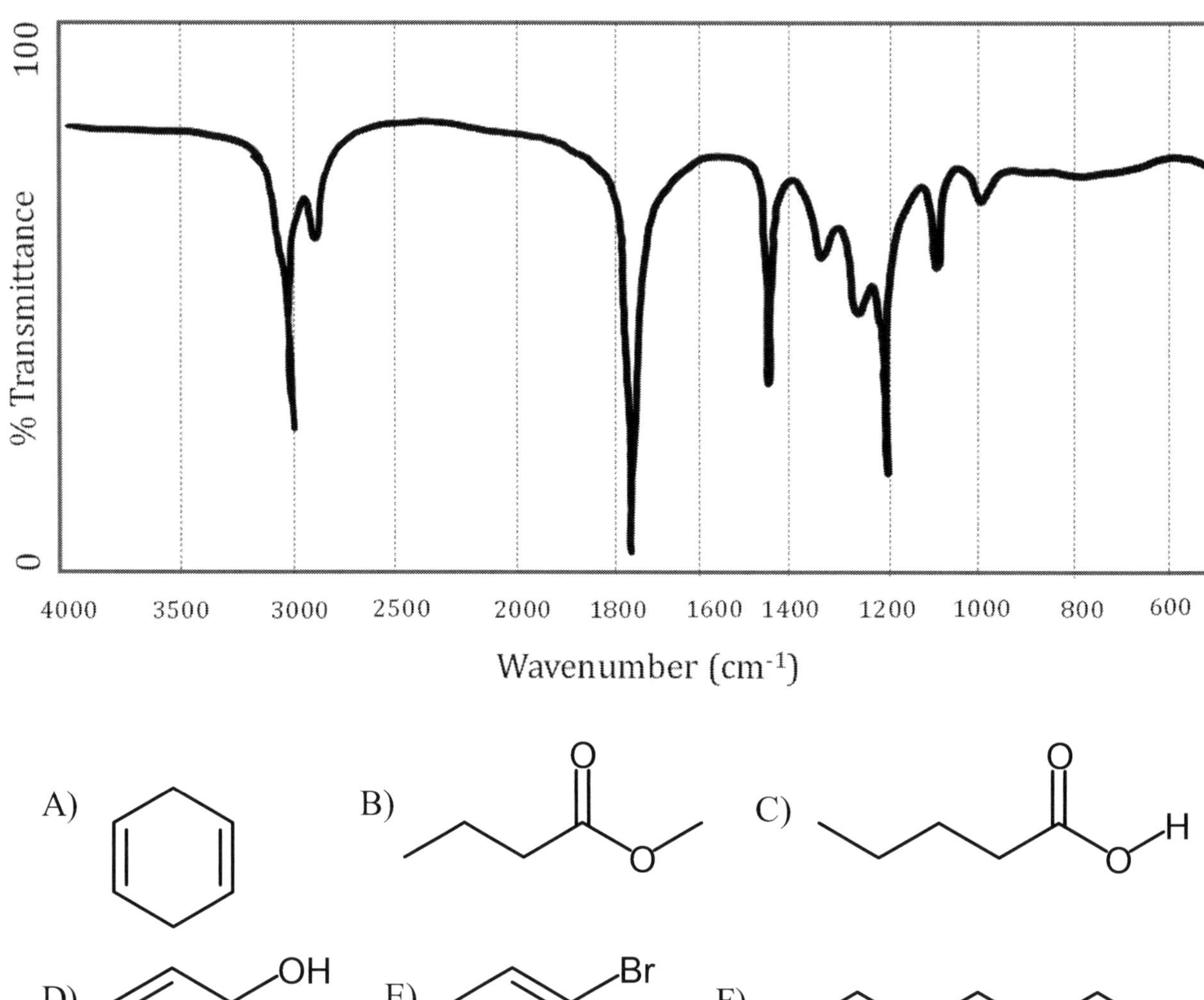

IR Spectroscopy Problem 7

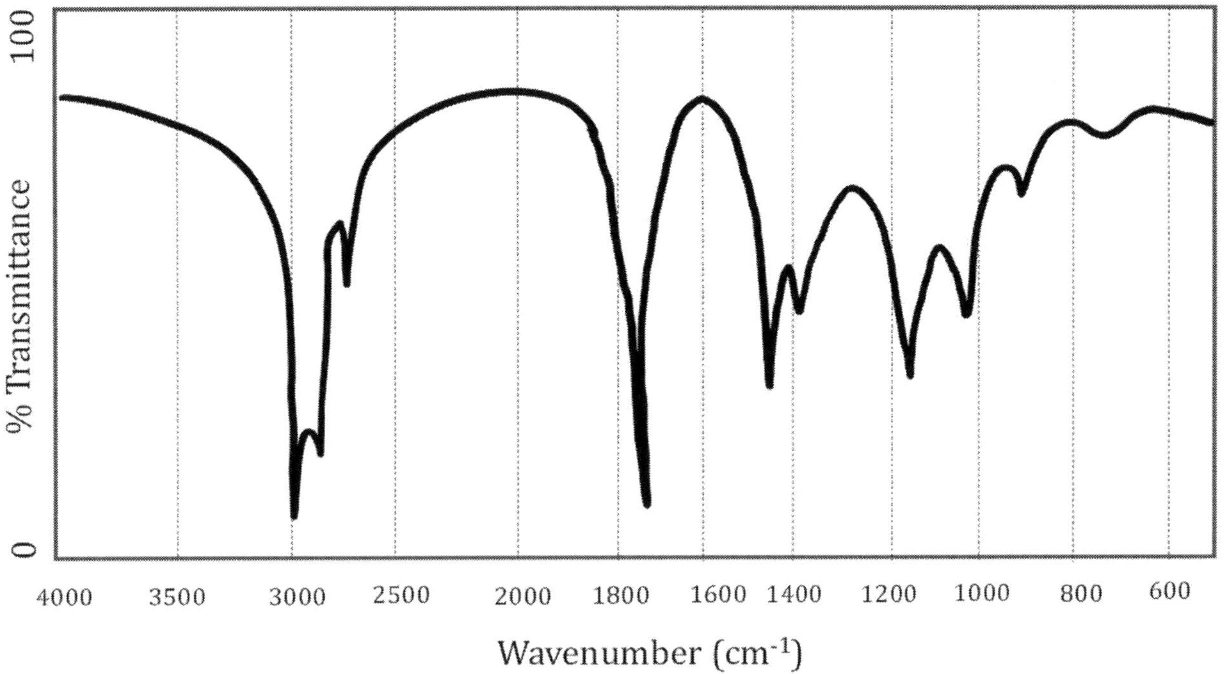


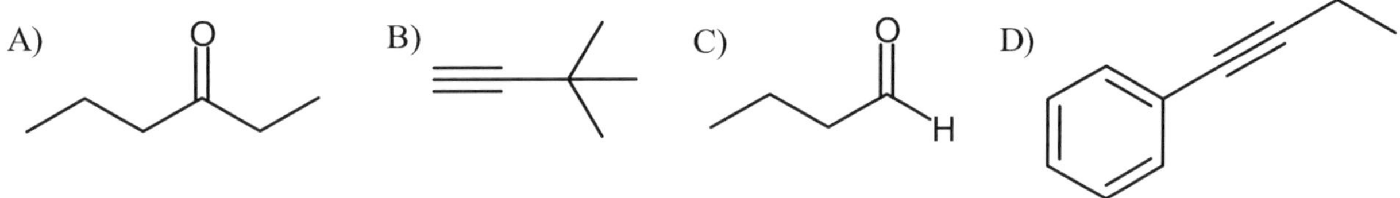

IR Spectroscopy Problem 8:

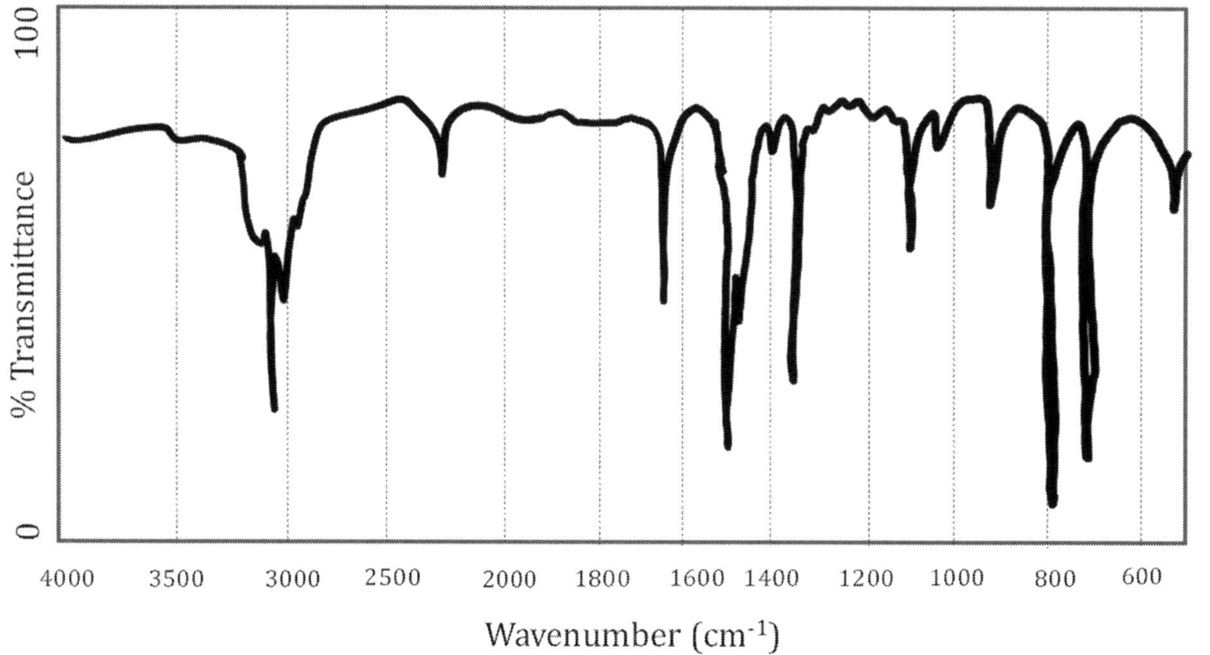


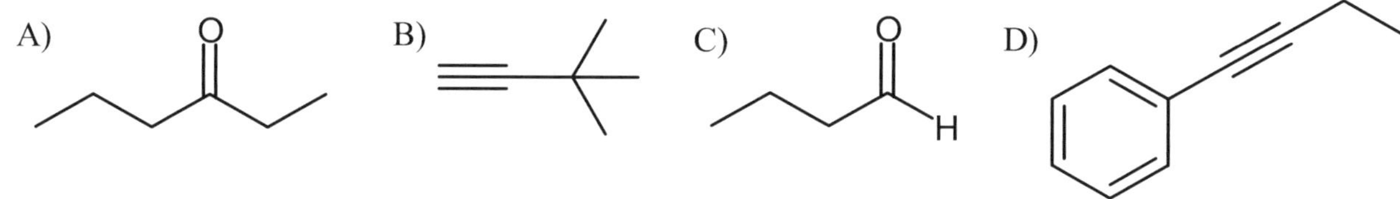

IR Spectroscopy Problem 9:

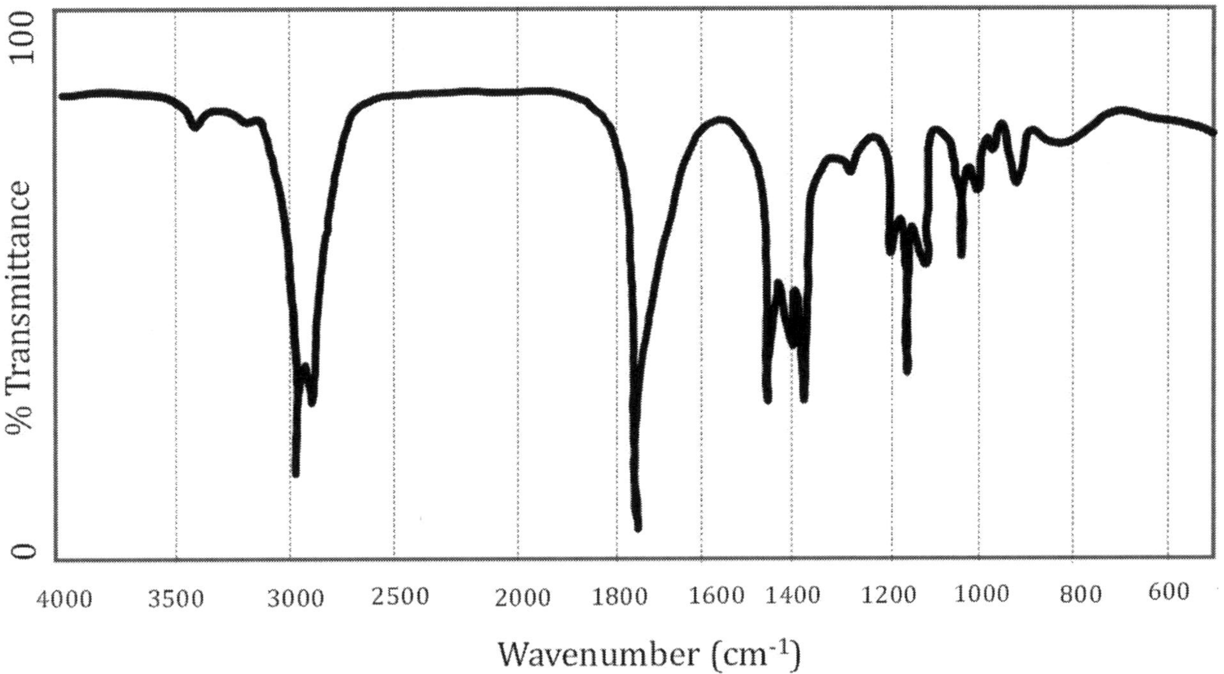


A)

B)

C)

D)

IR Spectroscopy Problem 10:

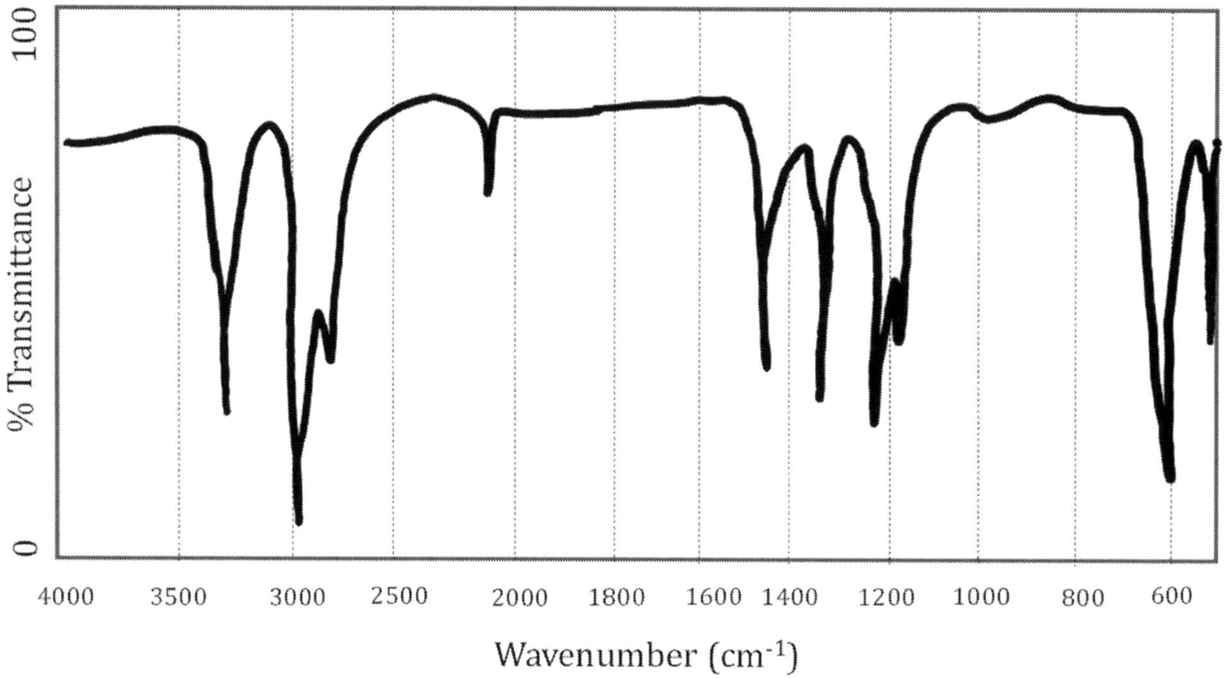


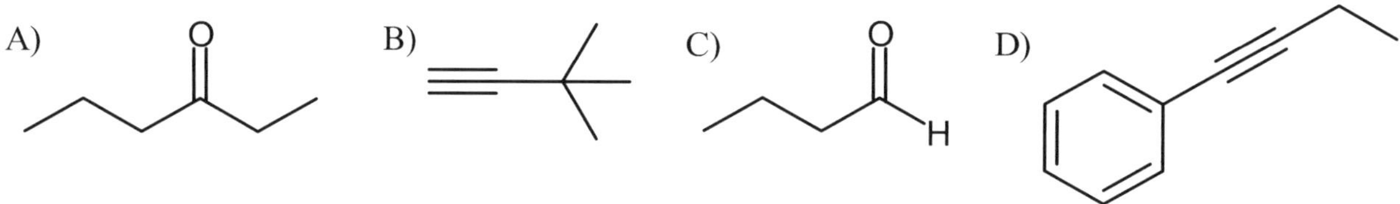

IR Spectroscopy Problem 11

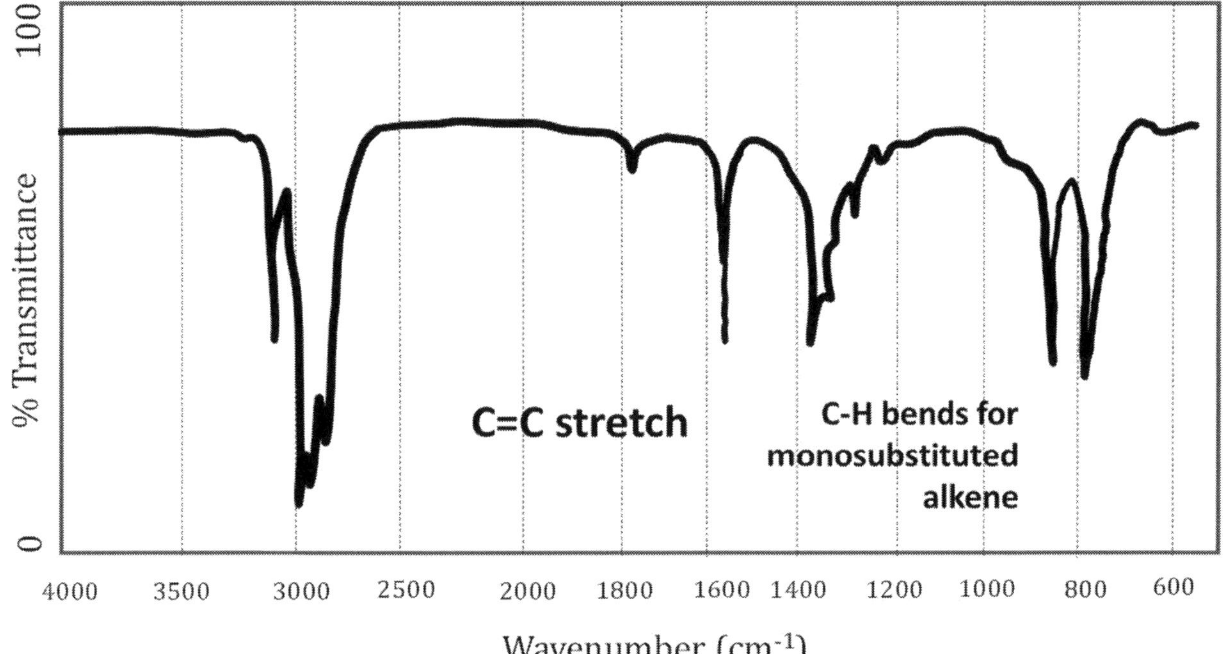


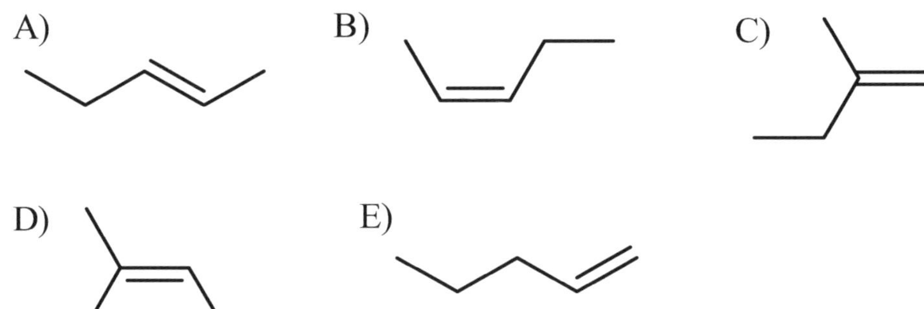

IR Spectroscopy Problem 12

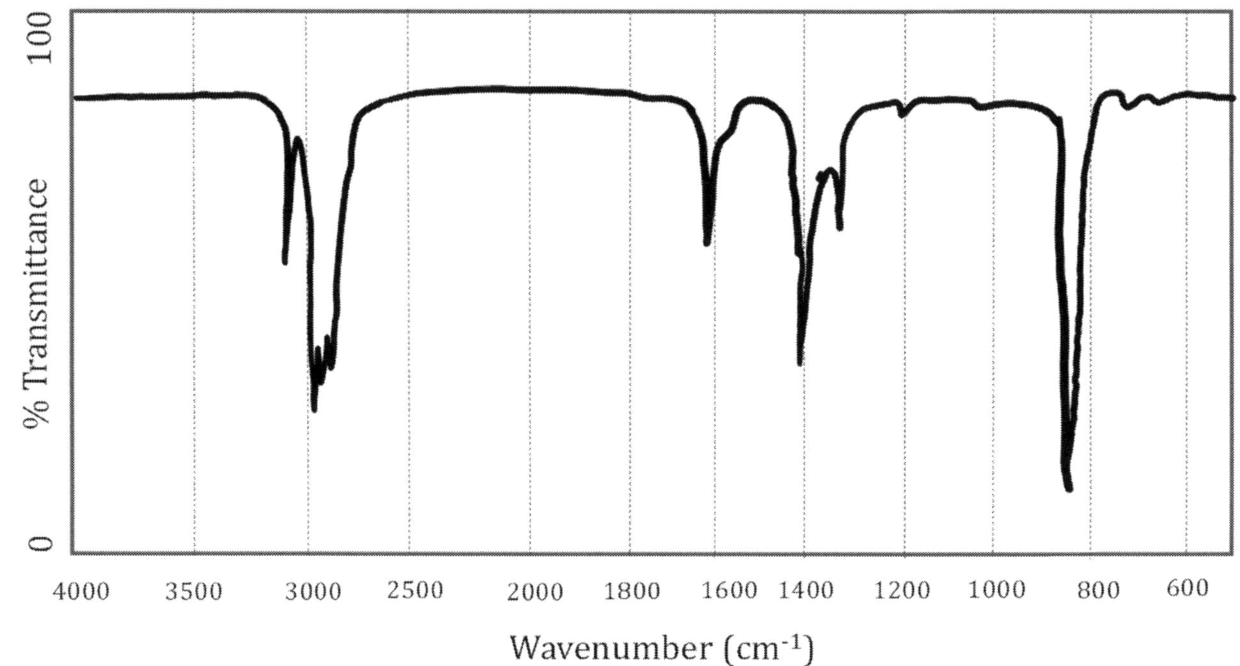


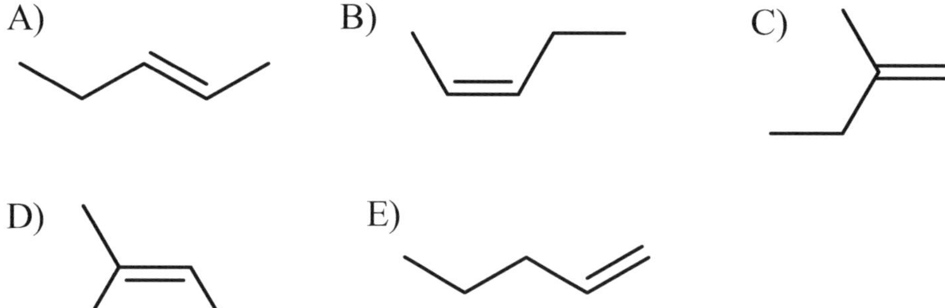

IR Spectroscopy Problem 13

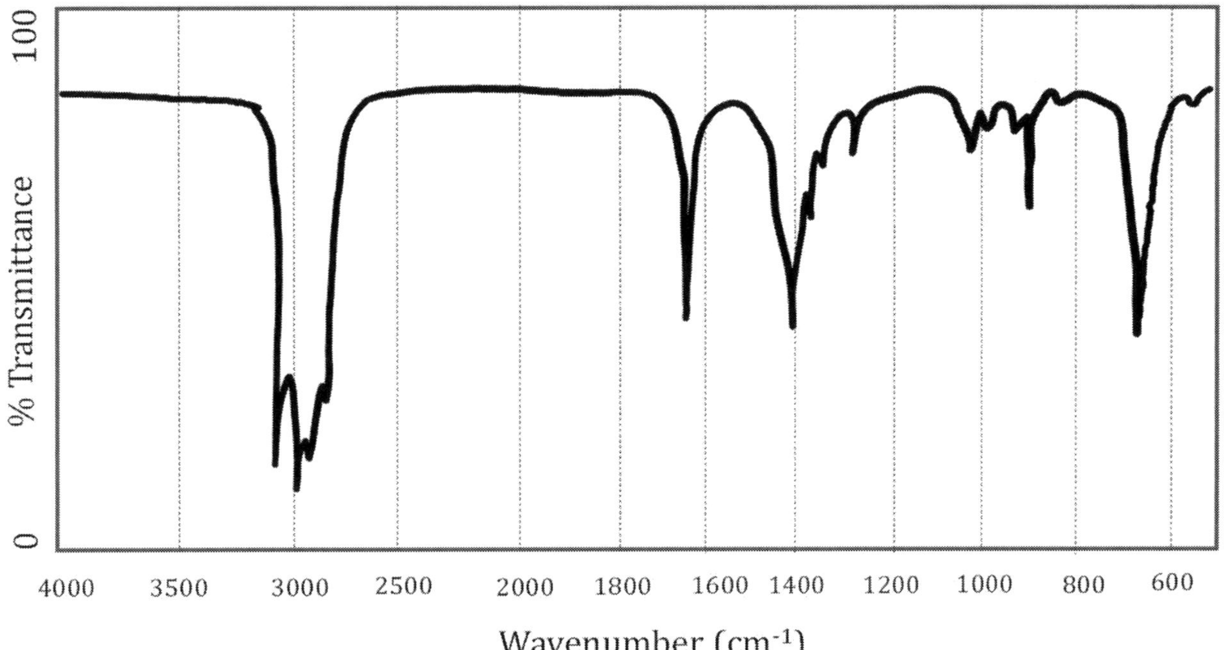


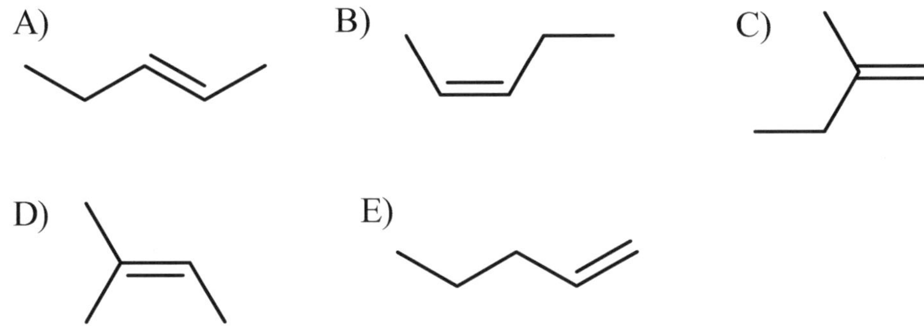

IR Spectroscopy Problem 14

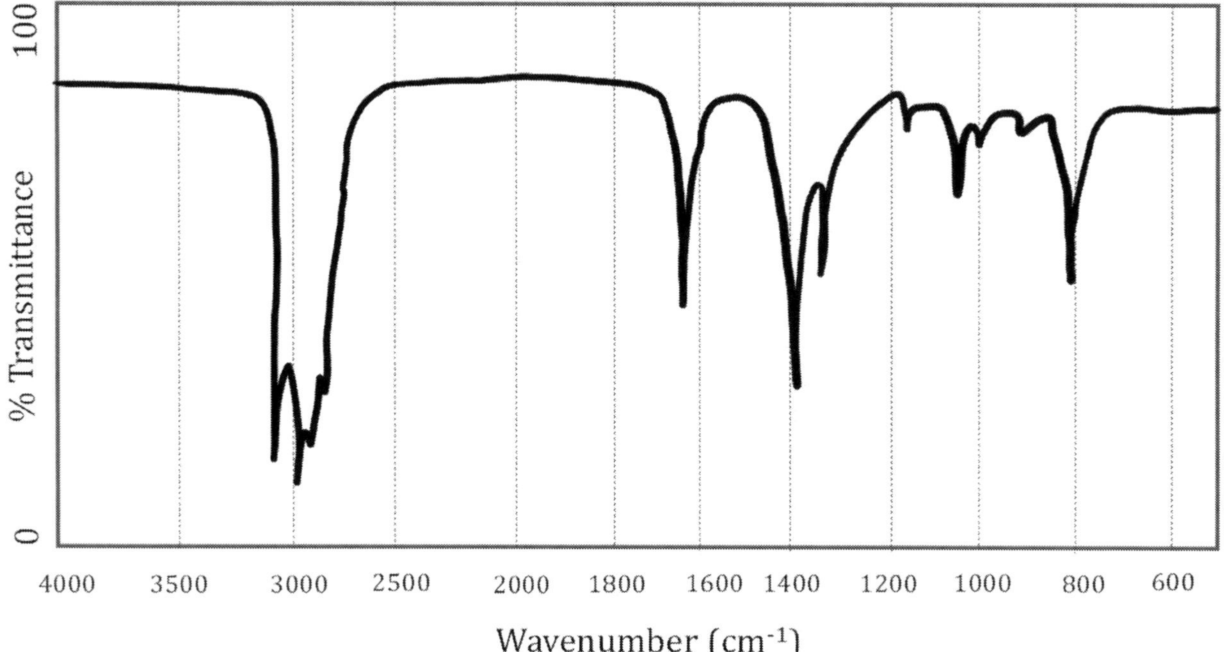


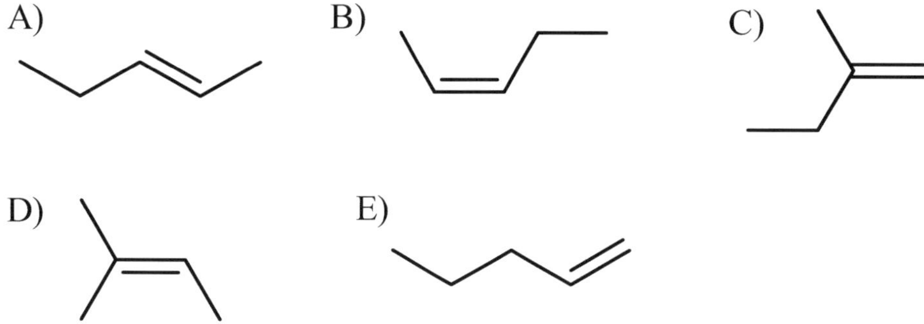

IR Spectroscopy Problem 15

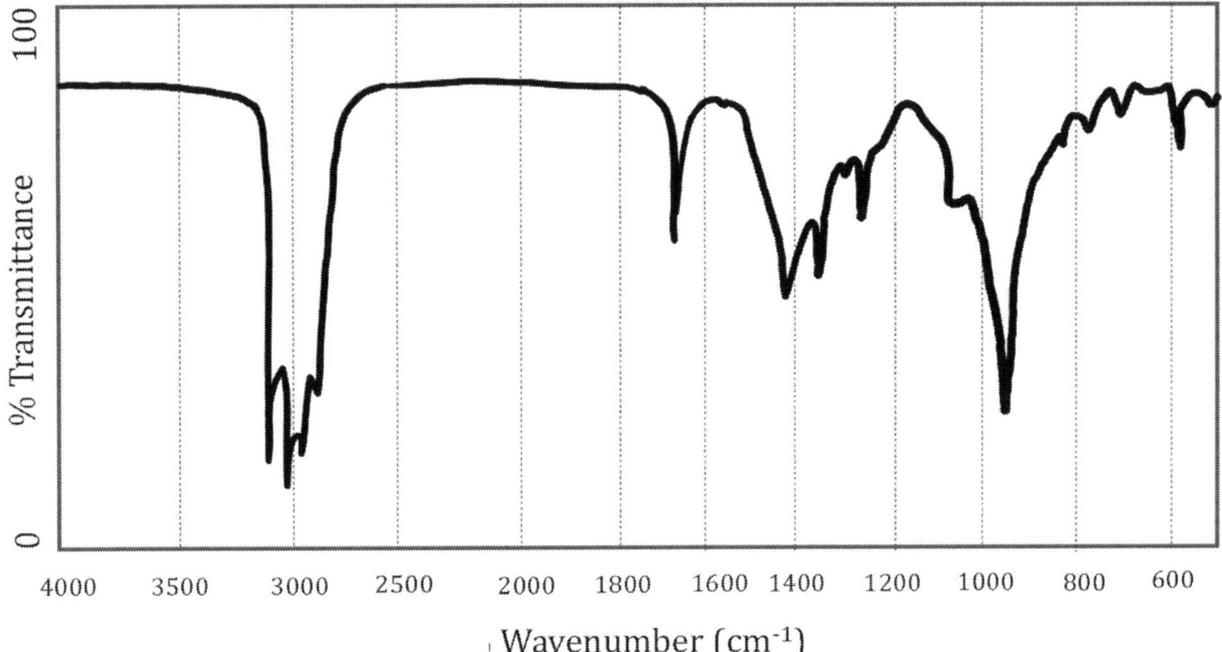


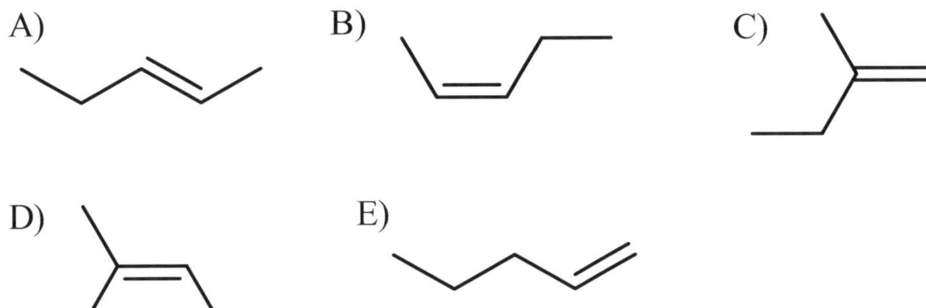

## Lesson VIII.3. Answers to Infrared Spectroscopy Practice IR Spectroscopy Problems

Answer to IR Spectroscopy Problem 1

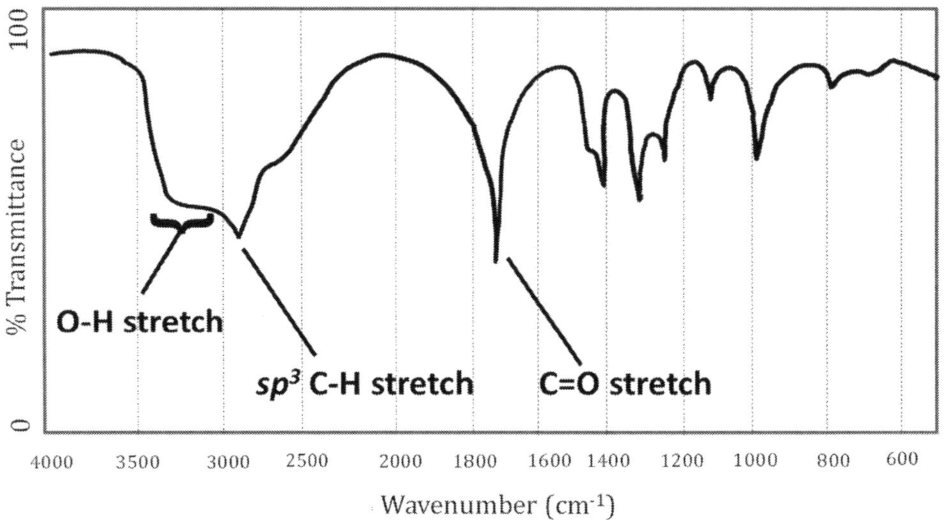


C) 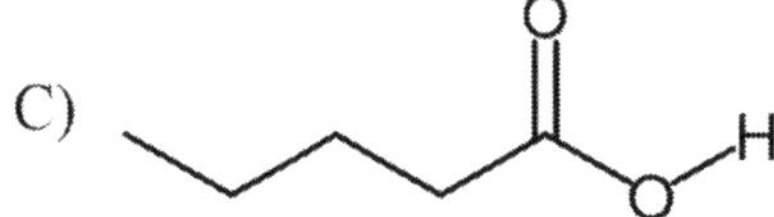


Answer to IR Spectroscopy Problem 2

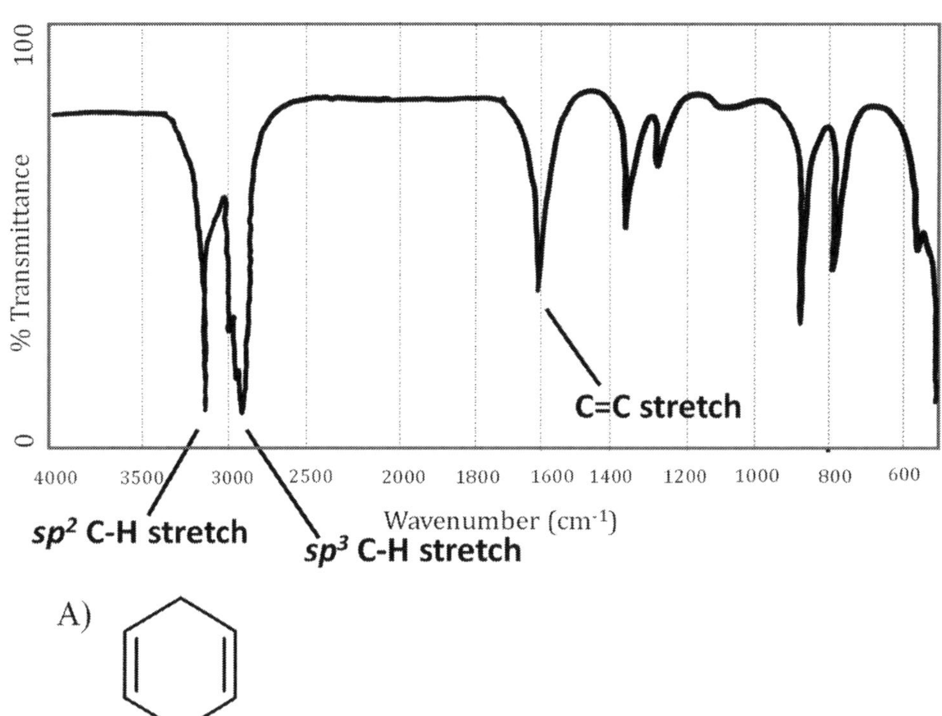

Answer to IR Spectroscopy Problem 3

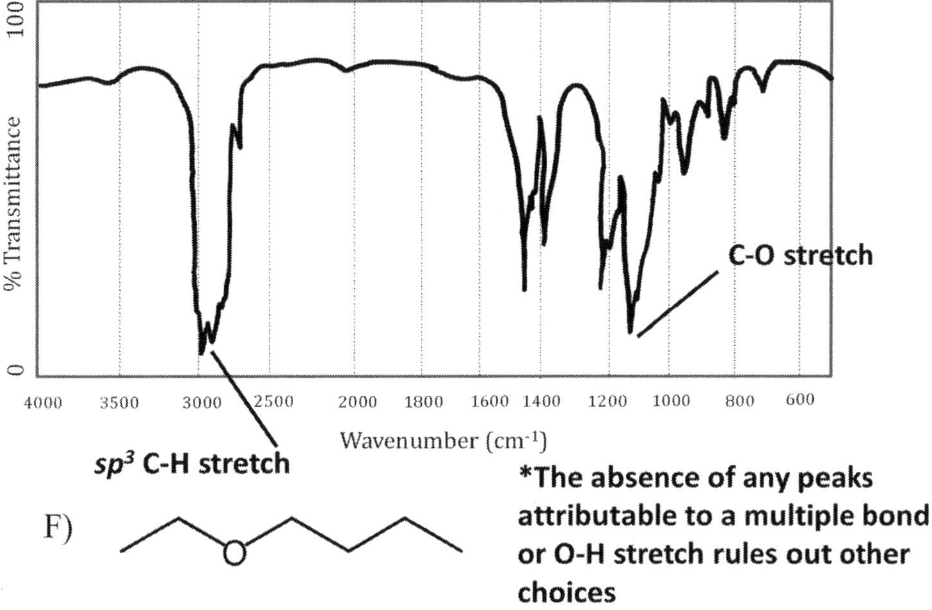


Answer to IR Spectroscopy Problem 4

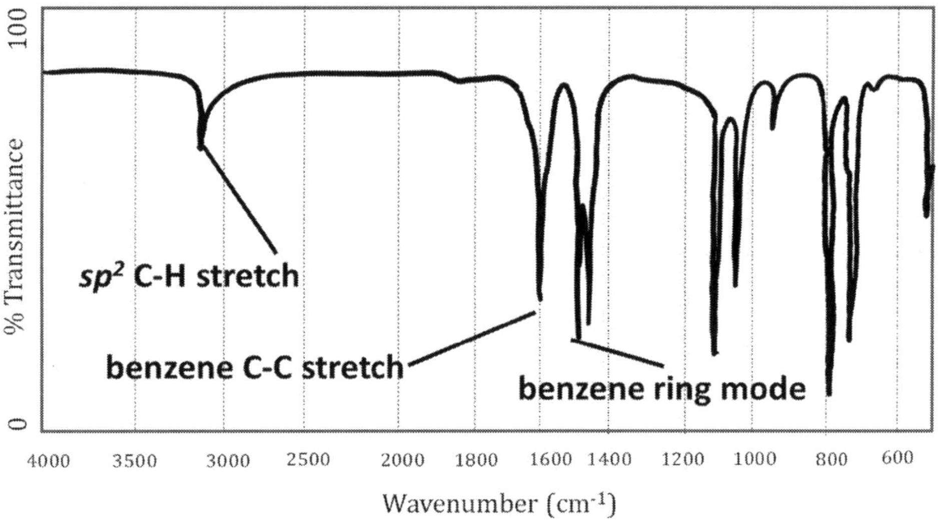


E) Br

Answer to IR Spectroscopy Problem 5

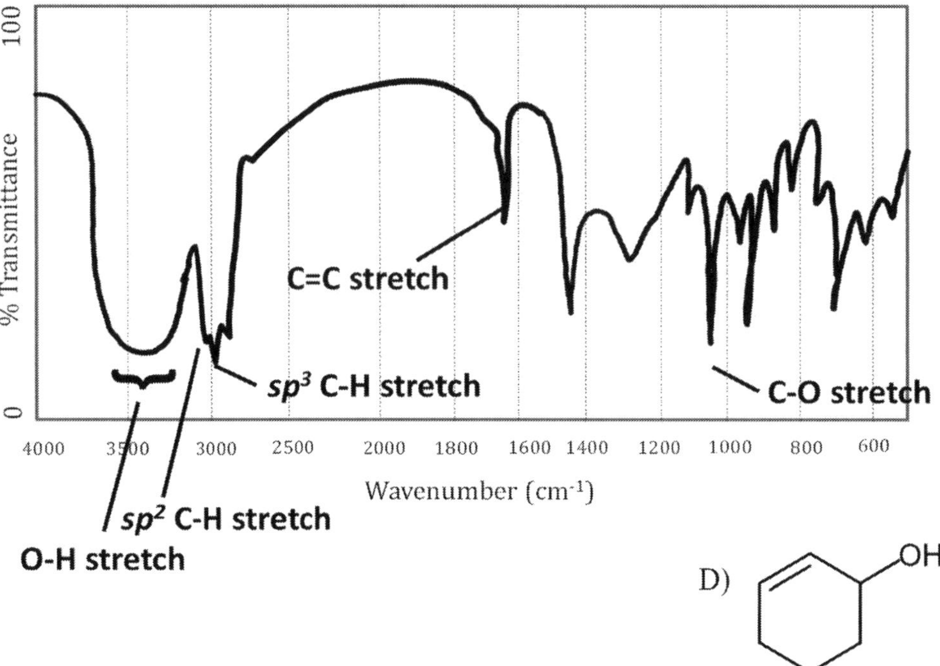


Answer to IR Spectroscopy Problem 6

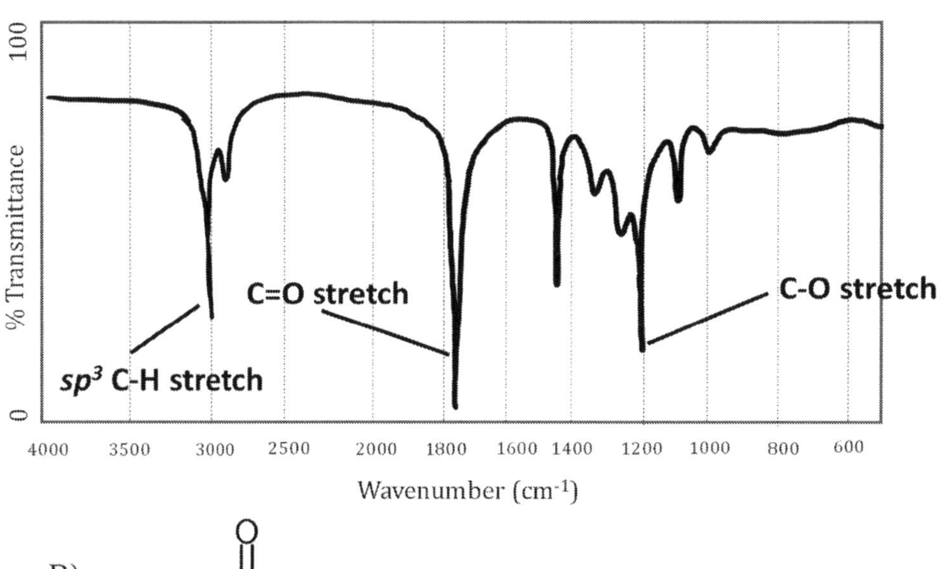

Answer to IR Spectroscopy Problem 7

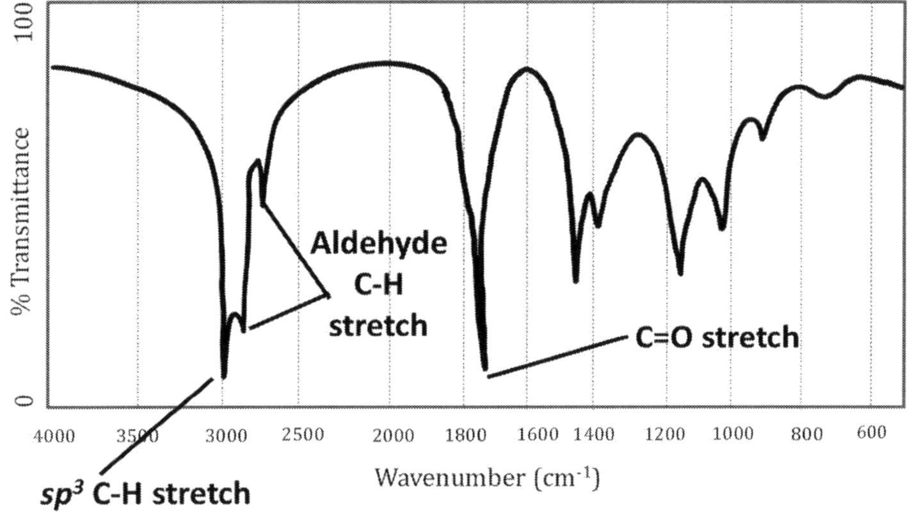


C)

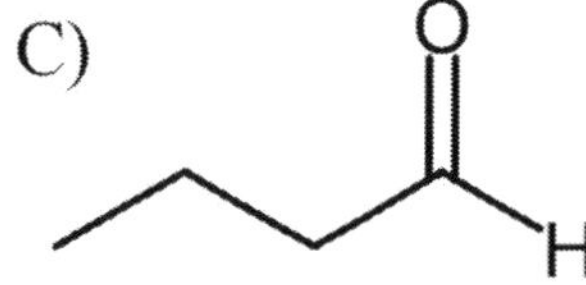


Answer to IR Spectroscopy Problem 8:

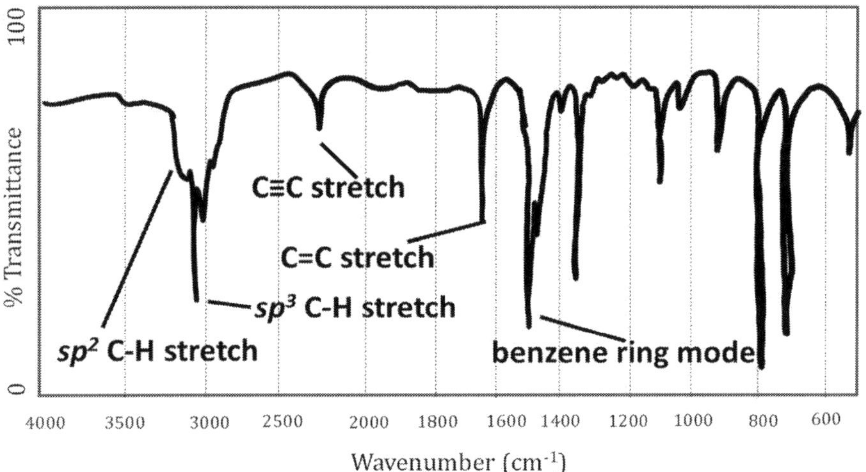


***The absence of an *sp* C-H stretch at 3300 cm$^{-1}$ rules out a terminal alkyne**

D)

Answer to IR Spectroscopy Problem 9:

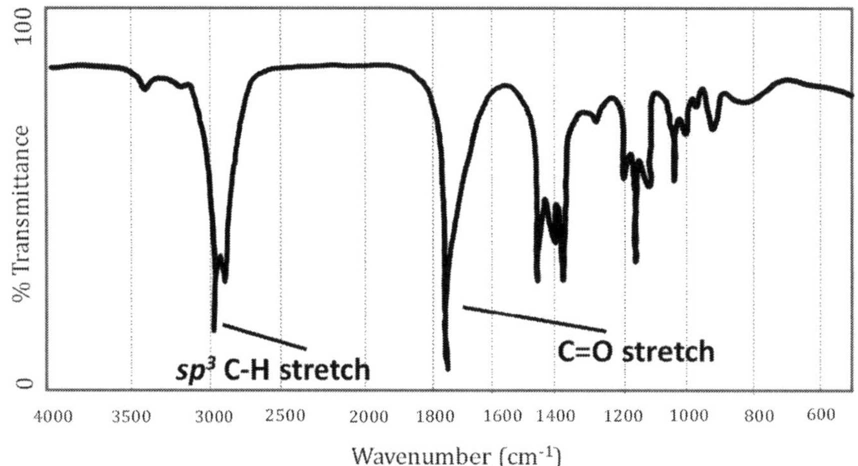


A)

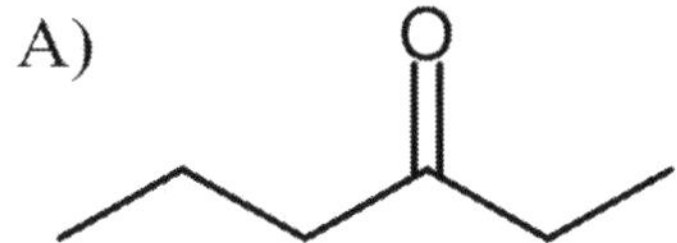


Answer to IR Spectroscopy Problem 10:

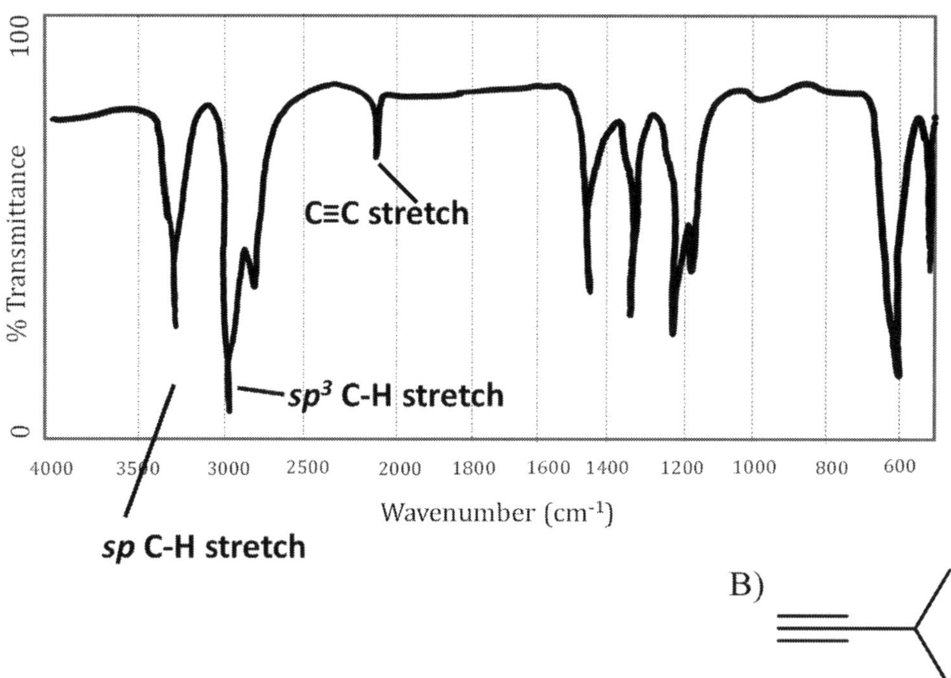


B)

Answer to IR Spectroscopy Problem 11

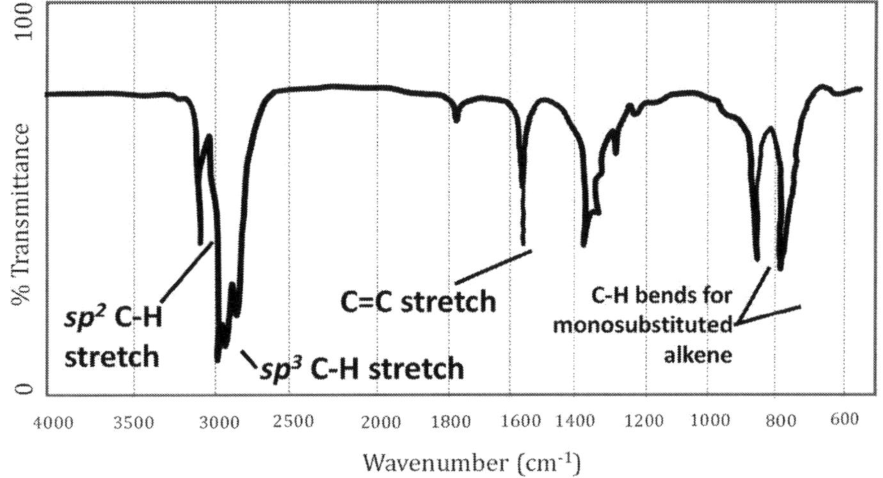


E)

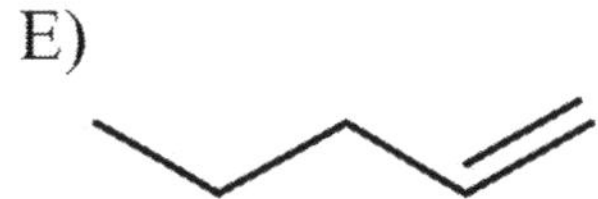

Answer to IR Spectroscopy Problem 12

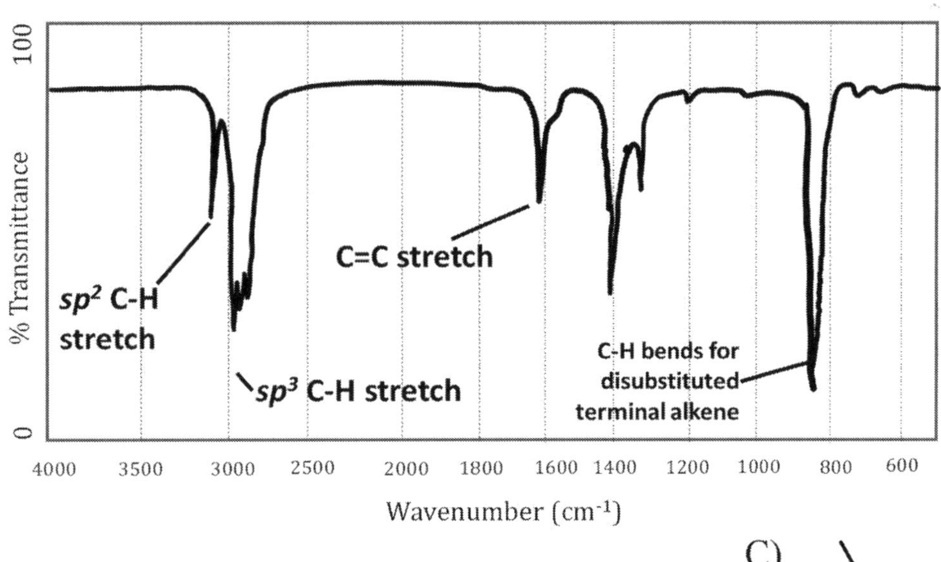


C)

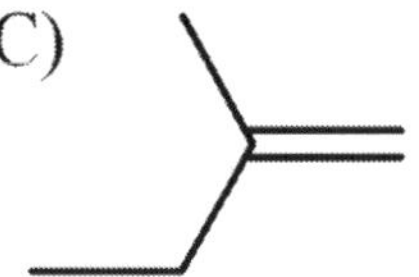

Answer to IR Spectroscopy Problem 13

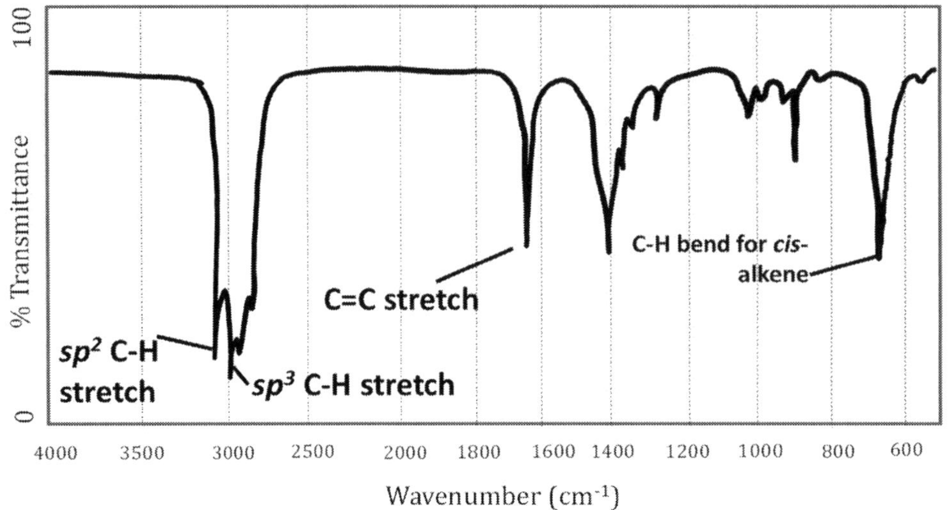


B) 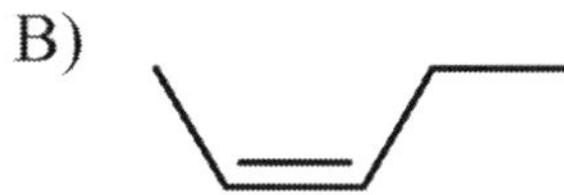

Answer to IR Spectroscopy Problem 14

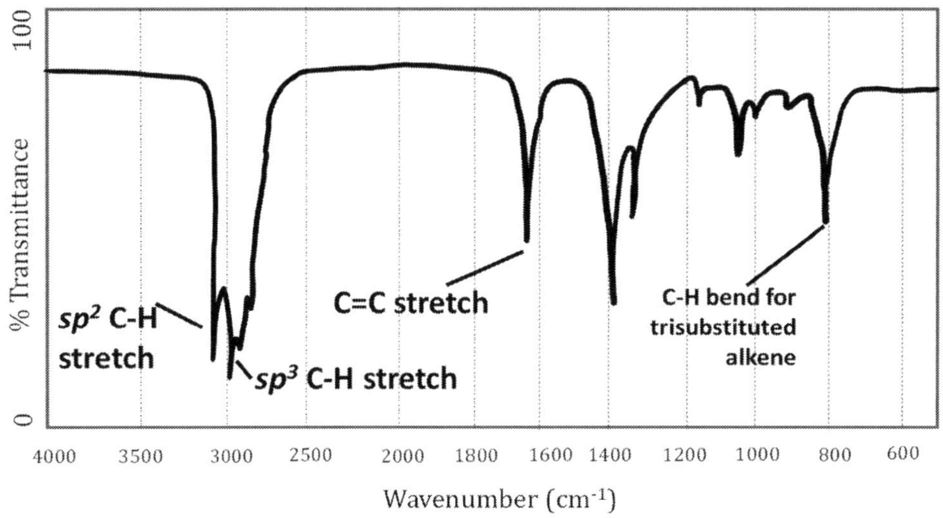


D) 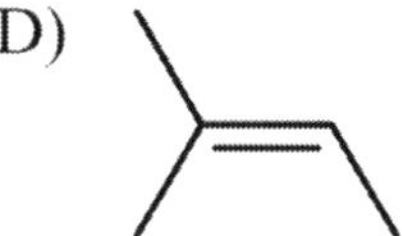

Answer to IR Spectroscopy Problem 15

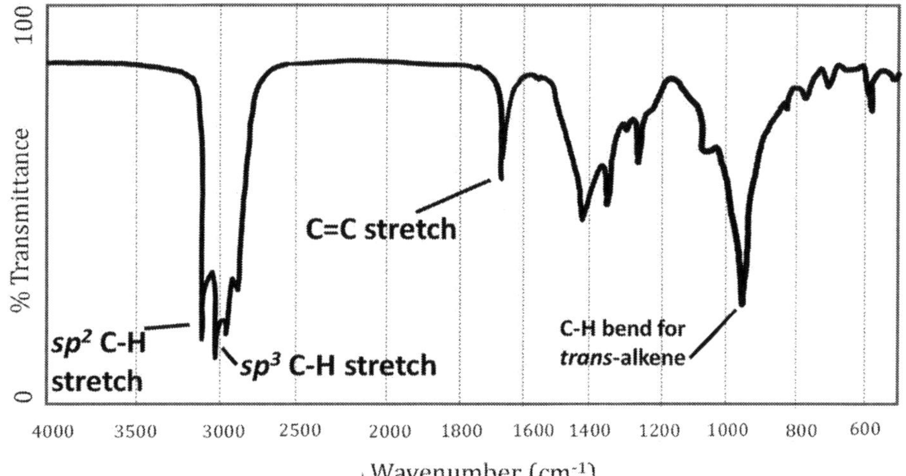


A)

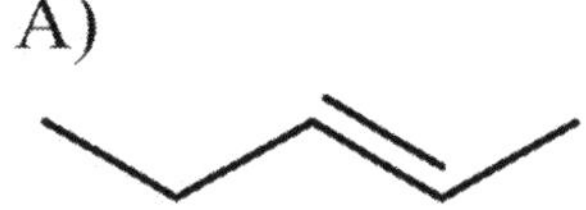

## Lesson VIII.4. Proton NMR Practice Problems

For each of the following, deduce the structure having the provided molecular formula that is most likely to have produced the accompanying $^1$H NMR spectrum.

Proton NMR Problem 1

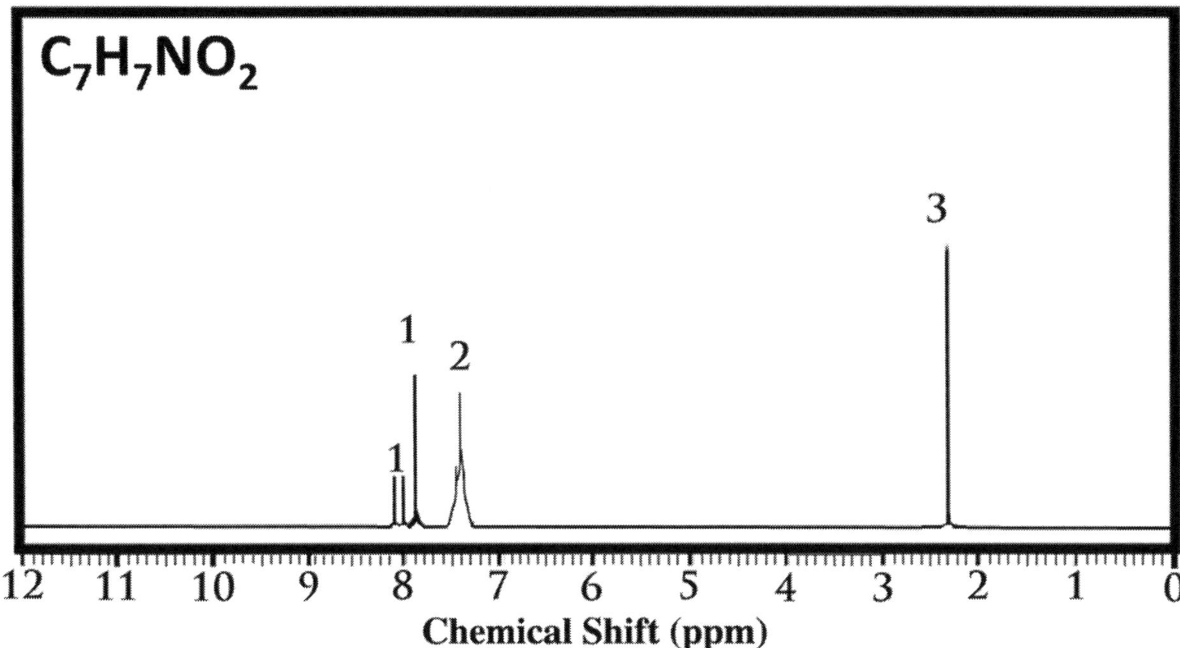


Proton NMR Problem 2

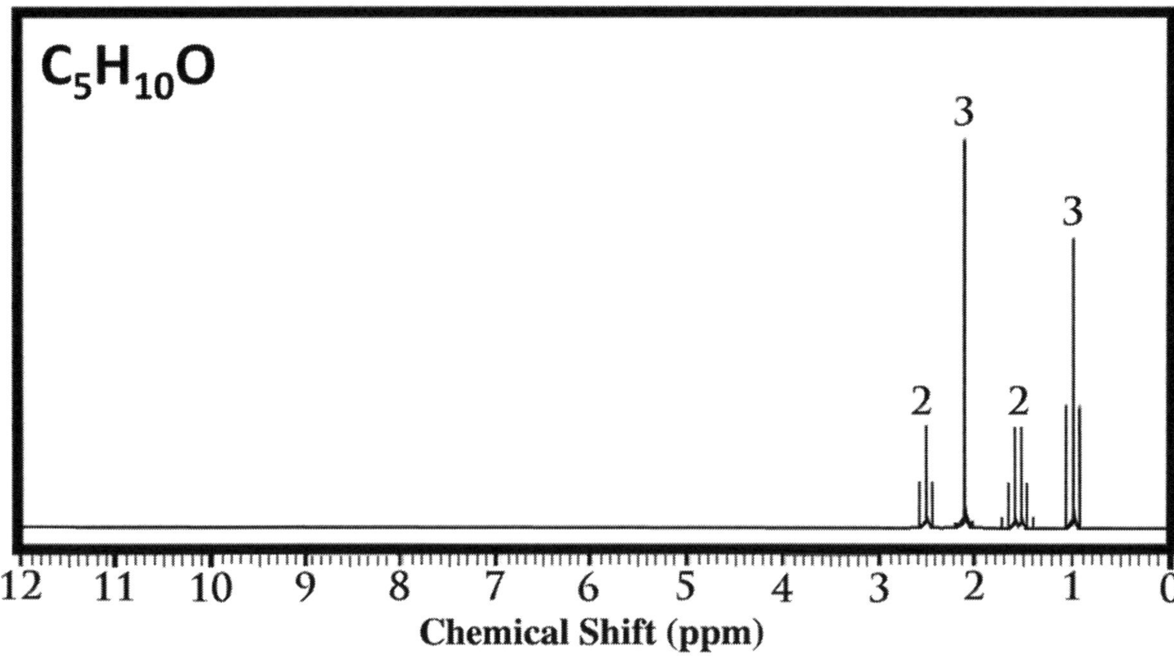

Proton NMR Problem 3

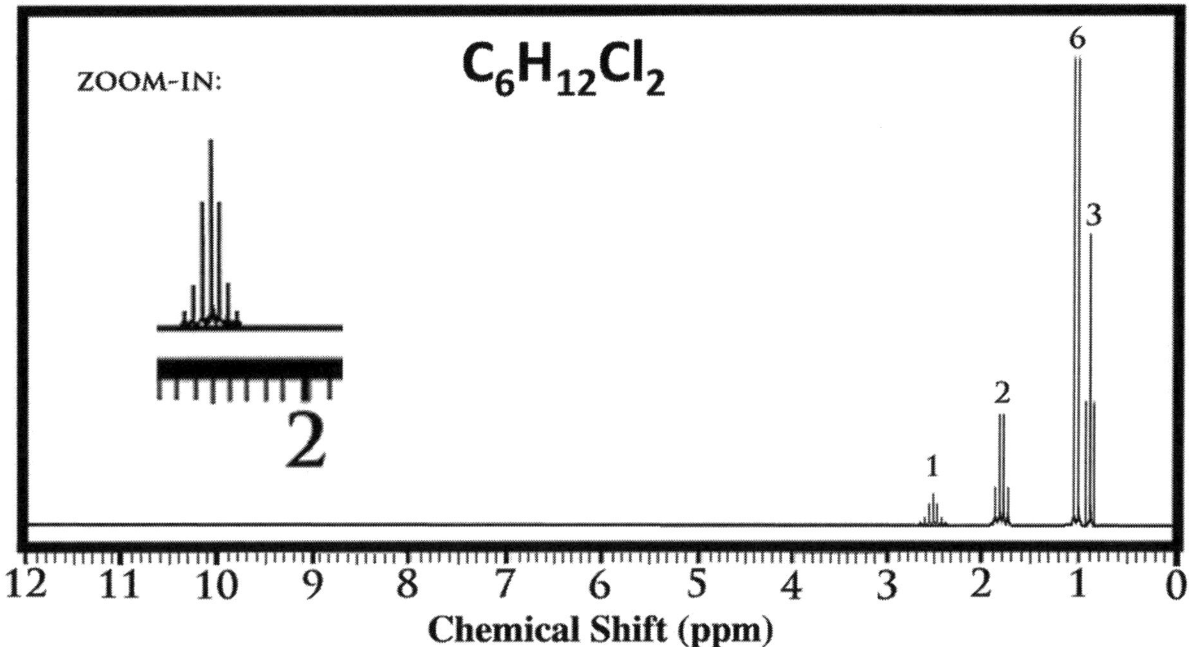


Proton NMR Problem 4

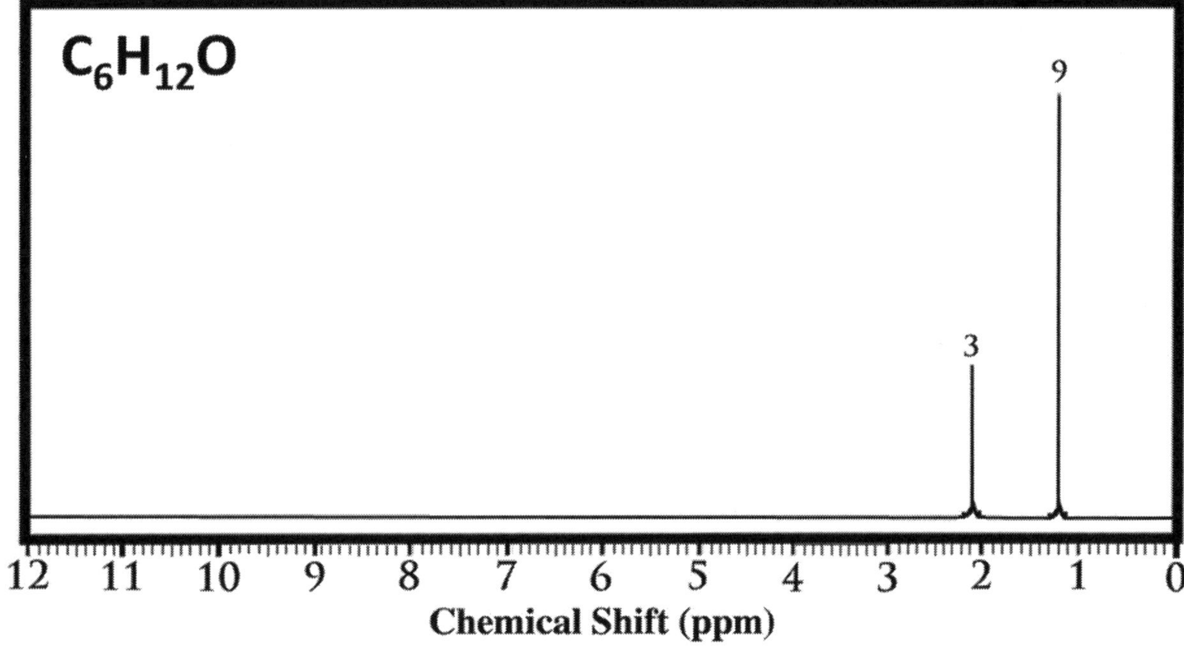

Proton NMR Problem 5

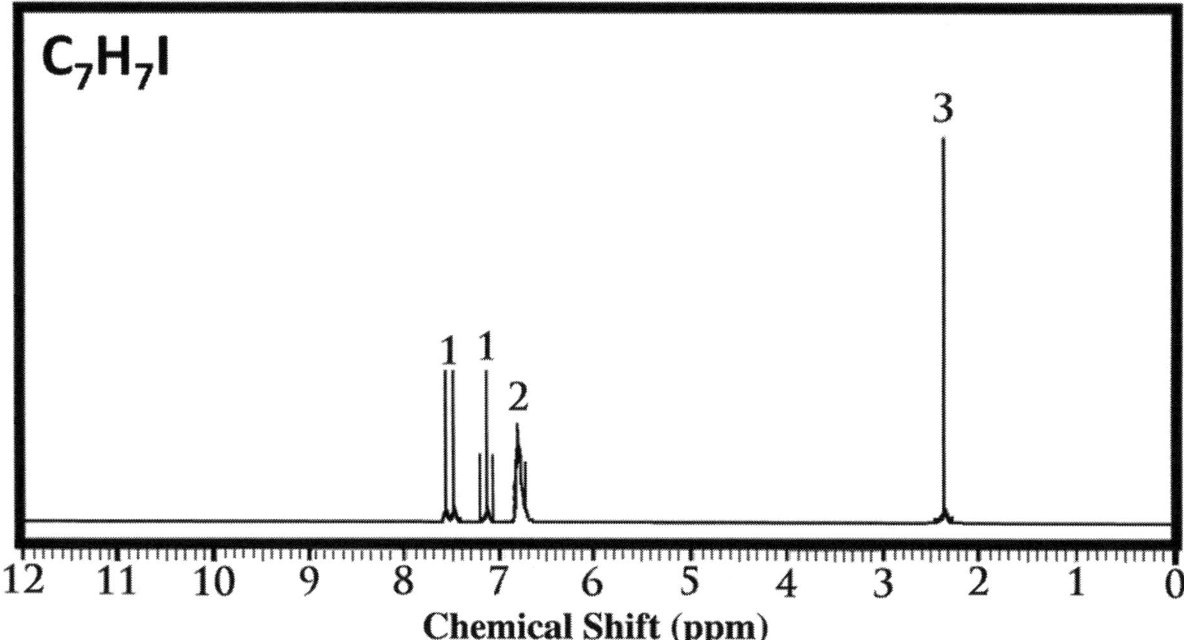


Proton NMR Problem 6

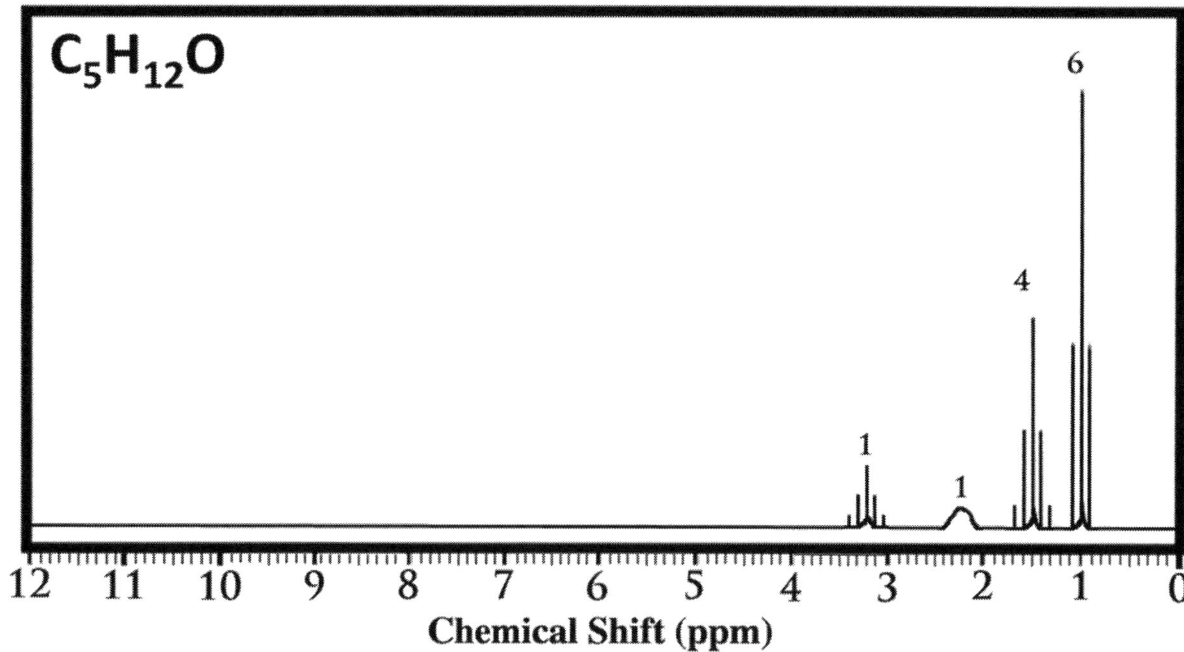

Proton NMR Problem 7

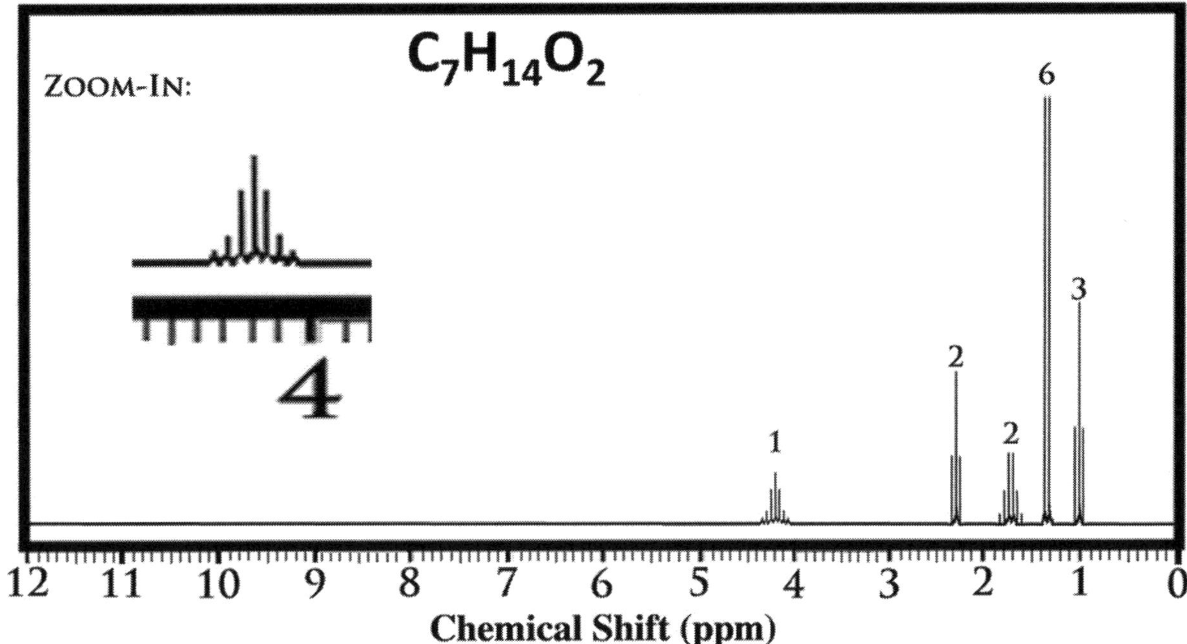


Proton NMR Problem 8

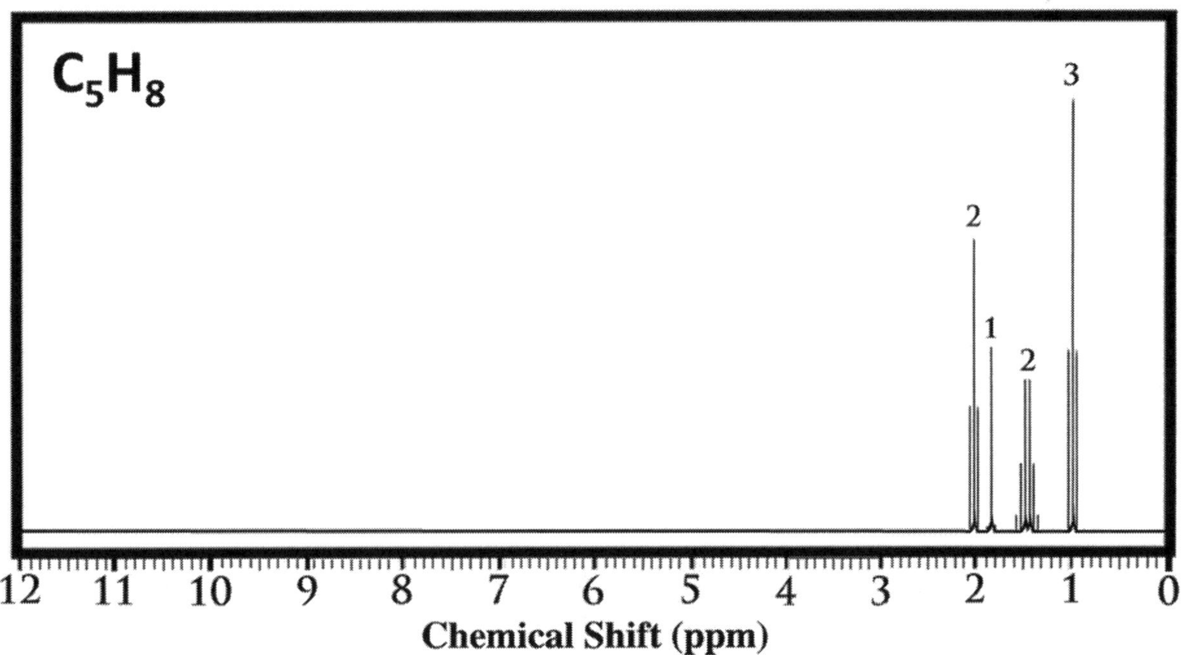

Proton NMR Problem 9

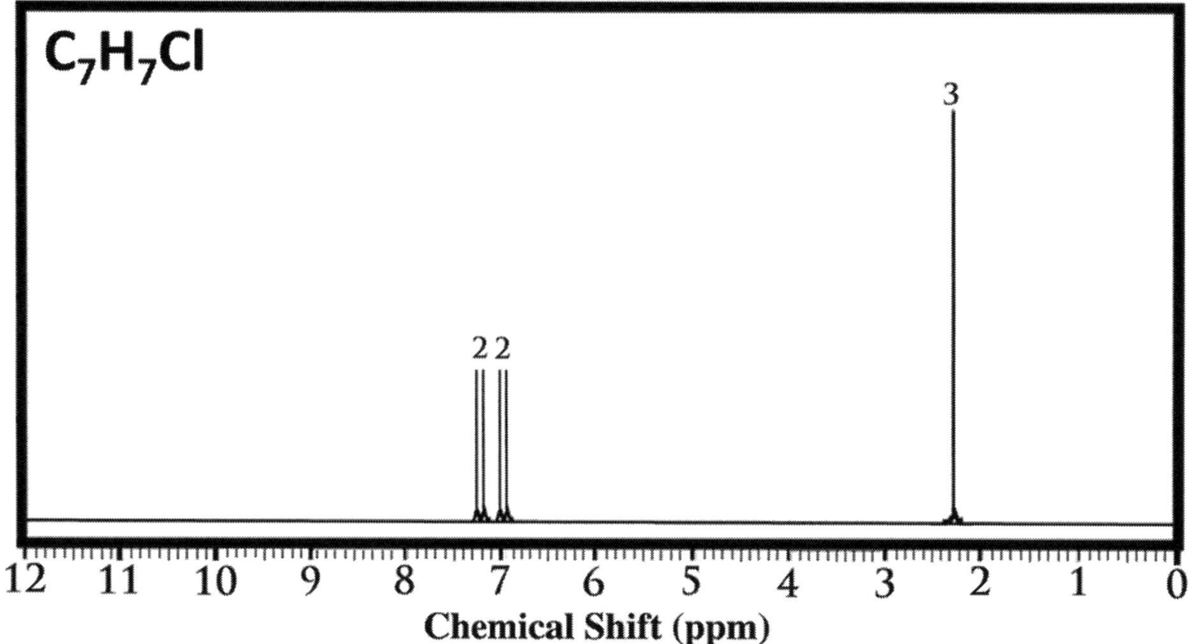


Proton NMR Problem 10

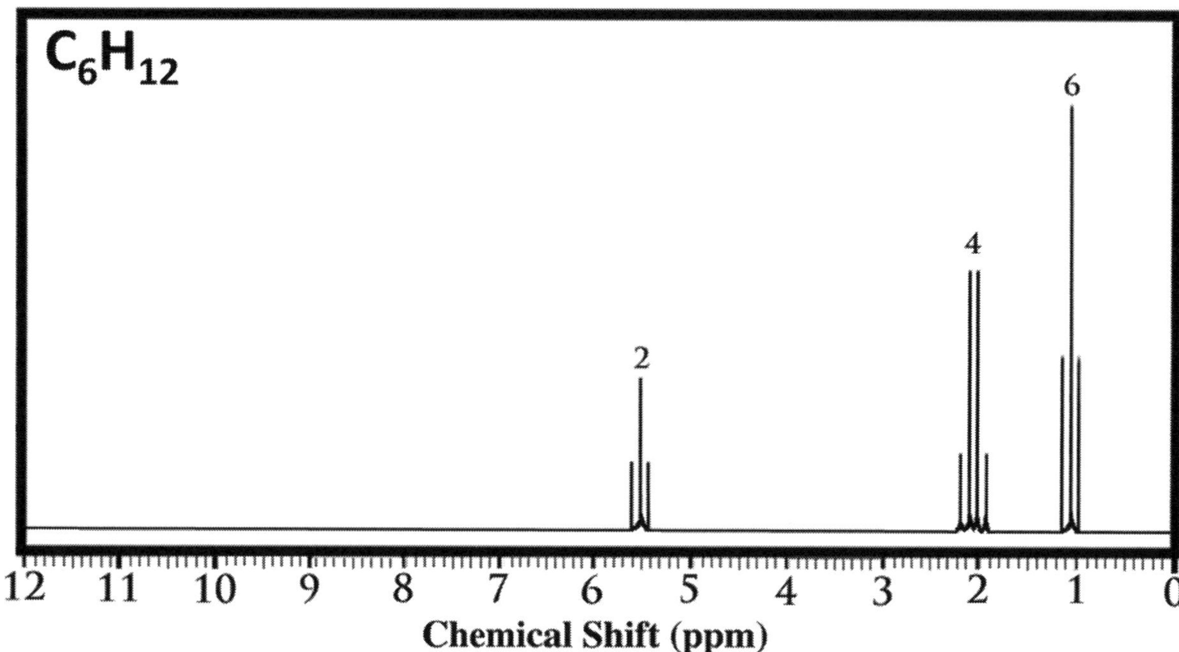

Proton NMR Problem 11

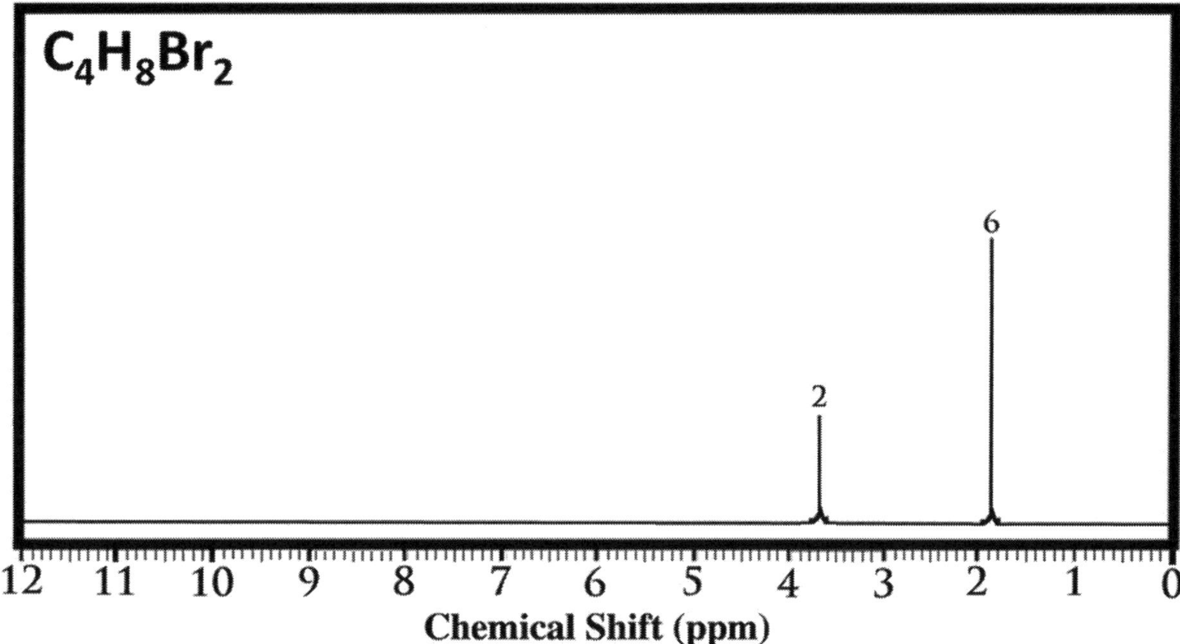


Proton NMR Problem 12

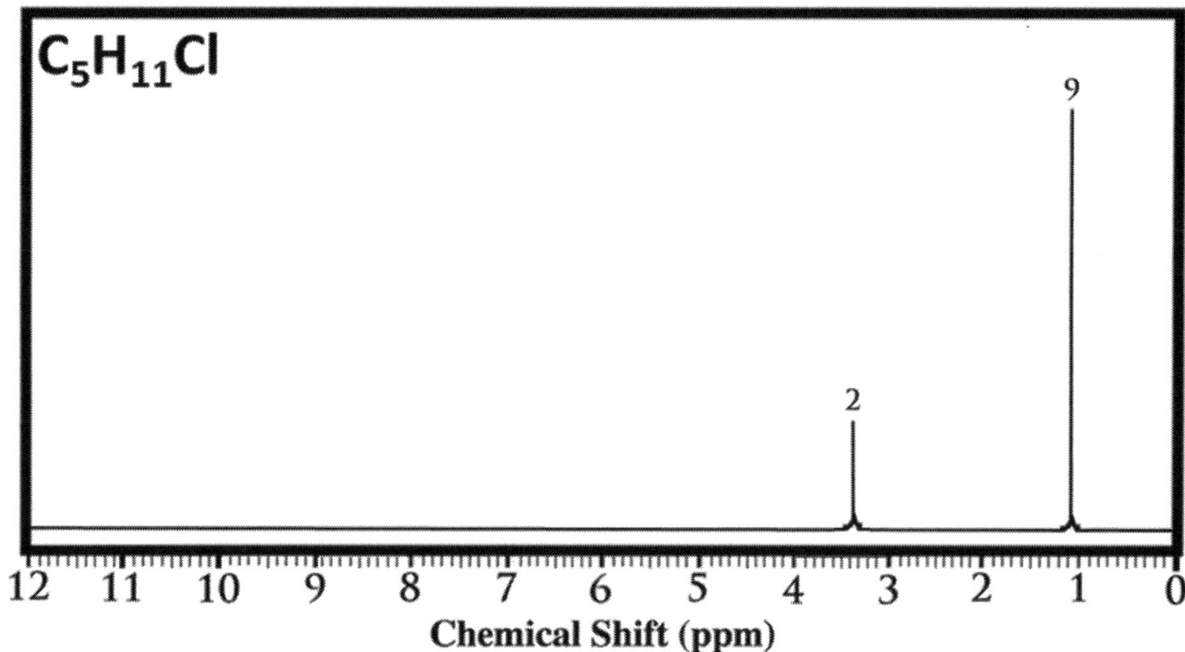

Proton NMR Problem 13

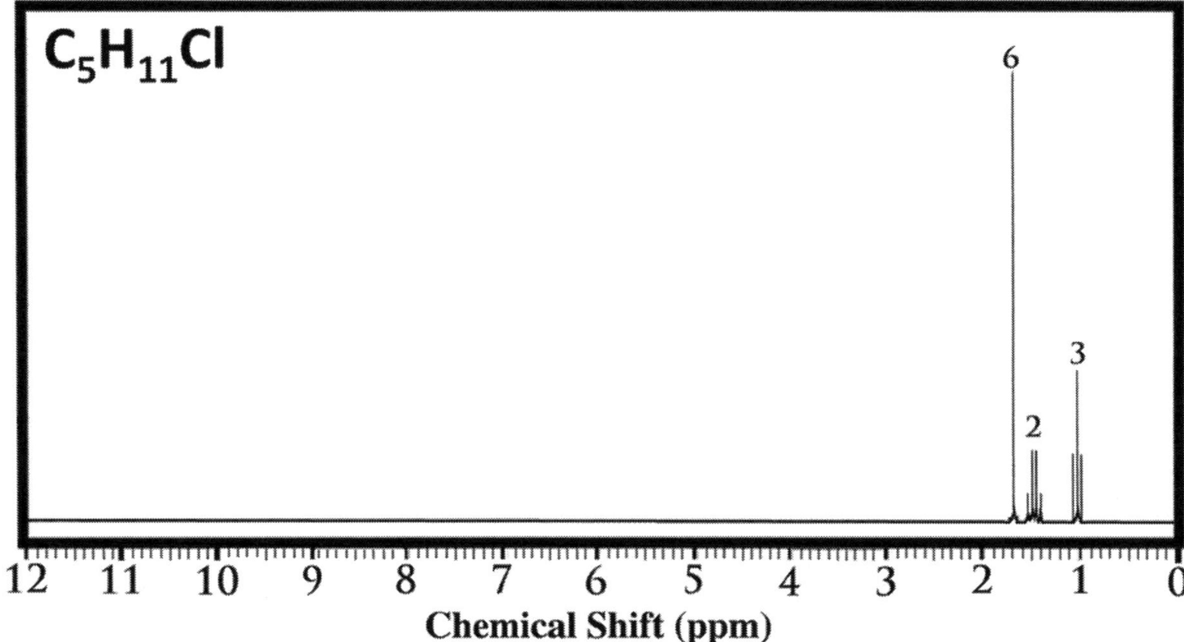


Proton NMR Problem 14

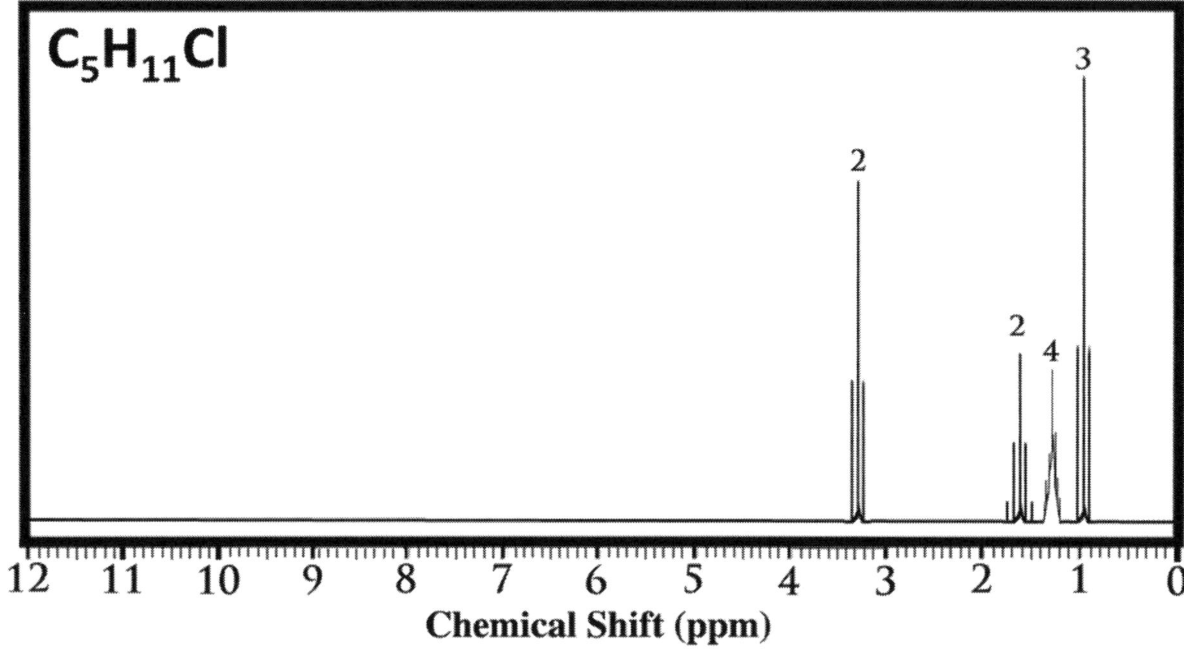

Proton NMR Problem 15

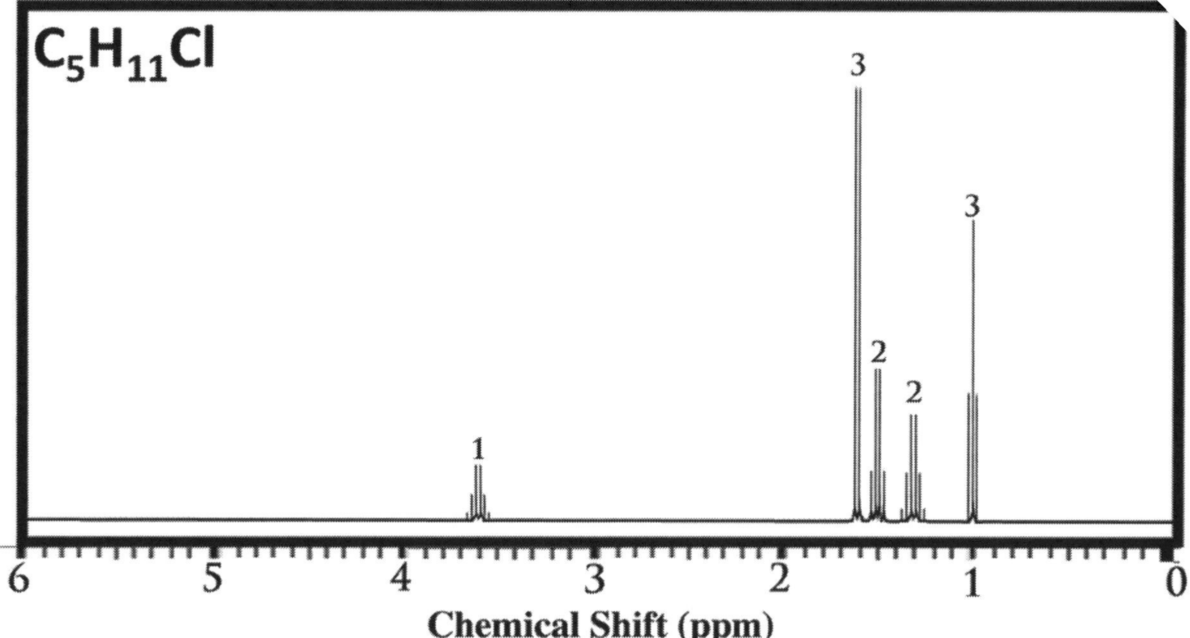

on NMR Problem 1

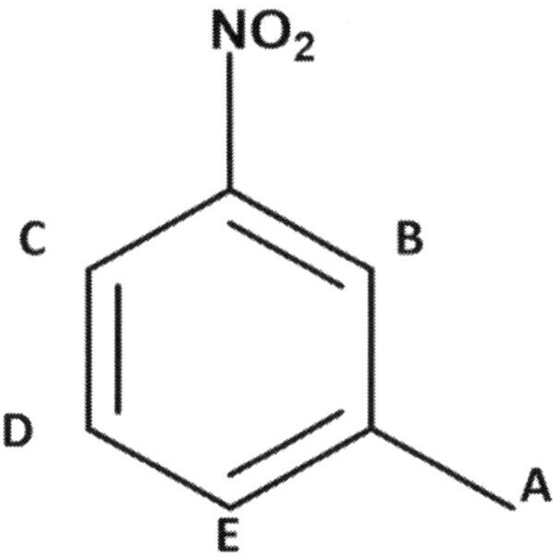


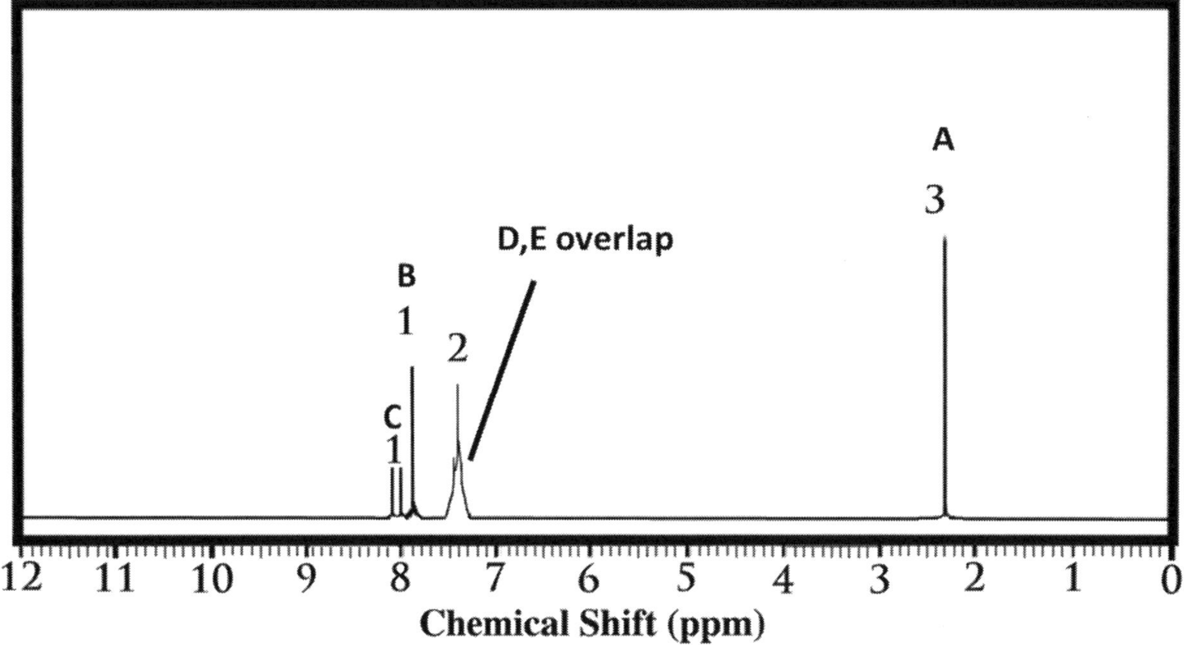

Answer to Proton NMR Problem 2

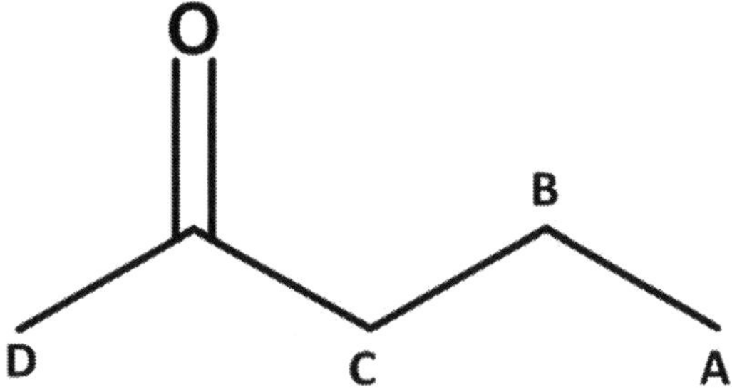


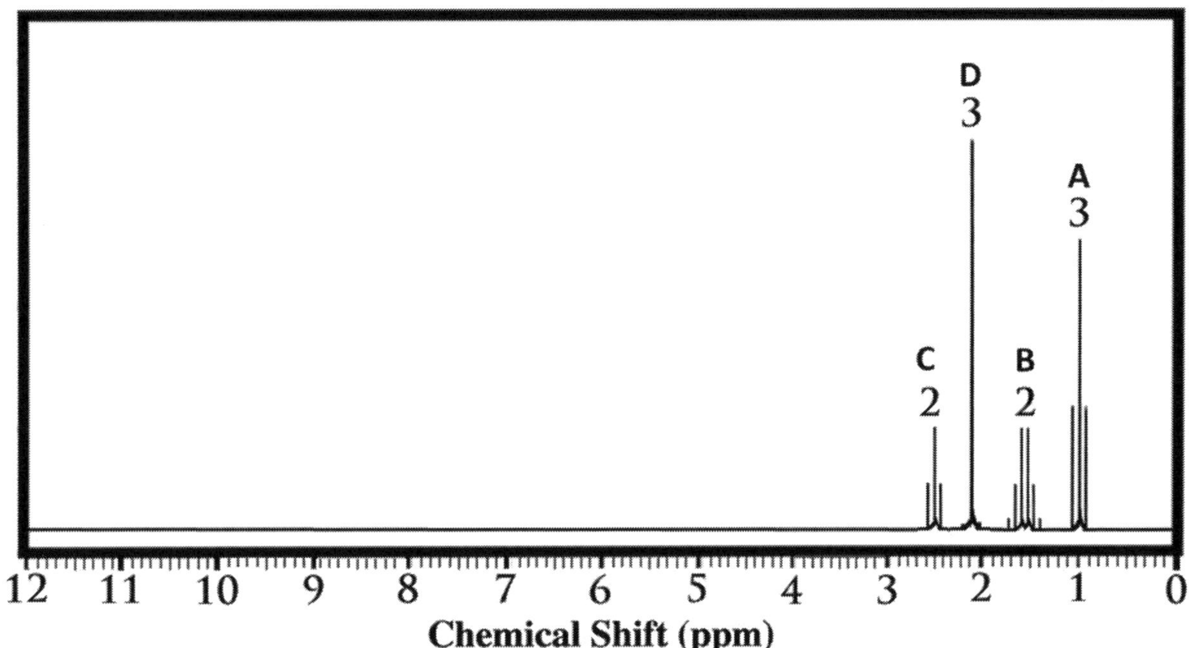

Answer to Proton NMR Problem 3

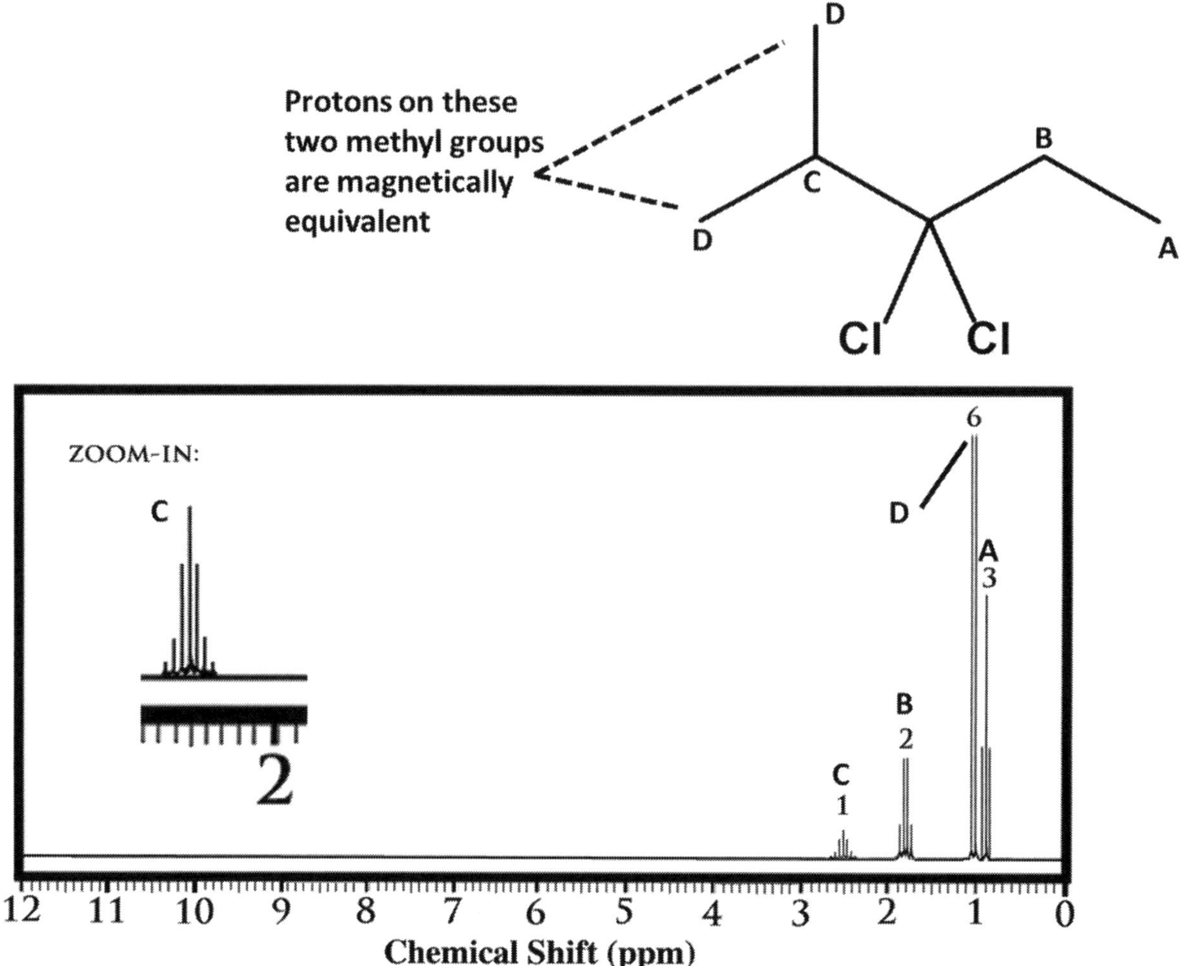

Answer to Proton NMR Problem 4

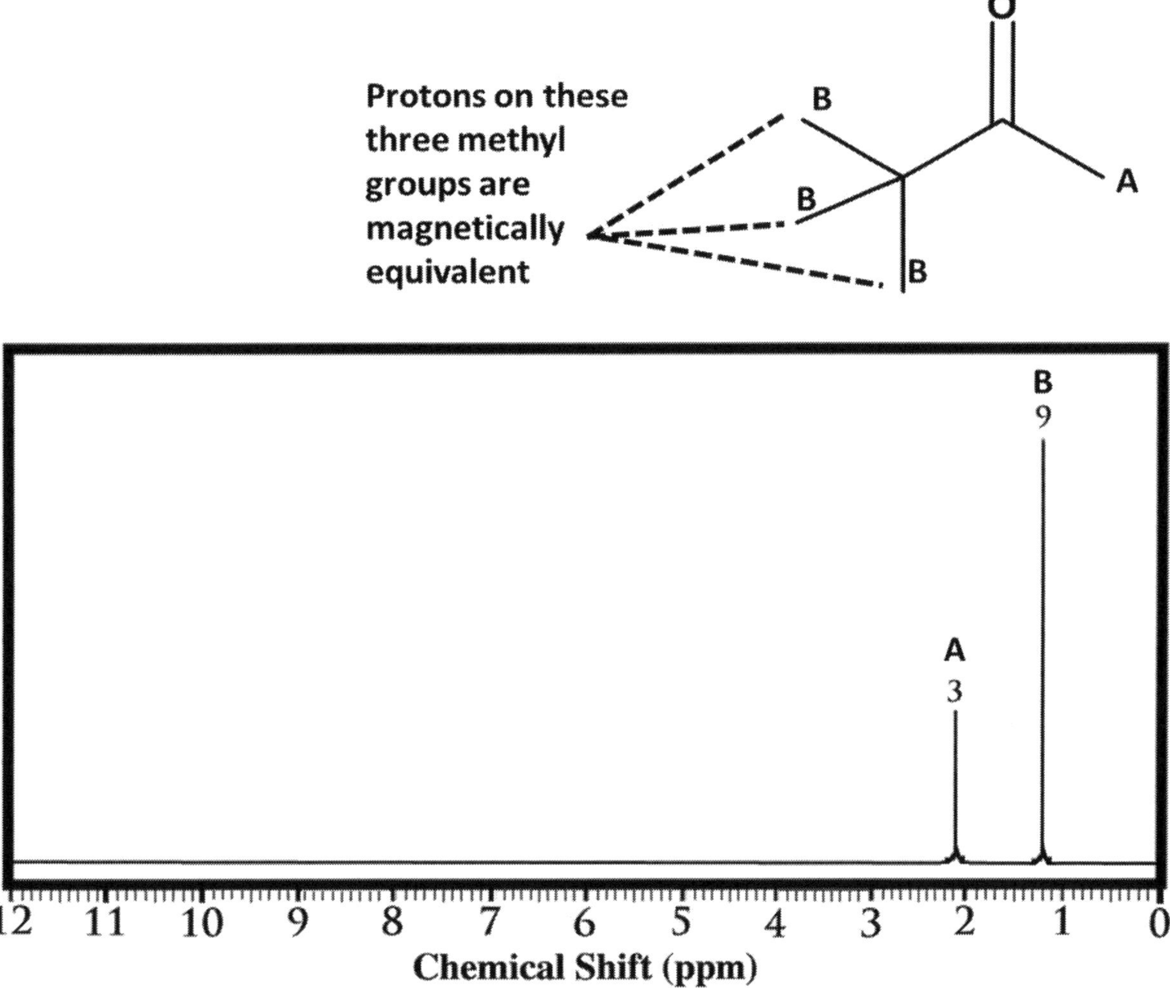

Answer to Proton NMR Problem 5

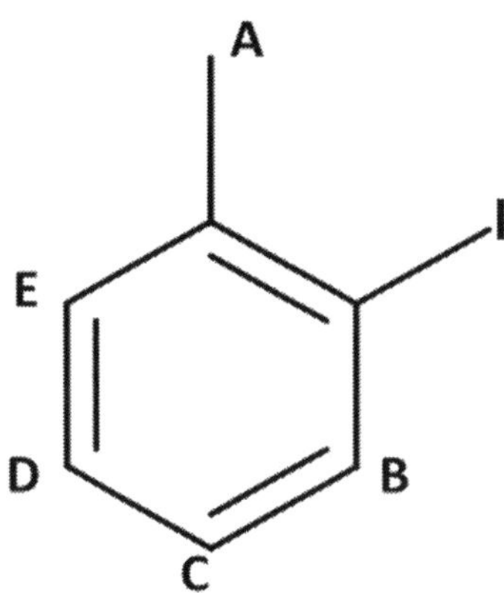


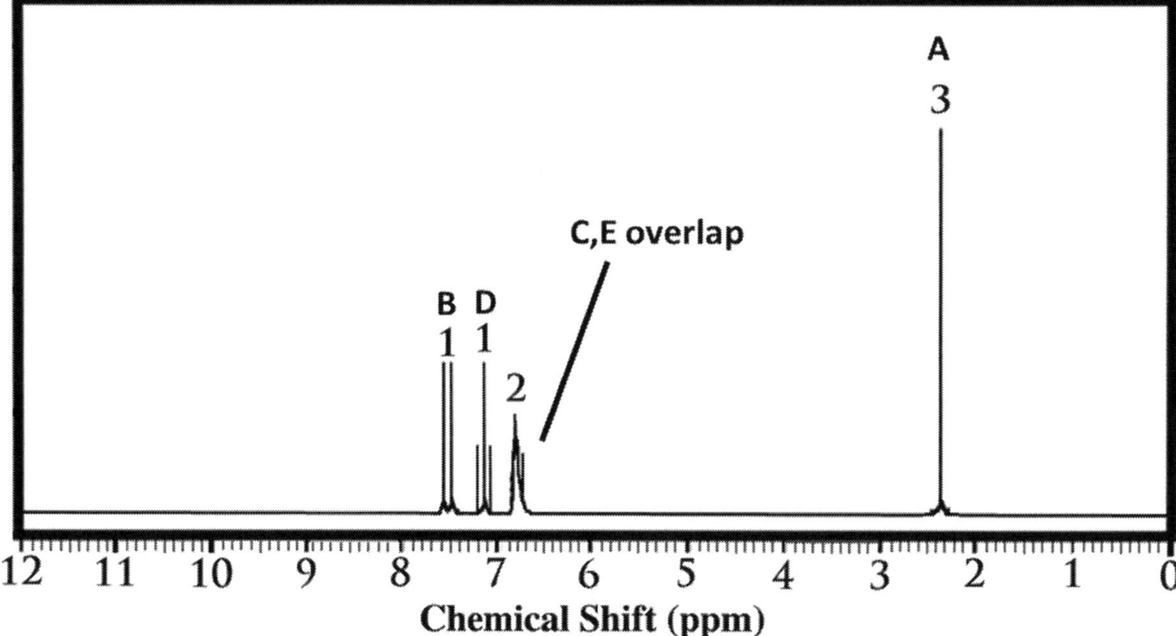

Answer to Proton NMR Problem 6

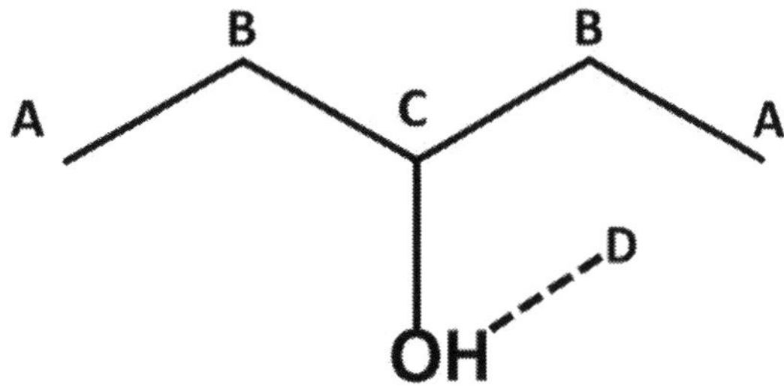


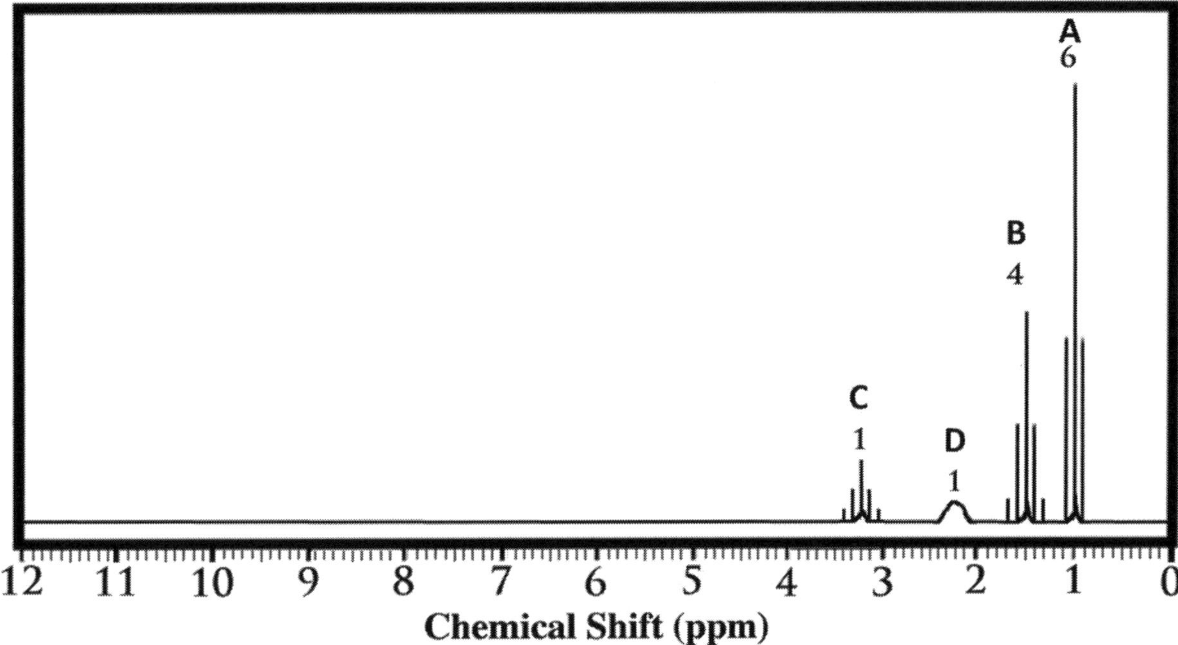

Answer to Proton NMR Problem 7

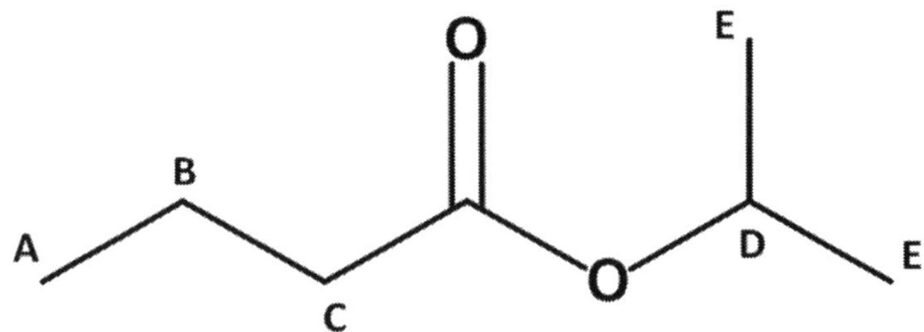


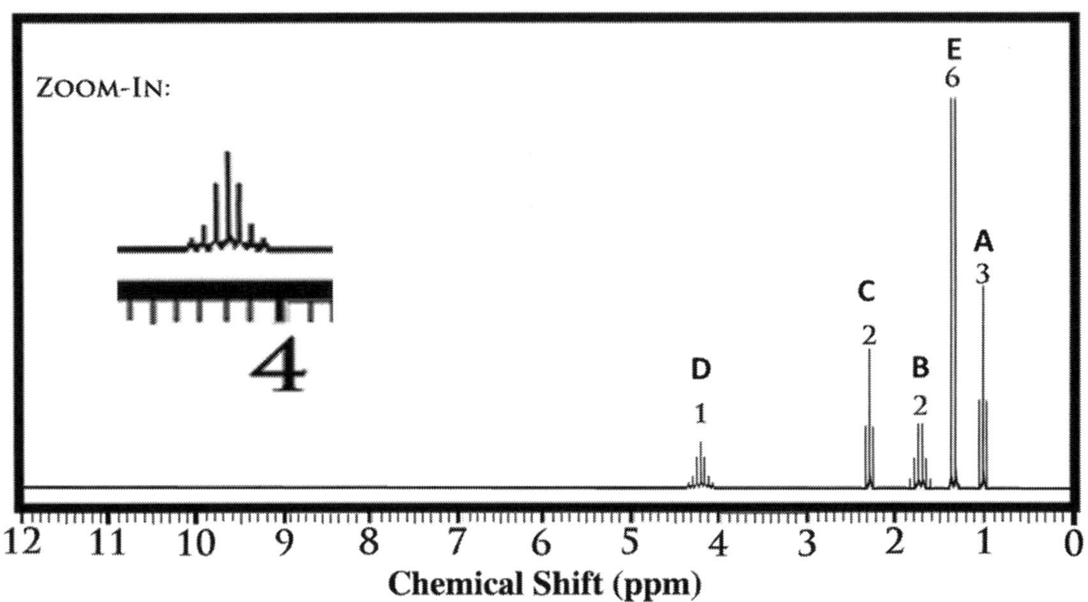

Answer to Proton NMR Problem 8

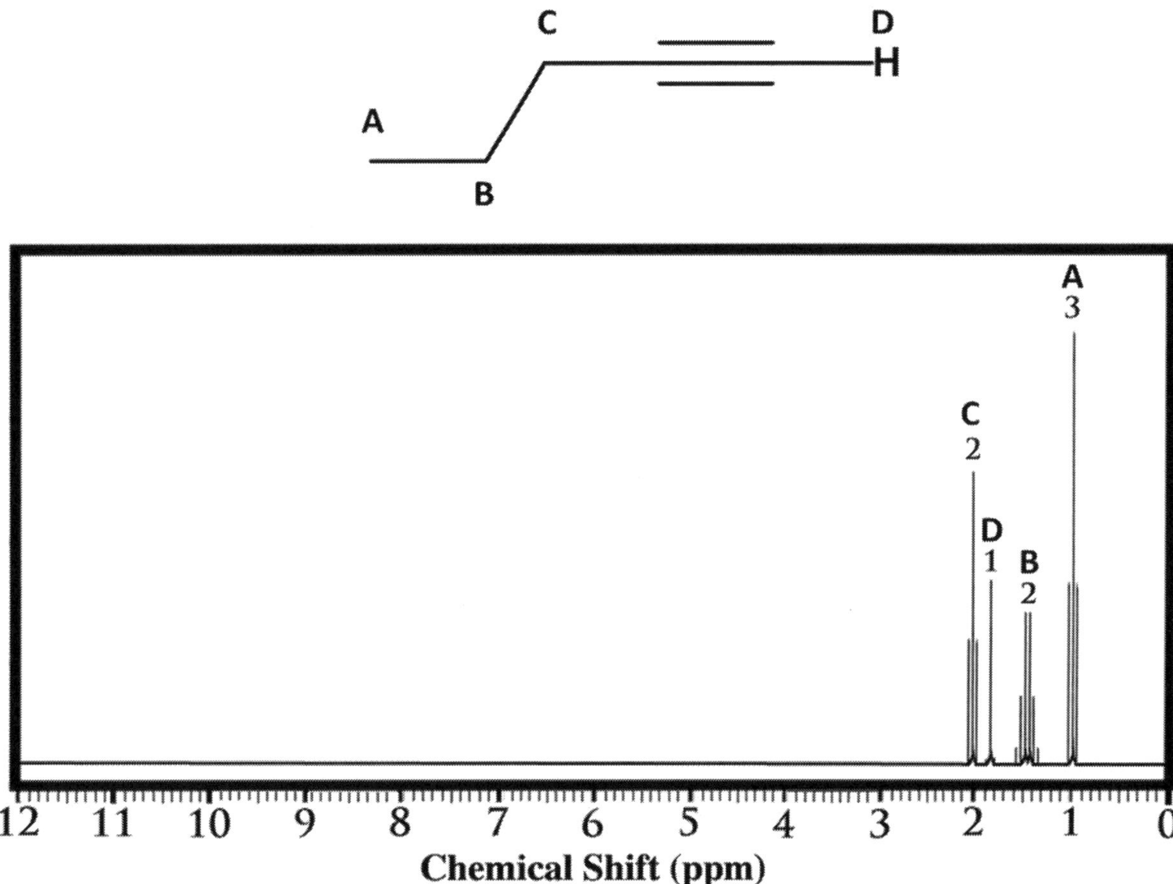

Answer to Proton NMR Problem 9

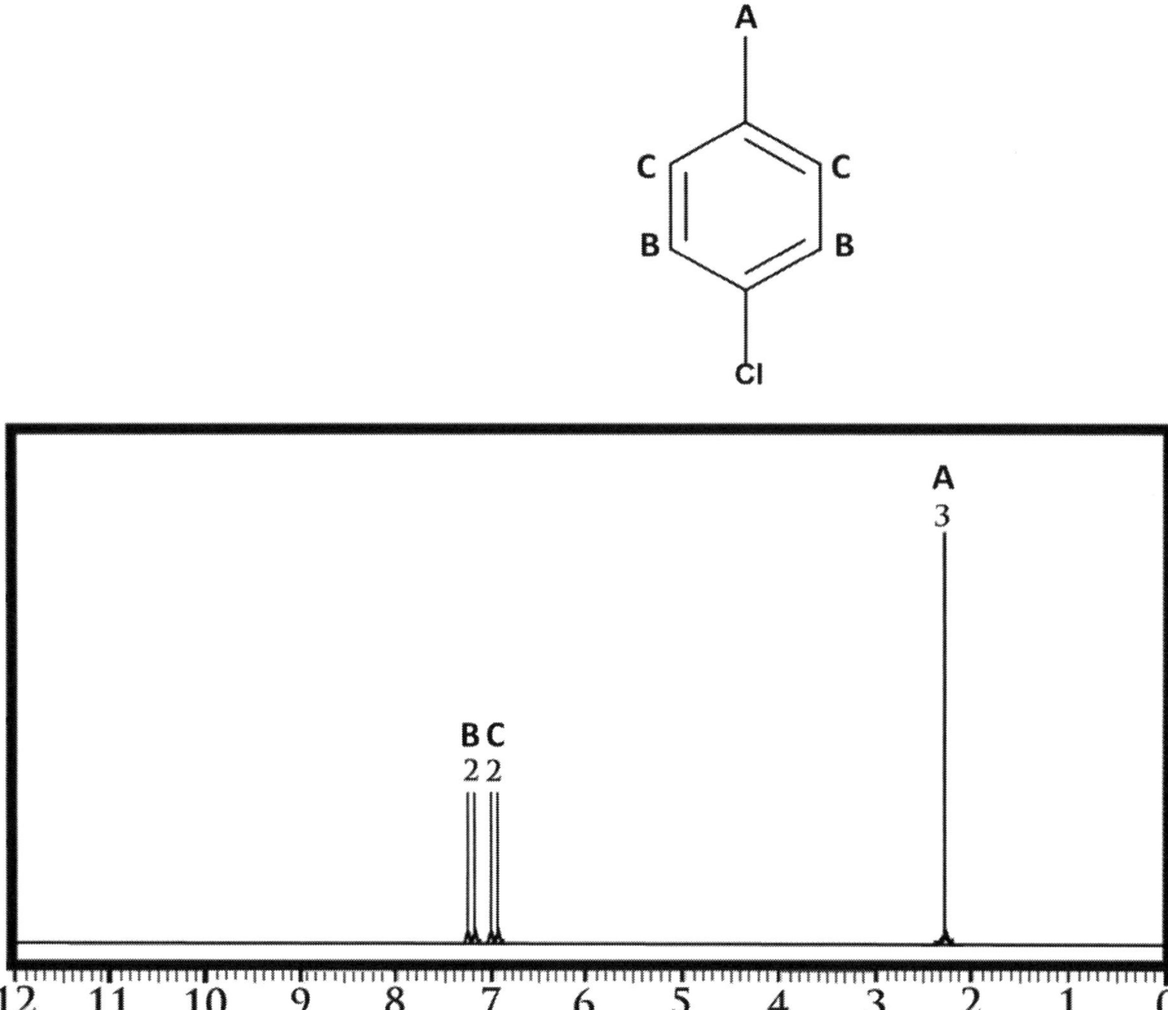

Answer to Proton NMR Problem 10

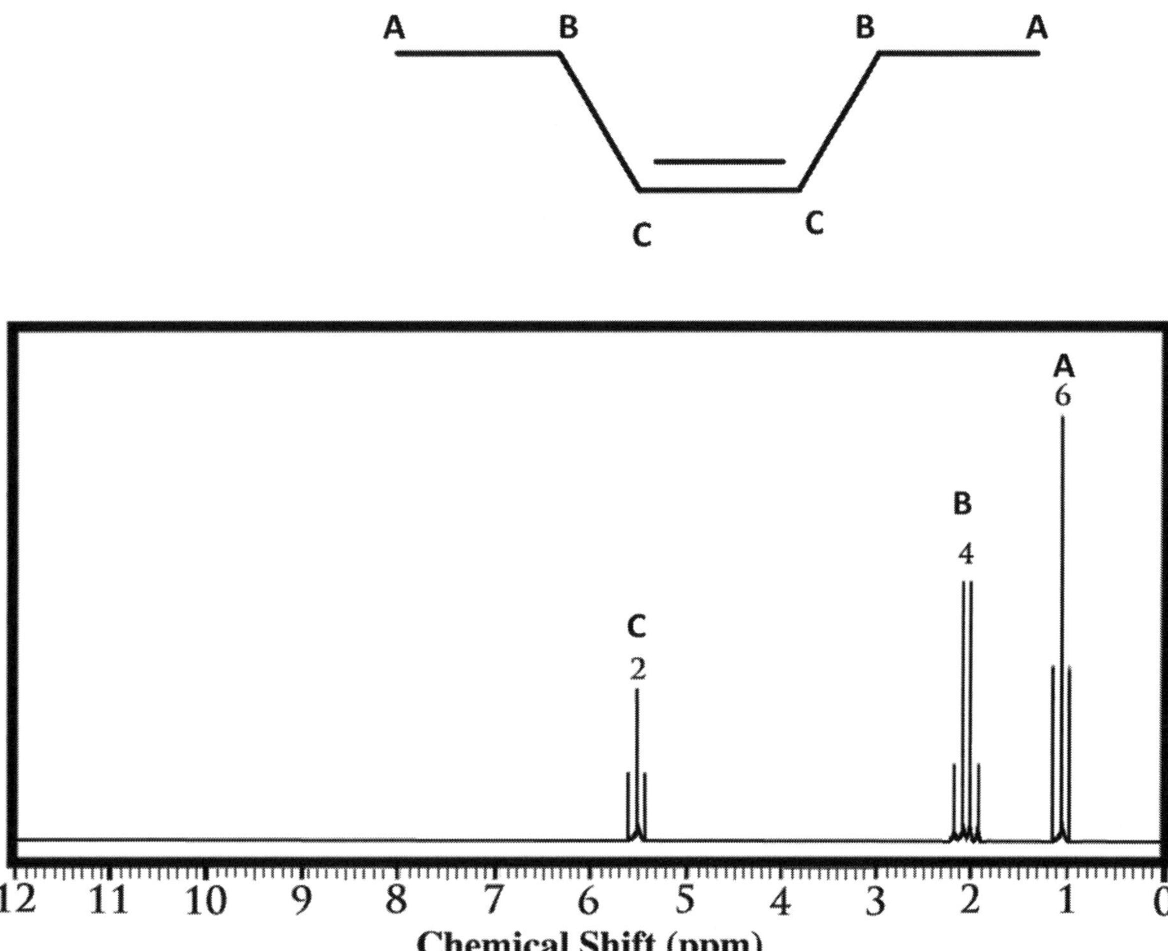

Answer to Proton NMR Problem 11

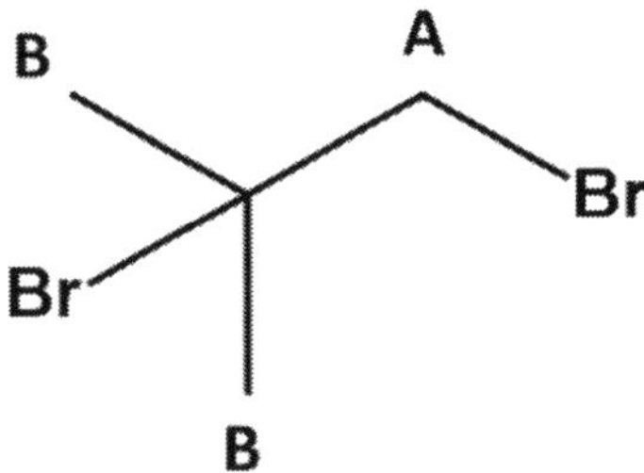


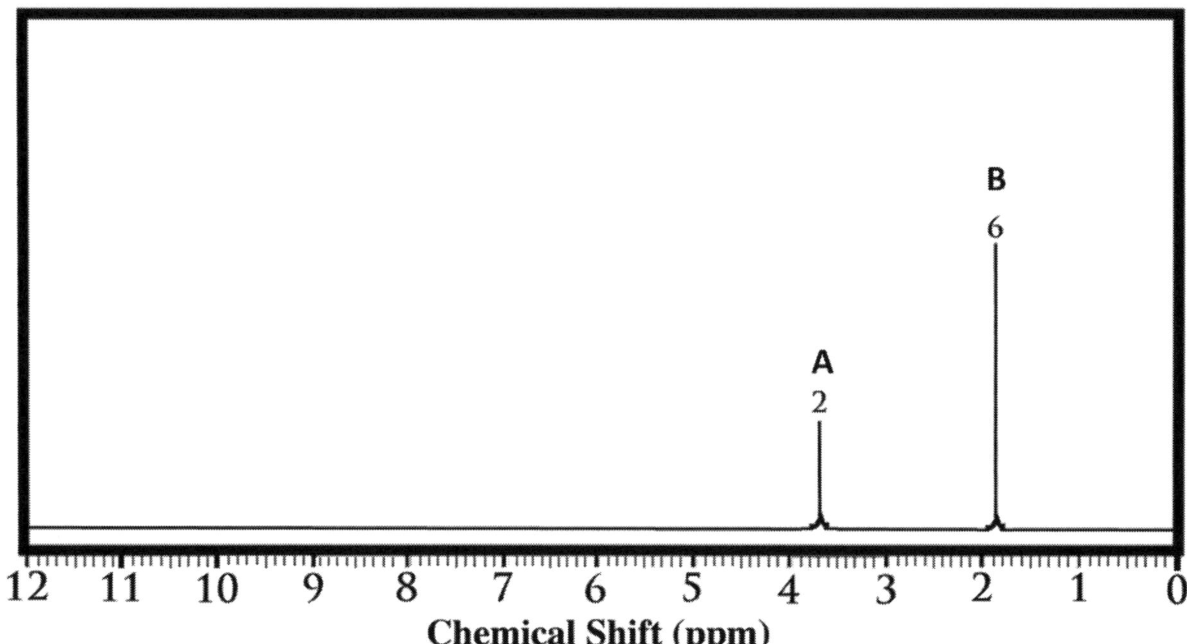

Answer to Proton NMR Problem 12

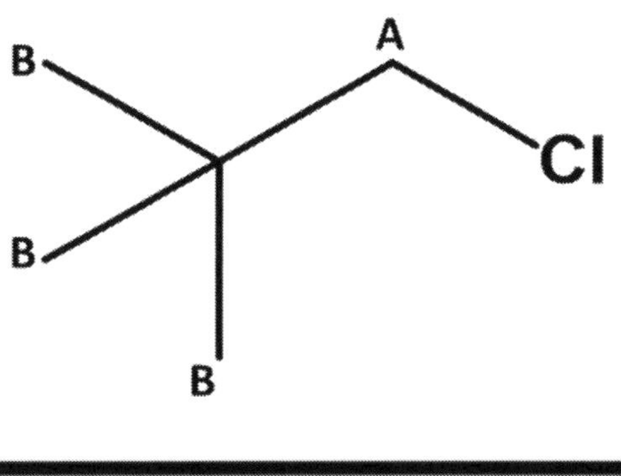


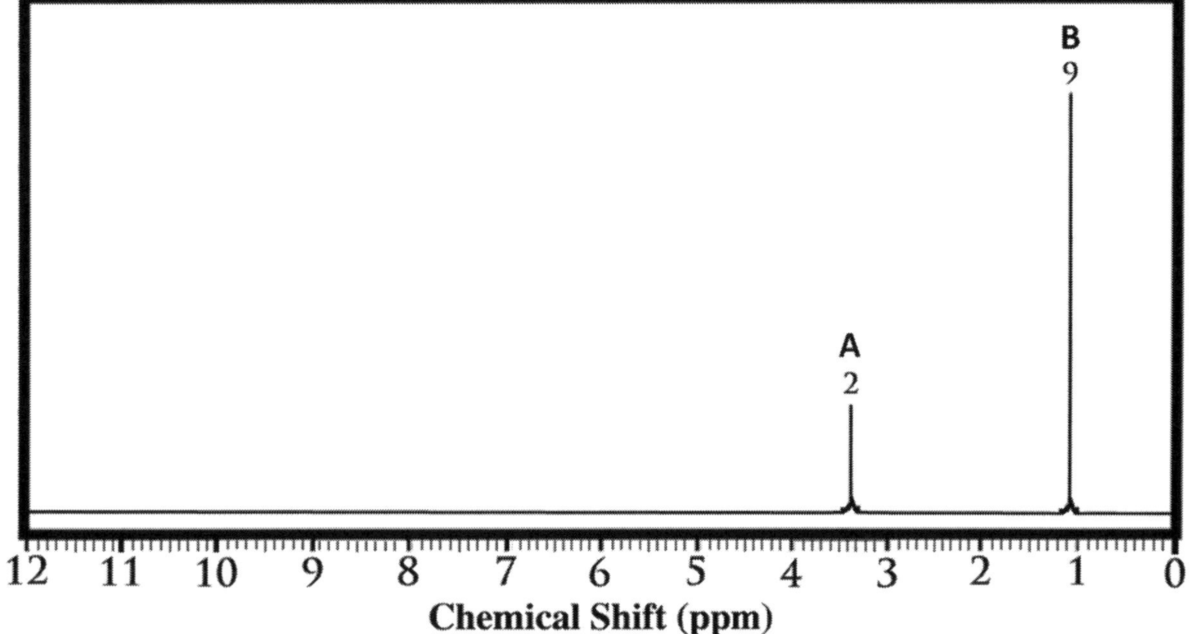

Answer to Proton NMR Problem 13

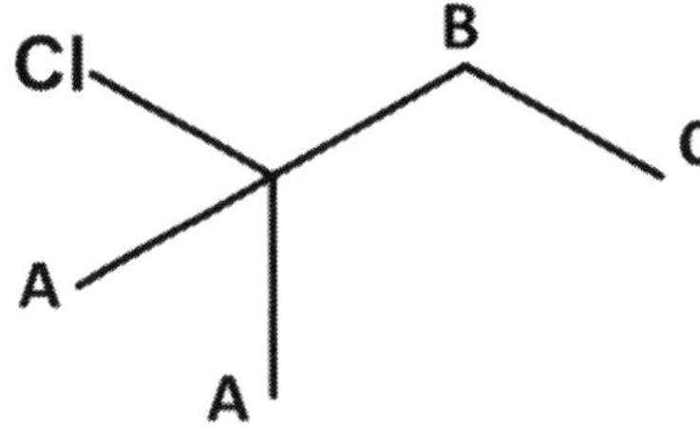


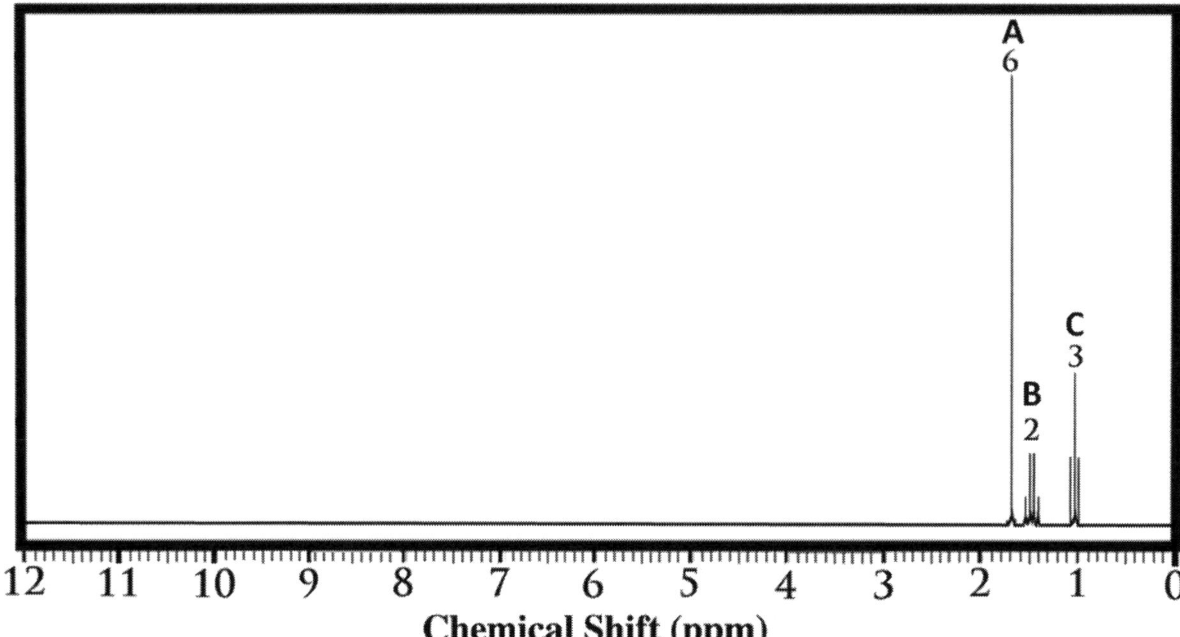

Answer to Proton NMR Problem 14

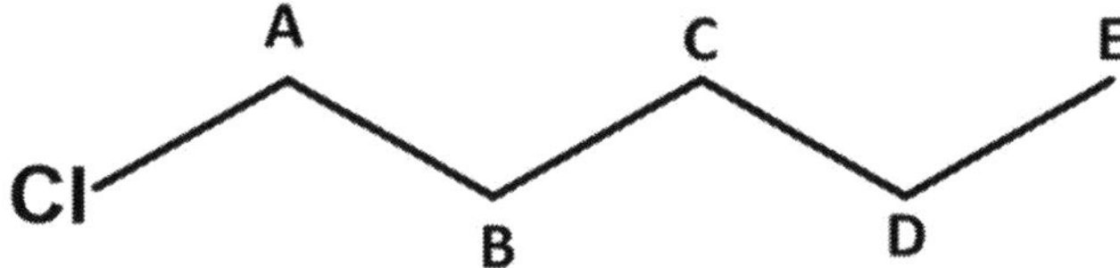


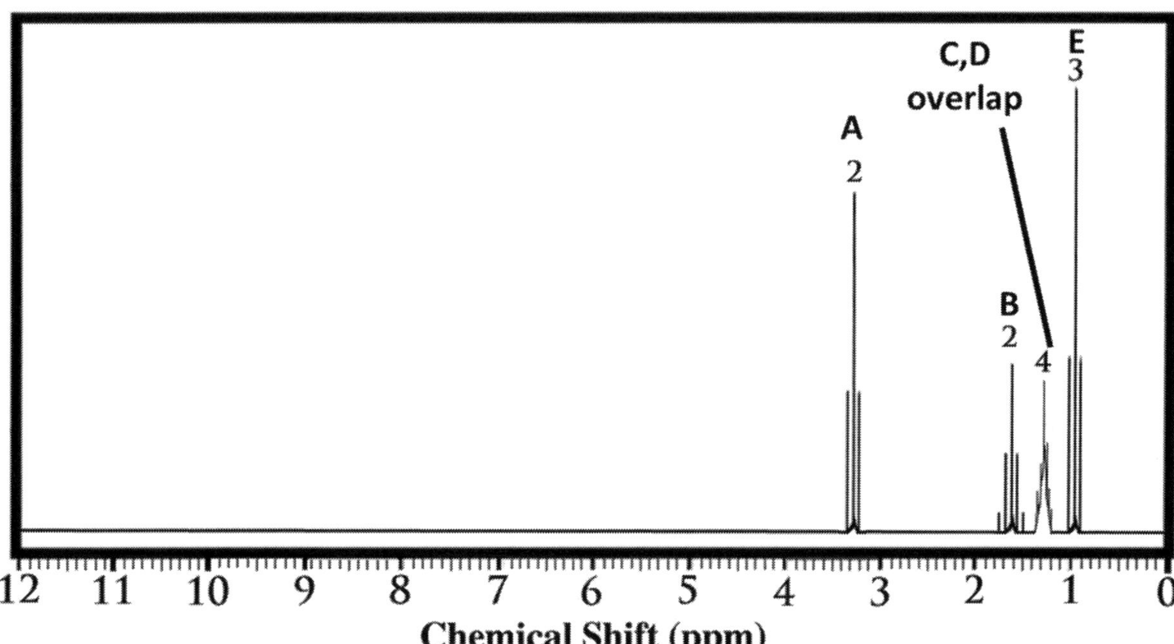

Answer to Proton NMR Problem 15

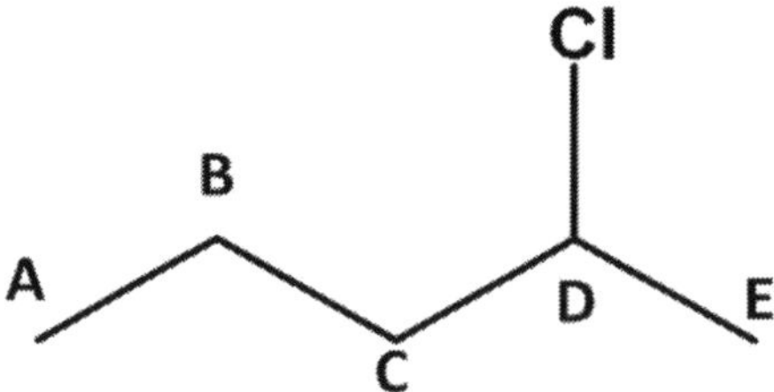


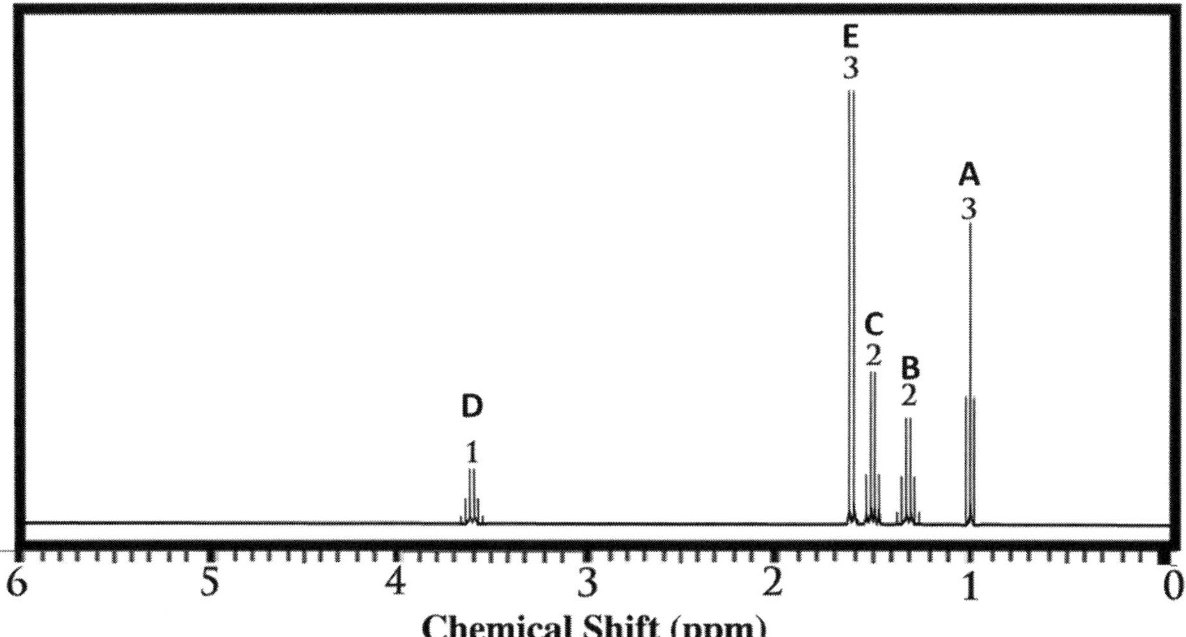

## Lesson VIII.6. Carbon-13 NMR Practice Problems with Solutions

Carbon-13 NMR Question 1.

Which of the molecules **I-IV** would give rise to the $^{13}C$ NMR spectrum shown below?

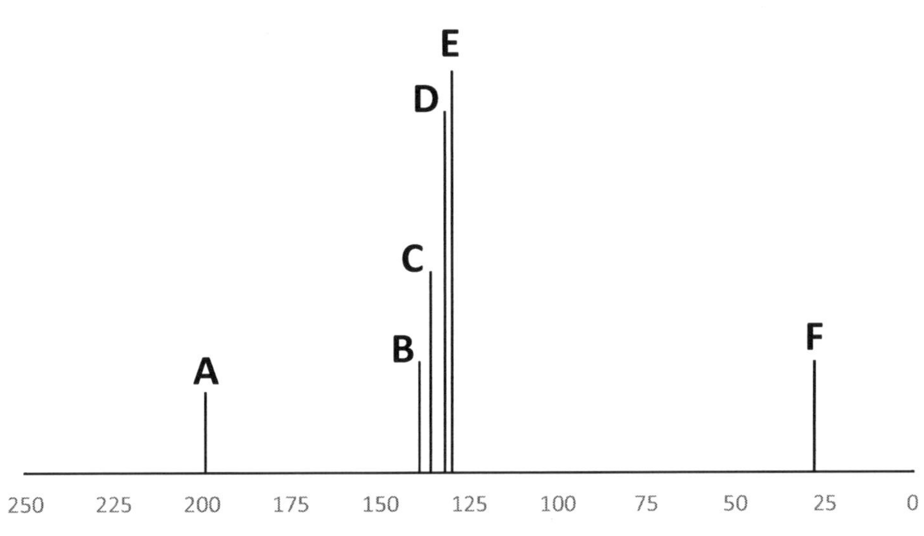


OH I

OH II

O III

O O IV

Carbon-13 NMR Answer 1.

**Peaks**

| | |
|---|---|
| **Carbonyls:** | 1 |
| **C=C:** | 4 |
| **Sp/highly E.N.:** | 0 |
| **Alkyl:** | 1 |

---

**Answer Choices:**

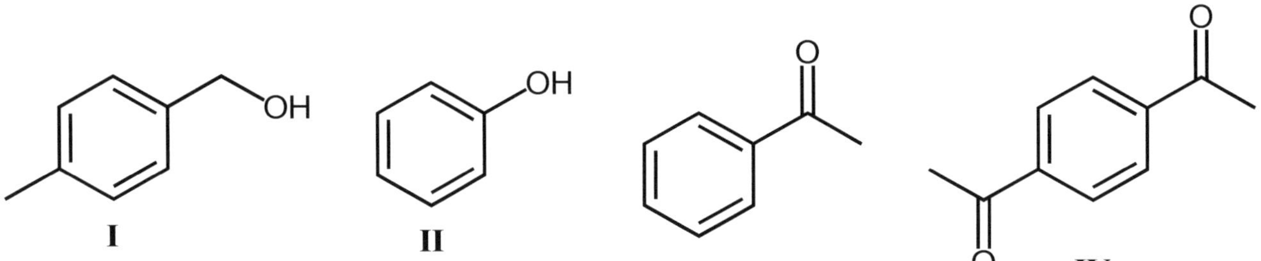


I II III IV

**I** and **II** do not have any carbonyl-based functional groups and would thus not exhibit any peaks in the **Carbonyls** region. **IV** has multiple mirror planes and would only exhibit 2 peaks in the **C=C** region. **III** is the only molecule consistent with the $^{13}$C NMR spectrum, and the peak assignments are shown below:

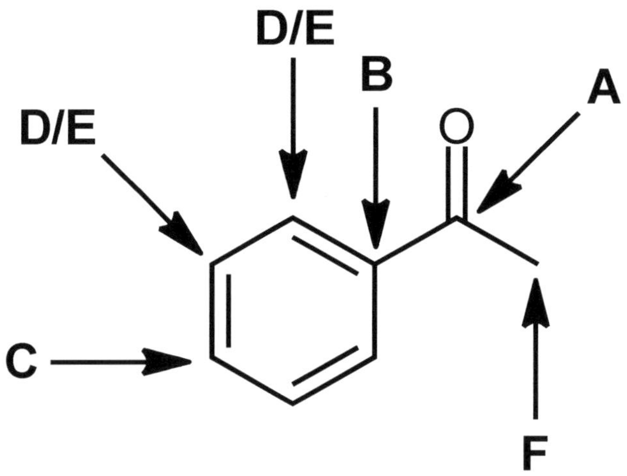

Carbon-13 NMR Question 2.

Which of the molecules **I-IV** would give rise to the $^{13}C$ NMR spectrum shown below?

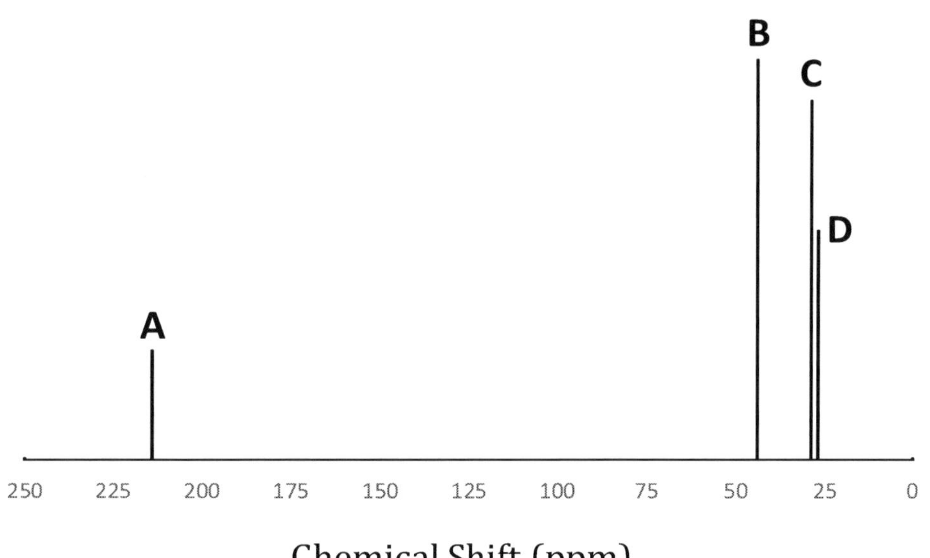


O OH O O

I II III IV

Carbon-13 NMR Answer 2.

**Peaks**

| | |
|---|---|
| **Carbonyls:** | 1 |
| **C=C:** | 0 |
| **Sp/highly E.N.:** | 1 (borderline) |
| **Alkyl:** | 2 |

---

**Answer Choices:**

O OH O O

I II III IV

**II** does not have any carbonyl-based functional groups and would thus not exhibit any peaks in the **Carbonyls** region. All of the carbons in **III** and **IV** are inequivalent, so both would exhibit 6 total peaks in the combined **Sp/highly E.N.** and **Alkyl** regions. **I** is the only molecule consistent with the $^{13}C$ NMR spectrum, and the peak assignments are shown below:

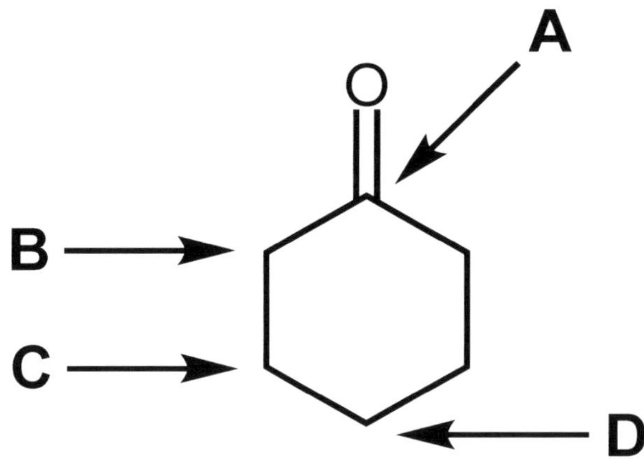

Carbon-13 NMR Question 3.

Which of the molecules **I-IV** would give rise to the $^{13}C$ NMR spectrum shown below?

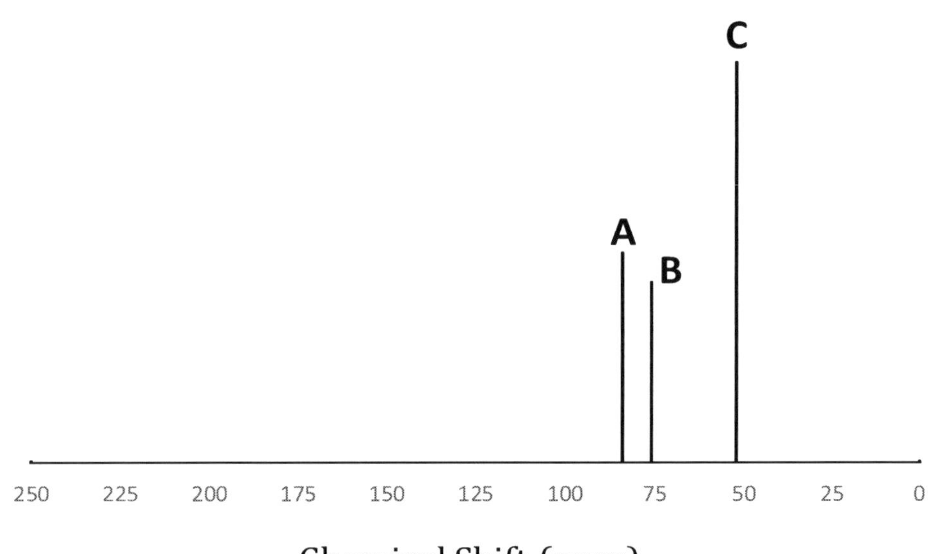


I

II

III

IV

Carbon-13 NMR Answer 3.

**Peaks**

**Carbonyls:** 0

**C=C:** 0

**Sp/highly E.N.:** 3

**Alkyl:** 0

---

**Answer Choices:**

I II III IV

**I** has a carbonyl-based functional group and would thus exhibit a peak in the **Carbonyls** region. **II** has an alkene and would thus exhibit 2 peaks in the **C=C** region. **III** contains a mirror plane and would thus only exhibit 2 peaks in the **sp/highly E.N.** region. **IV** is the only molecule consistent with the $^{13}C$ NMR spectrum, and the peak assignments are shown below:

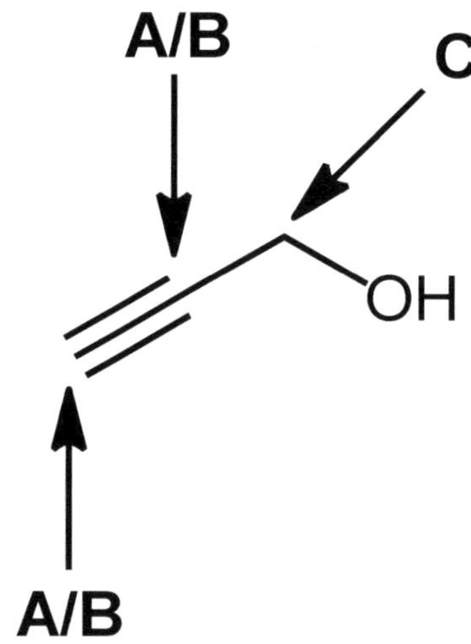

Carbon-13 NMR Question 4.

Which of the molecules **I-IV** would give rise to the $^{13}C$ NMR spectrum shown below?

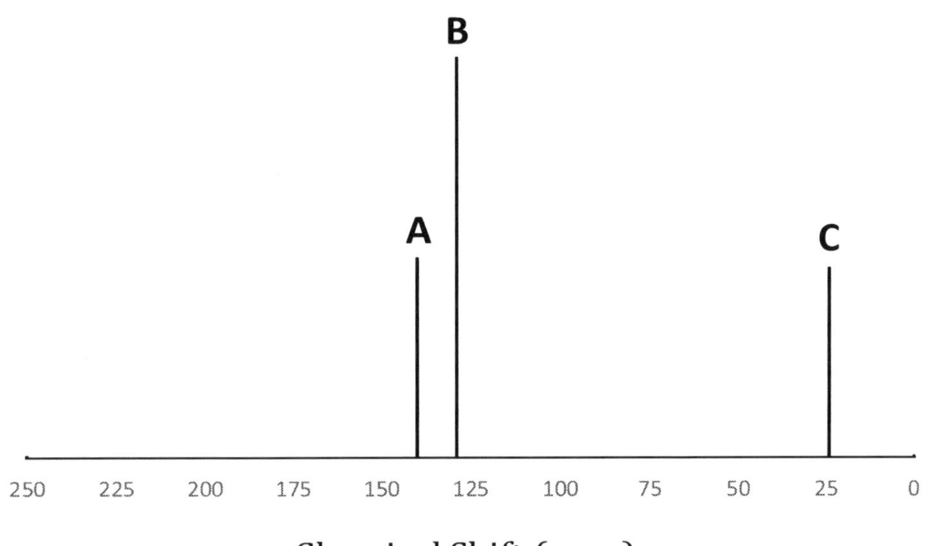


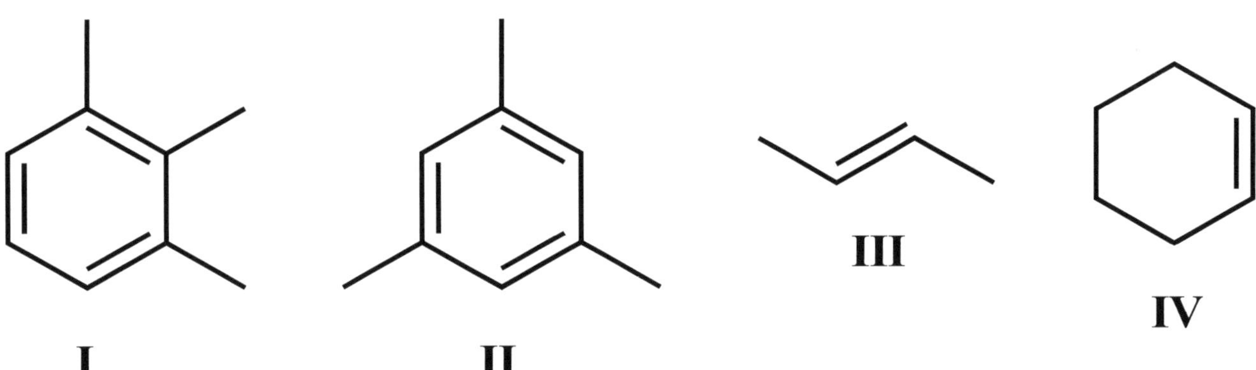

Carbon-13 NMR Answer 4.

**Peaks**

| | |
|---|---|
| **Carbonyls:** | 0 |
| **C=C:** | 2 |
| **Sp/highly E.N.:** | 0 |
| **Alkyl:** | 1 |

---

**Answer Choices:**

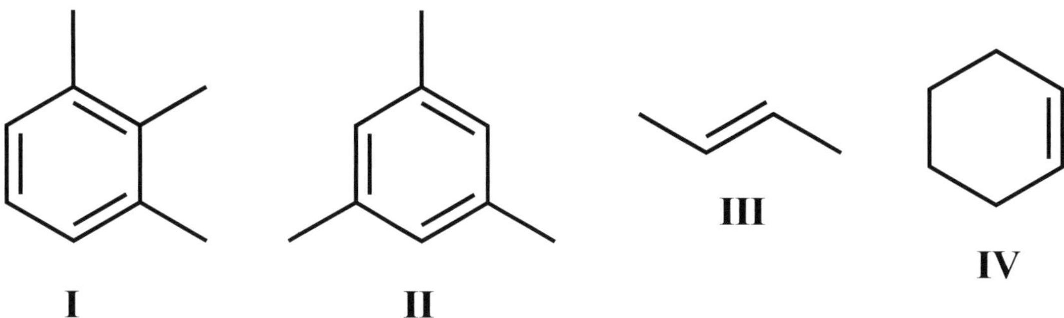


**I** and **IV** have a mirror planes but each would still have 2 inequivalent *sp*$^3$-carbons (without highly electronegative substituents), so each would exhibit 2 peaks in the **Alkyl** region. **III** has a mirror plane that would make the 2 alkene carbons equivalent, so it would exhibit only 1 peak in the **C=C** region. **II** is the only molecule consistent with the $^{13}$C NMR spectrum, and the peak assignments are shown below (the carbons with H-substituents will relax faster):

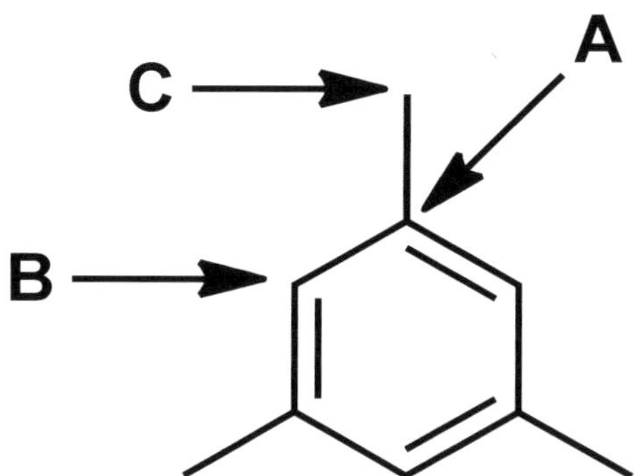

Carbon-13 NMR Question 5.

Which of the molecules **I-IV** would give rise to the $^{13}C$ NMR spectrum shown below?

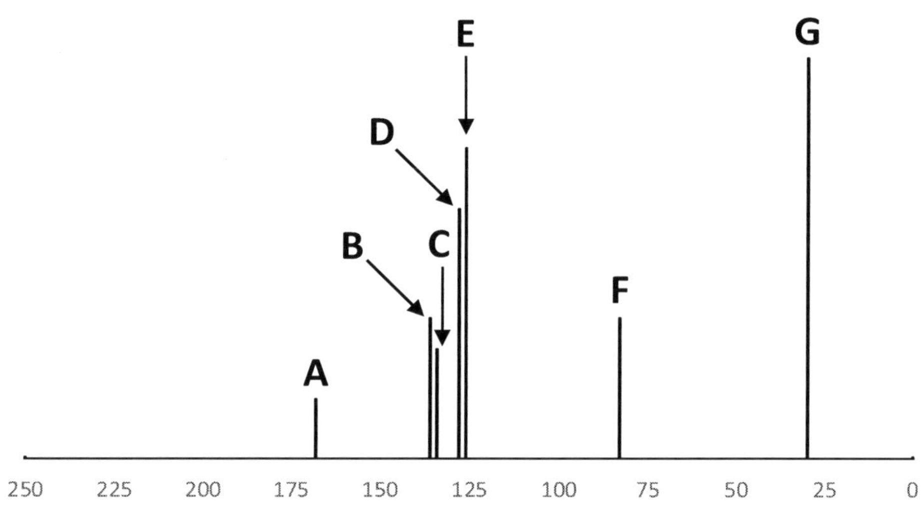


Chemical Shift (ppm)

I

II

III

IV

Carbon-13 NMR Answer 5.

**Peaks**

**Carbonyls:** 1

**C=C:** 4

**Sp/highly E.N.:** 1

**Alkyl:** 1

---

**Answer Choices:**

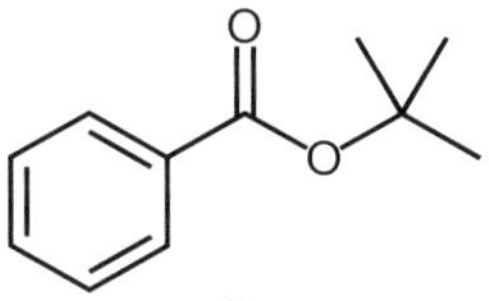

I 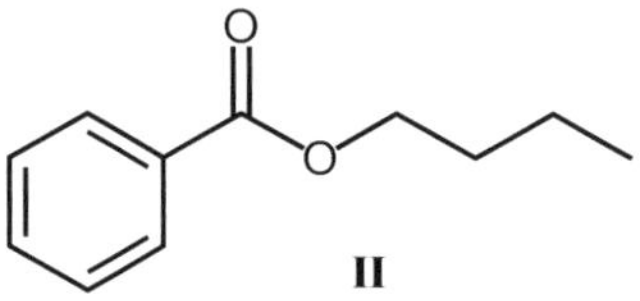 II 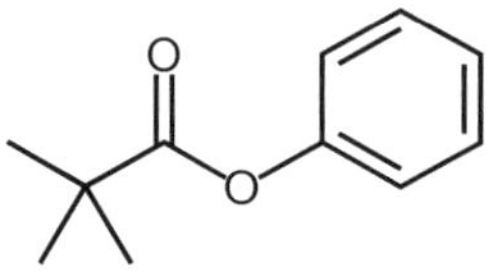 III 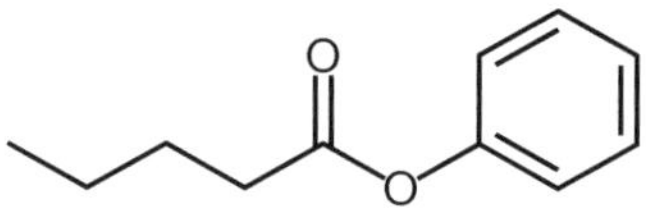 IV

**III** and **IV** do not have any $sp^3$-carbons with highly electronegative substituents, so neither molecule would exhibit any peaks in the **Sp/highly E.N.** region. **IV** has 1 $sp^3$-carbon with a highly electronegative substituent and 3 $sp^3$-carbons without highly electronegative substituents, so it would exhibit 1 and 3 peaks in the **Sp/highly E.N.** and **Alkyl** regions, respectively. **I** is the only molecule consistent with the $^{13}$C NMR spectrum, and the peak assignments are shown below (remember that carbons with H-substituents will relax faster and more equivalent carbons will give stronger peak intensities):

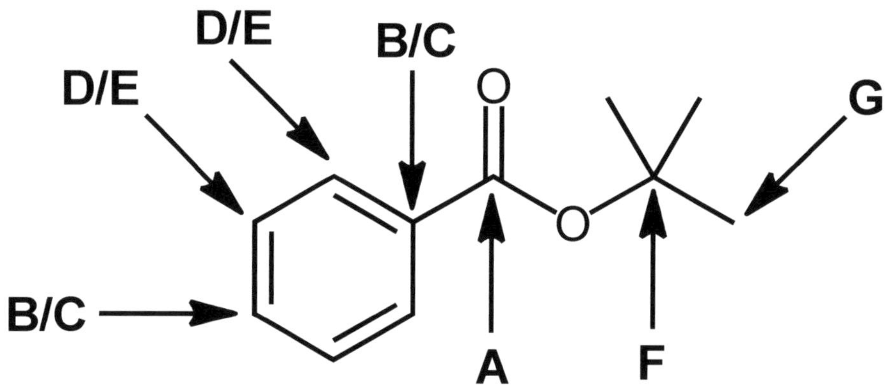

D/E D/E B/C G B/C A F

Carbon-13 NMR Question 6.

Which of the molecules **I-IV** would give rise to the $^{13}C$ NMR spectrum shown below?

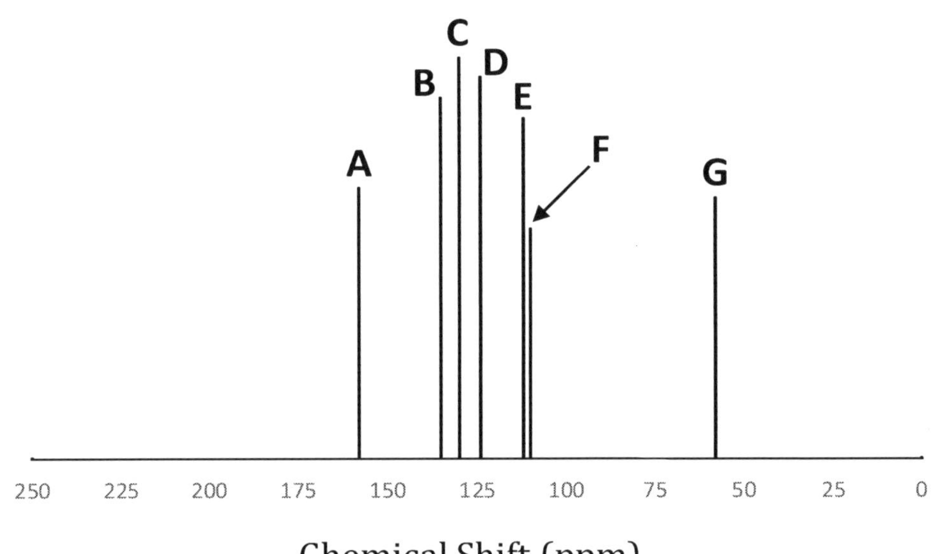


I: O (methyl ether), Br

II: O (methyl ether), Br

III: O (methyl ether), Br, Br, Br

IV: N, $NH_2$, Br, Br

Carbon-13 NMR Answer 6.

**Peaks**

**Carbonyls:** 0

**C=C:** 6 (1 is borderline with **Carbonyls**)

**Sp/highly E.N.:** 1

**Alkyl:** 0

---

**Answer Choices:**

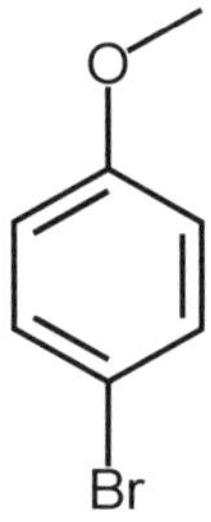

I 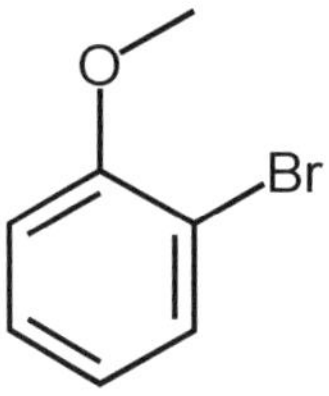 II 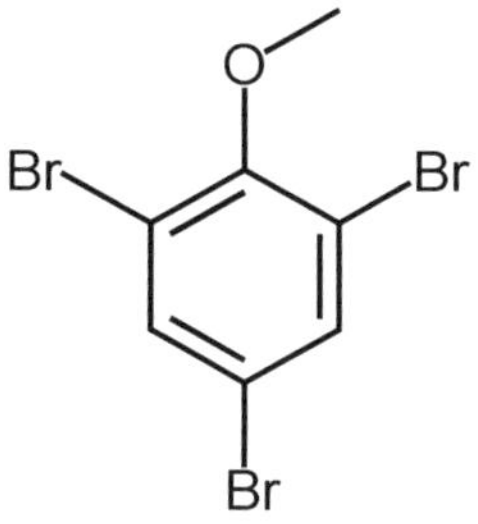 III 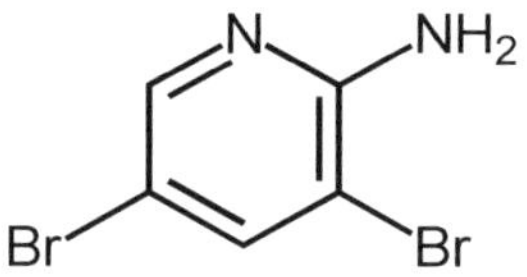 IV

**I** contains a mirror plane and would thus exhibit only 5 peaks in total, and only 4 peaks in the **C=C** region. Likewise, **III** contains a mirror plane and would thus exhibit only 5 peaks in total, and only 4 peaks in the **C=C** region (although its peaks would be shifted more downfield). **IV** only has 5 carbons in total and no heteroatoms that commonly cause peak splitting, so there is no way it could exhibit 7 total peaks in its $^{13}$C NMR spectrum. II is the only molecule consistent with this $^{13}$C NMR spectrum, and the peak assignments are shown below (remember that oxygen is a strong *ortho/para*-director, so it increases electron density at the positions *ortho* and, to a lesser extent, *para* relative to it!):

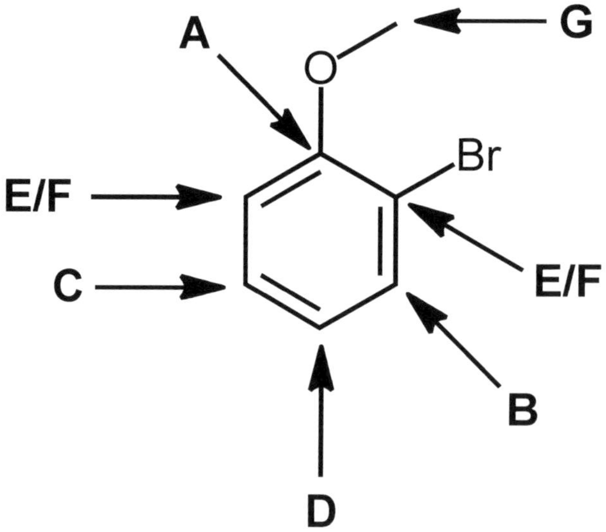

A G E/F E/F C B D

Carbon-13 NMR Question 7. Which of the molecules **I-IV** would give rise to the $^{13}C$ NMR spectrum shown below?

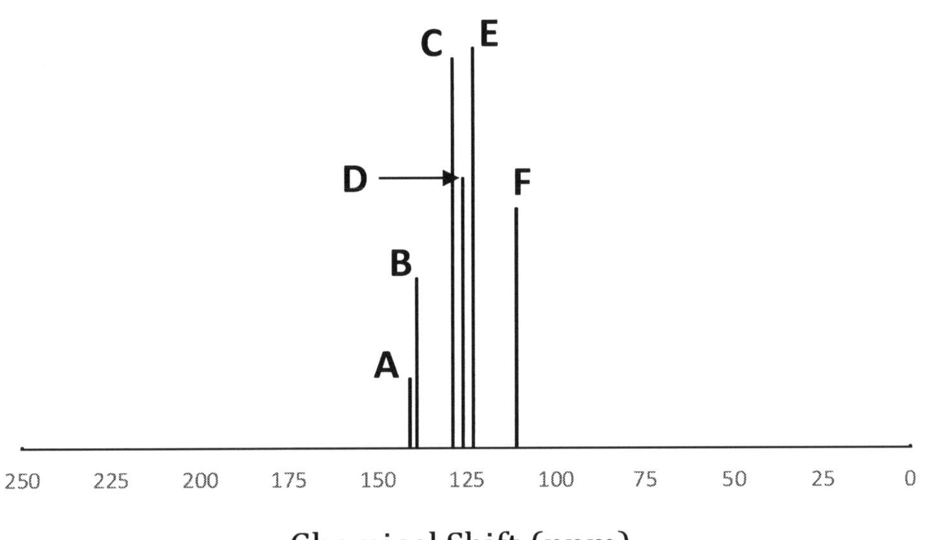


I

II

III

IV

Carbon-13 NMR Answer 7.

**Peaks**

| | |
|---|---|
| **Carbonyls:** | 0 |
| **C=C:** | 6 |
| **Sp/highly E.N.:** | 0 |
| **Alkyl:** | 0 |

---

**Answer Choices:**

I 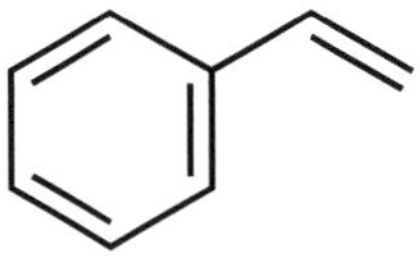 II 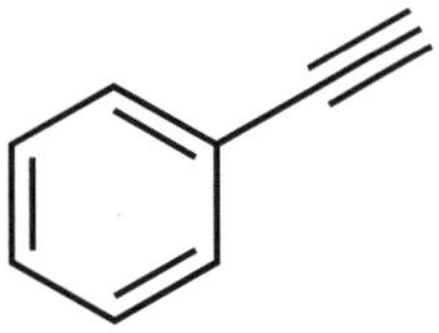 III 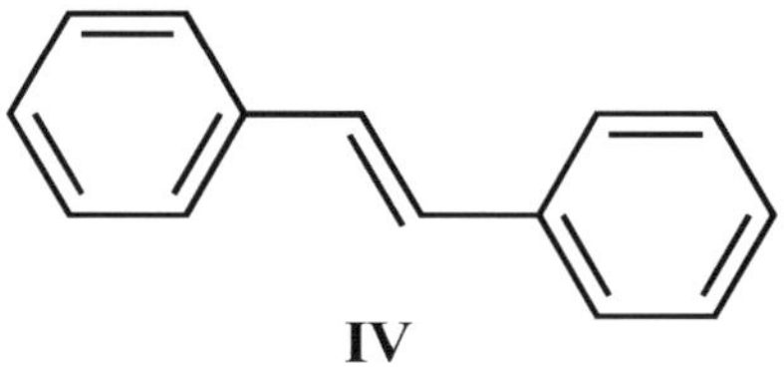 IV

**III** has 2 inequivalent *sp*-carbons and would thus exhibit 2 peaks in the **Sp/highly E.N.** region. **I** has only 3 inequivalent $sp^2$-carbons (rotate molecule by 180º perpendicular to plane of page to see why), so it would only exhibit 3 peaks in the **C=C** region. Similarly, **IV** has only 5 inequivalent $sp^2$-carbons (rotate molecule by 180º perpendicular to plane of page to see why, and remember C–C single bonds can rotate), so it would only exhibit 5 peaks in the **C=C** region. **II** is the only molecule consistent with the $^{13}$C NMR spectrum. Because all of the carbons are $sp^2$-hybridized with no heteroatom substituents, there is not enough information to be able to make any peak assignments.

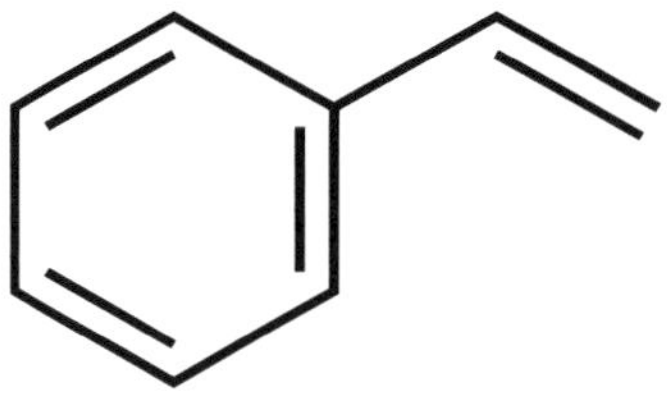

Carbon-13 NMR Question 8.

Which of the molecules **I-IV** would give rise to the $^{13}C$ NMR spectrum shown below?

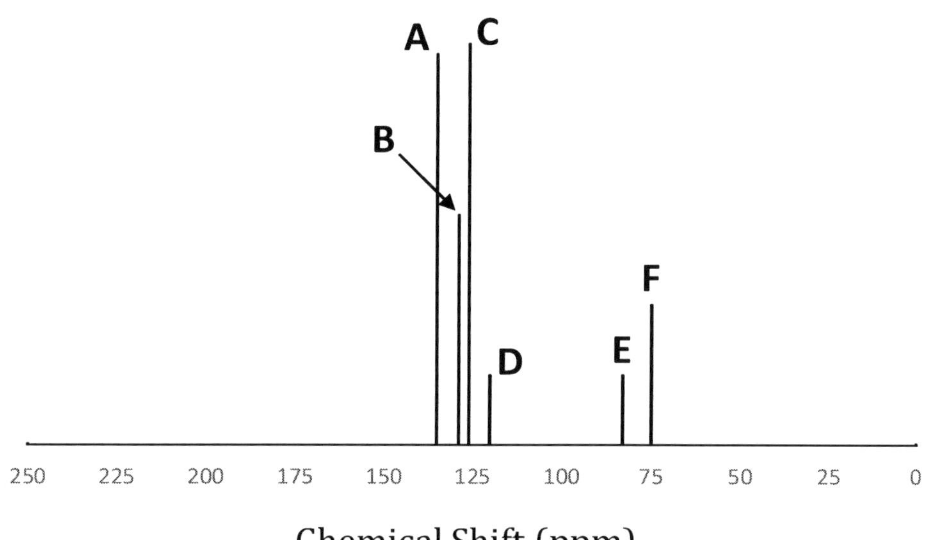


I

II

III

IV

Carbon-13 NMR Answer 8.

**Peaks**

| | |
|---|---|
| **Carbonyls:** | 0 |
| **C=C:** | 4 |
| **Sp/highly E.N.:** | 2 |
| **Alkyl:** | 0 |

---

**Answer Choices:**

I II III IV

**II** only has $sp^2$-carbons, so it would not exhibit any peaks in the **Sp/highly E.N.** region. **I** has multiple mirror planes and would thus only exhibit 2 peaks in the **C=C** region. **IV** also has multiple mirror planes and would only exhibit 1 peak in the **Sp/highly E.N.** region. **III** is the only molecule consistent with the $^{13}C$ NMR spectrum, and the peak assignments are shown below (remember that carbons with H-substituents will relax faster and more equivalent carbons will give stronger peak intensities):

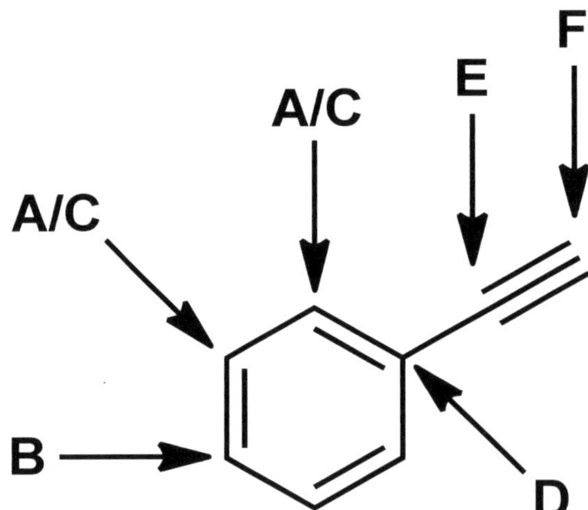

Carbon-13 NMR Question 9.

Which of the molecules **I-IV** would give rise to the $^{13}C$ NMR spectrum shown below?

***Hint:*** **Peak B is in the region for a C directly attached to a highly electronegative atom.**

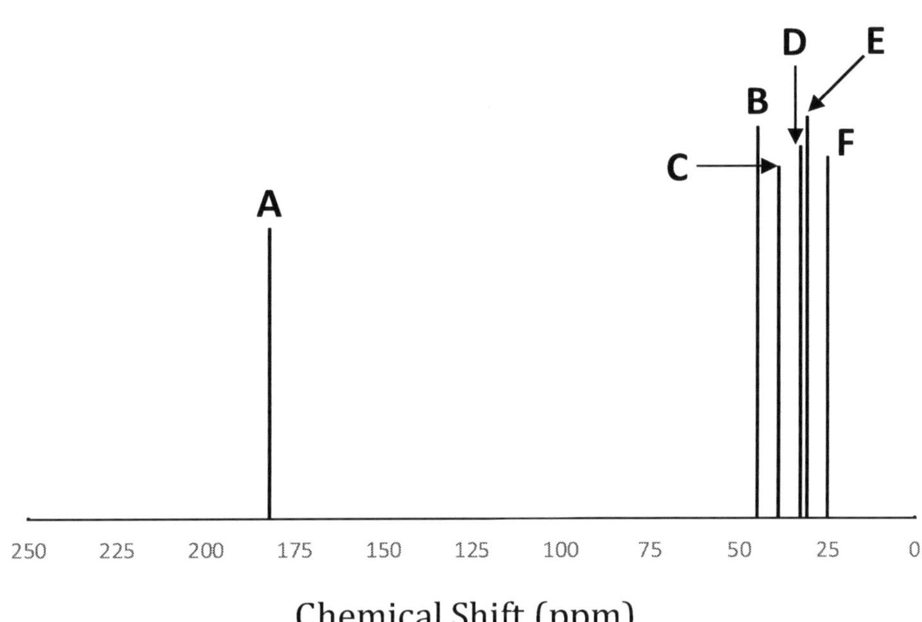


O, $NH_2$ — I

O, NH — II

O, NH — III

O, $NH_2$ — IV

Carbon-13 NMR Answer 9.

**Peaks**

**Carbonyls:** 1

**C=C:** 0

**Sp/highly E.N.:** 1

**Alkyl:** 4

---

**Answer Choices:**

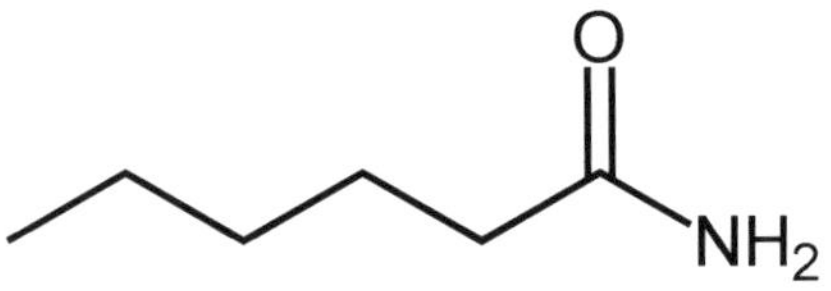

**I** has no substituents on the nitrogen (other than the carbonyl C that we will count in the carbonyl region peaks), so it would not exhibit any peaks in the **Sp/highly E.N.** region. **IV** has a phenyl ring, so it would exhibit 4 peaks in the **C=C** region. **III** only has 5 carbons, so it could not give rise to more than 5 peaks total in the spectrum. **II** is the only molecule consistent with the $^{13}$C NMR spectrum. We can easily assign peaks **A** (only 1 carbonyl in the structure) and **B** (only one carbon attached to a highly-electronegative substituent), but we do not have enough information to assign peaks **C-F** (this is a non-trivial exercise even for advanced graduate students).

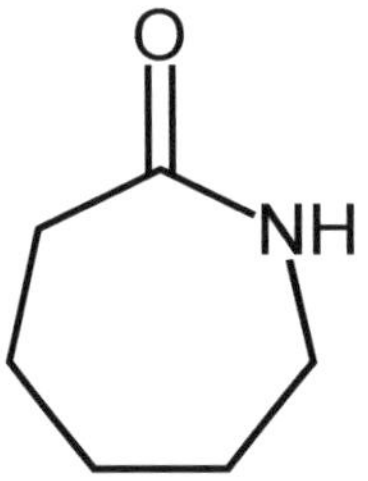

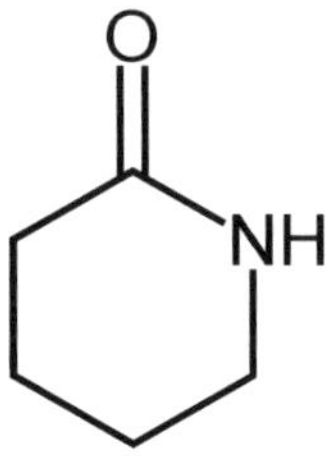

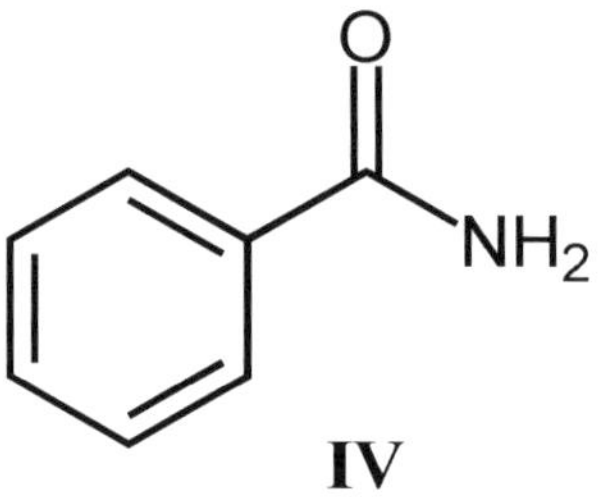

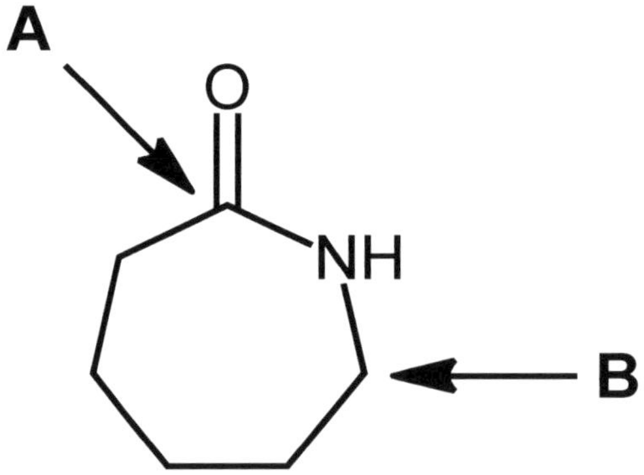

Carbon-13 NMR Question 10.

Which of the molecules **I-IV** would give rise to the $^{13}C$ NMR spectrum shown below?

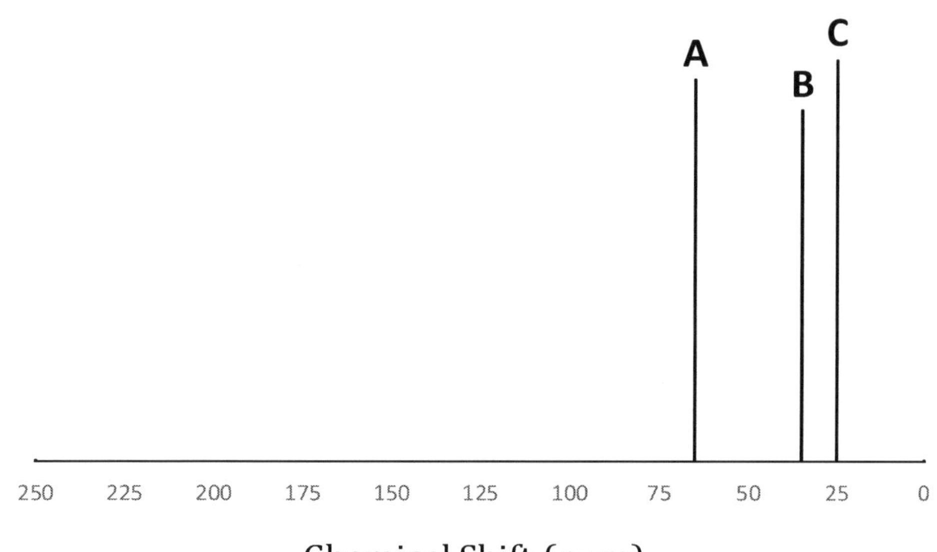


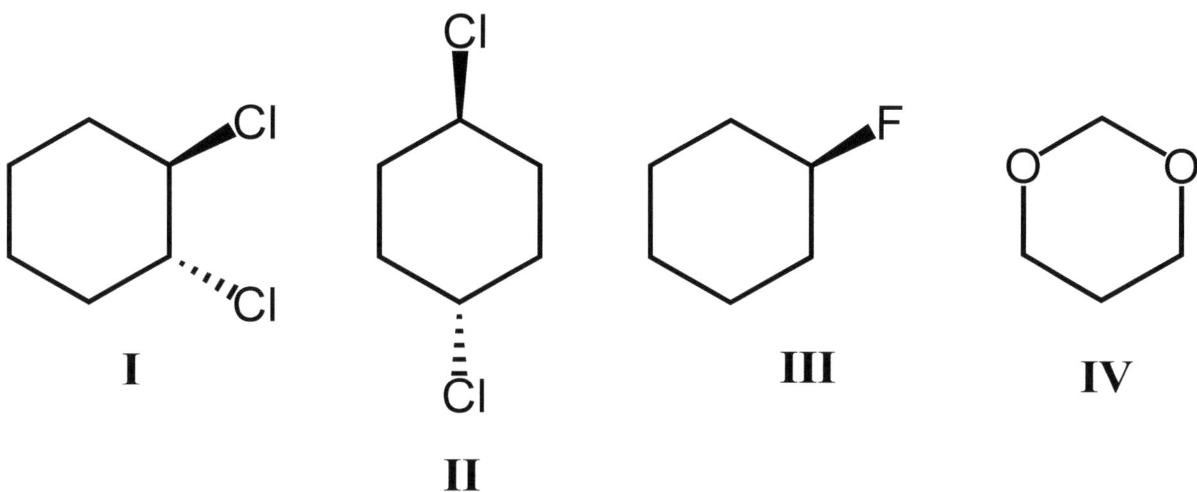

Carbon-13 NMR Answer 10.

**Peaks**

**Carbonyls:** 0

**C=C:** 0

**Sp/highly E.N.:** 1

**Alkyl:** 2

---

**Answer Choices:**

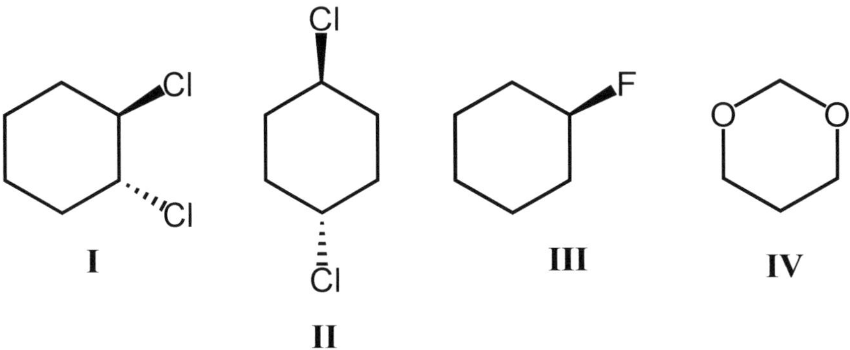


**II** only contains 2 unique carbons total (there is a mirror plane bisecting the carbons with the Cl substituents, and you can imagine rotating 180º on an axis containing the plane of the page to see why), and would thus exhibit only 1 peak in the **Sp/highly E.N.** region and 1 peak in the **Alkyl** region. **III** contains 4 inequivalent carbons. The mirror plane in **IV** leaves 3 inequivalent carbons, one of which is between two oxygens (and gets shifted up to the **C=C** region), so **IV** would exhibit 1 peak in the **C=C** region, 1 peak in the **Sp/highly E.N.** region, and 1 peak in the **Alkyl** region. **I** is the only molecule consistent with this $^{13}C$ NMR spectrum, and the peak assignments are shown below:

Carbon-13 NMR Question 11.

Which of the molecules **I-IV** would give rise to the $^{13}C$ NMR spectrum shown below?

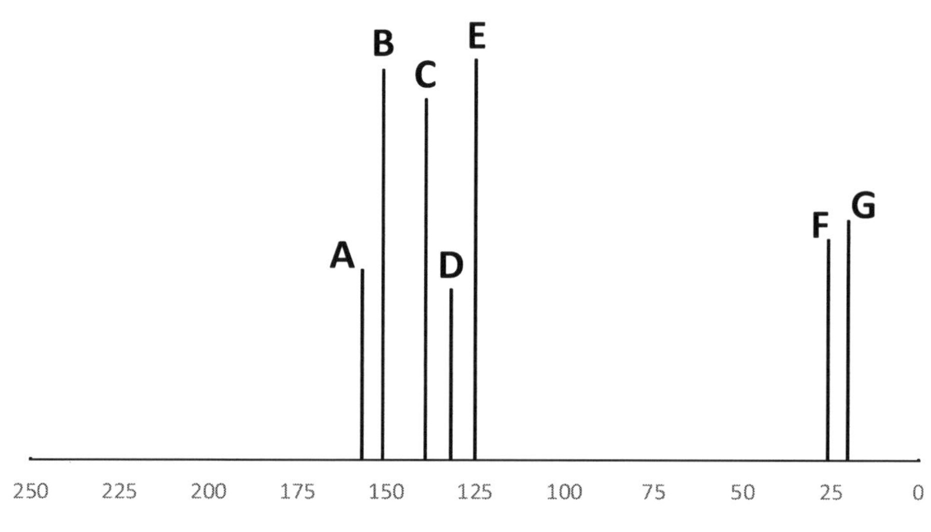


I II III IV

Carbon-13 NMR Answer 11.

**Peaks**

| | |
|---|---|
| **Carbonyls:** | 0 |
| **C=C:** | 5 (2 are borderline with **Carbonyls**) |
| **Sp/highly E.N.:** | 0 |
| **Alkyl:** | 2 |

---

**Answer Choices:**

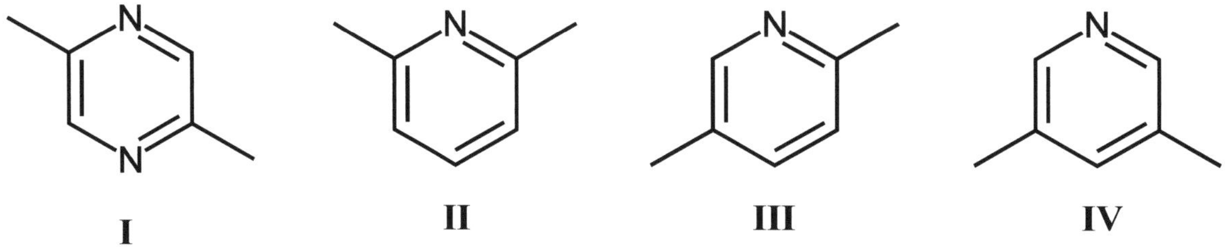

I II III IV

**I** only contains 3 inequivalent carbons (imagine rotating the molecule 180º perpendicular to the plane of the page to see why), so it would only exhibit 2 peaks in the **C=C** region and 1 in the **Alkyl** region. **II** and **IV** each contain mirror planes which means that each molecule only has 4 inequivalent carbons in total. **II** and **IV** would thus exhibit only 3 peaks in the **C=C** region and 1 peak in the **Alkyl** region. **III** is the only molecule consistent with this $^{13}$C NMR spectrum, and the peak assignments are shown below (carbons that have H substituents will give more intense peaks than carbons without H substituents, and carbons next to a highly-electronegative element will be more downfield than those not next to one):

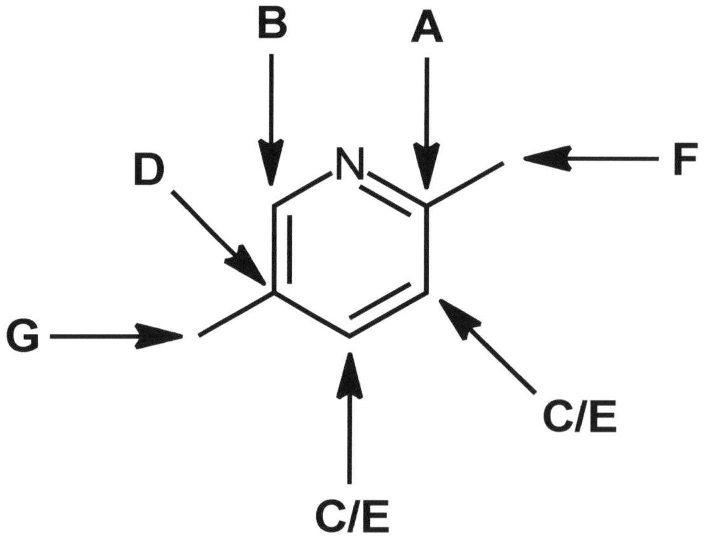

B A D F G C/E C/E

Carbon-13 NMR Question 12.

Which of the molecules **I-IV** would give rise to the $^{13}C$ NMR spectrum shown below?

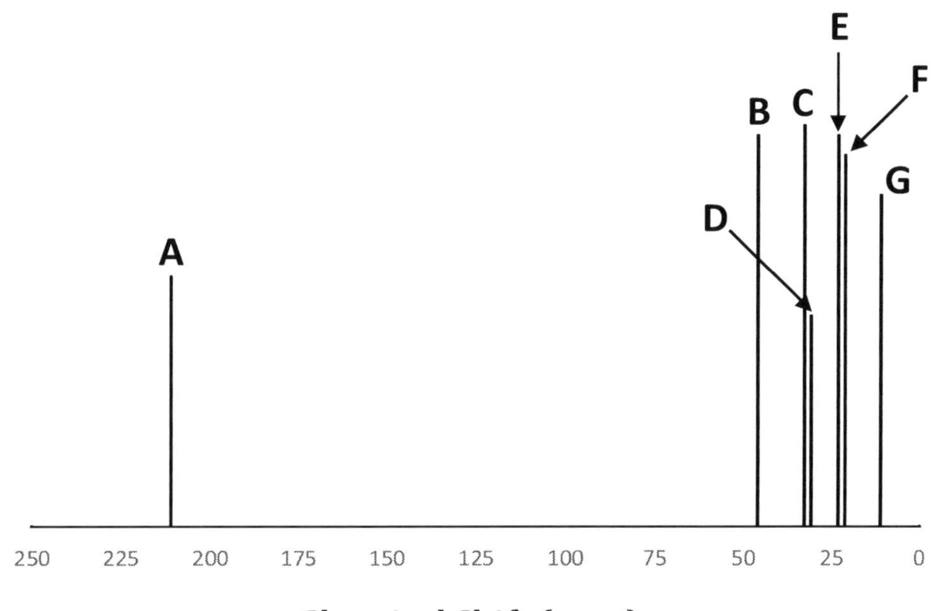


I II III IV

Carbon-13 NMR Answer 12.

**Peaks**

| | |
|---|---|
| **Carbonyls:** | 1 |
| **C=C:** | 0 |
| **Sp/highly E.N.:** | 0 |
| **Alkyl:** | 6 (1 is borderline with **Sp/highly E.N.**) |

---

**Answer Choices:**

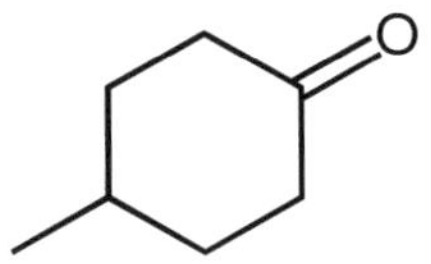

I

II

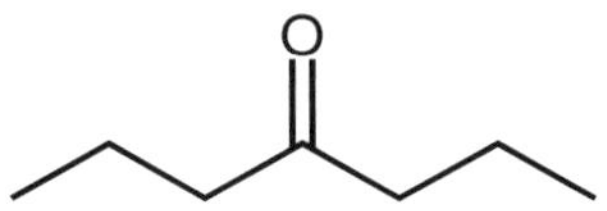

III

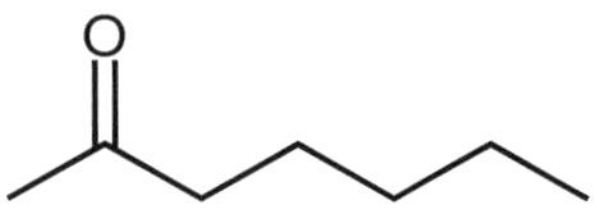

IV

**II** has 7 inequivalent carbons, but it would not exhibit any peaks in the **Carbonyls** region. **I** and **III** would each exhibit 1 peak in the **Carbonyls** region, but the presence of a mirror plane in each molecule would cause **I** to only exhibit 5 peaks in the **Alkyl** region, and **III** to only exhibit 4 peaks in the **Alkyl** region. **IV** is the only molecule consistent with the $^{13}C$ NMR spectrum. We can easily assign peak **A** (only 1 carbonyl in the structure), but we do not have enough information to assign peaks **B**-**G** (this is a non-trivial exercise even for advanced graduate students).

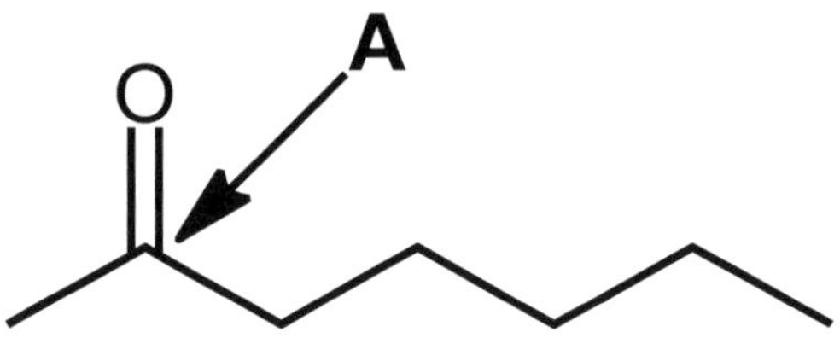

A

Carbon-13 NMR Question 13.

Which of the molecules **I-IV** would give rise to the $^{13}C$ NMR spectrum shown below?

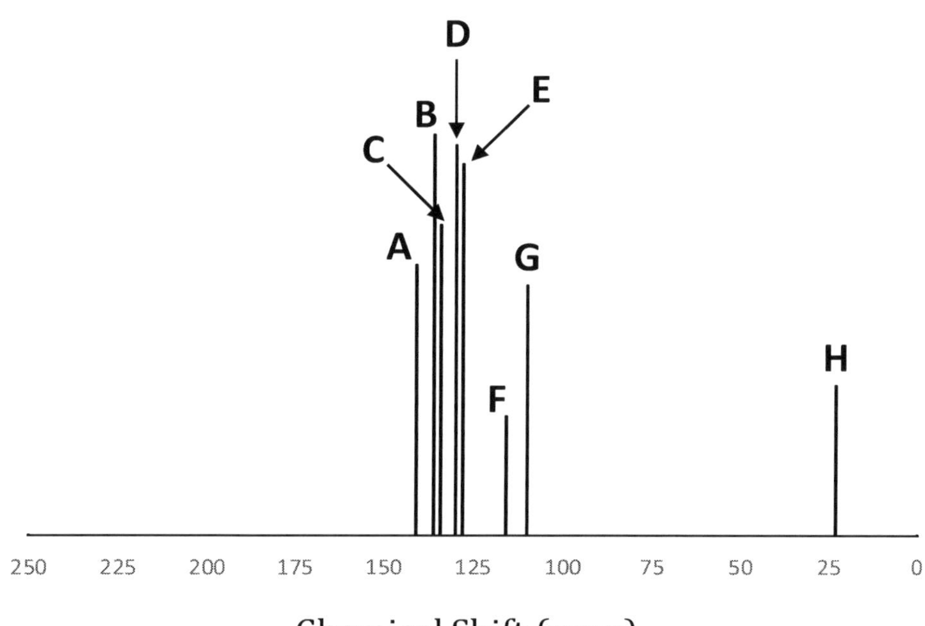


O

C≡N

C≡N

I II III IV

Carbon-13 NMR Answer 13.

**Peaks**

**Carbonyls:** 0

**C=C:** 7

**Sp/highly E.N.:** 0

**Alkyl:** 1

---

**Answer Choices:**

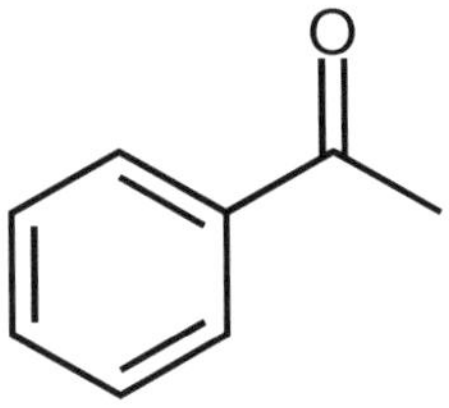

I 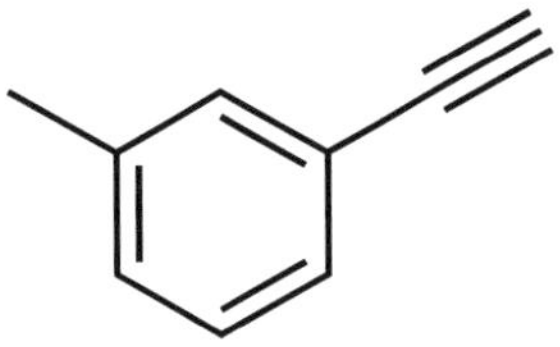 II 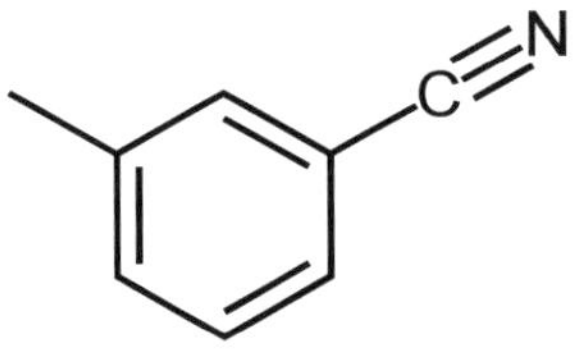 III 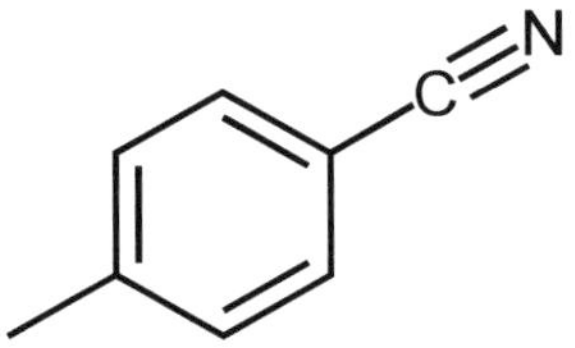 IV

**I** would exhibit 1 peak in the **Carbonyls** region and can be eliminated. **II** would exhibit 2 peaks in the **Sp/highly E.N.** region and can be eliminated. **III** and **IV** each have an *sp*-carbon attached to a highly-electronegative element, which shifts its signal into the **C=C** region. However, **IV** has a mirror plane and would thus have only 4 inequivalent aryl carbons, so it would exhibit only 5 peaks (4 aryl + 1 nitrile) in the **C=C** region. **III** is the only molecule consistent with the $^{13}C$ NMR spectrum. We can easily assign peak **A** (only 1 $sp^3$-carbon with no highly-electronegative substituents in this molecule), but we do not have enough information to assign peaks **B**-**H** (this is a non-trivial exercise even for advanced graduate students).

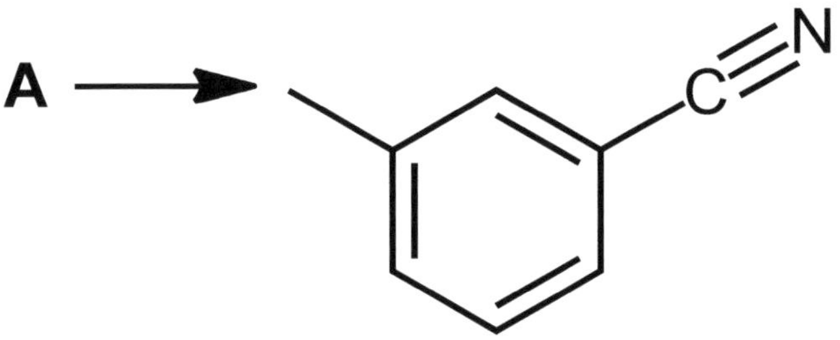

A

Carbon-13 NMR Question 14.

Which of the molecules **I-IV** would give rise to the $^{13}C$ NMR spectrum shown below?

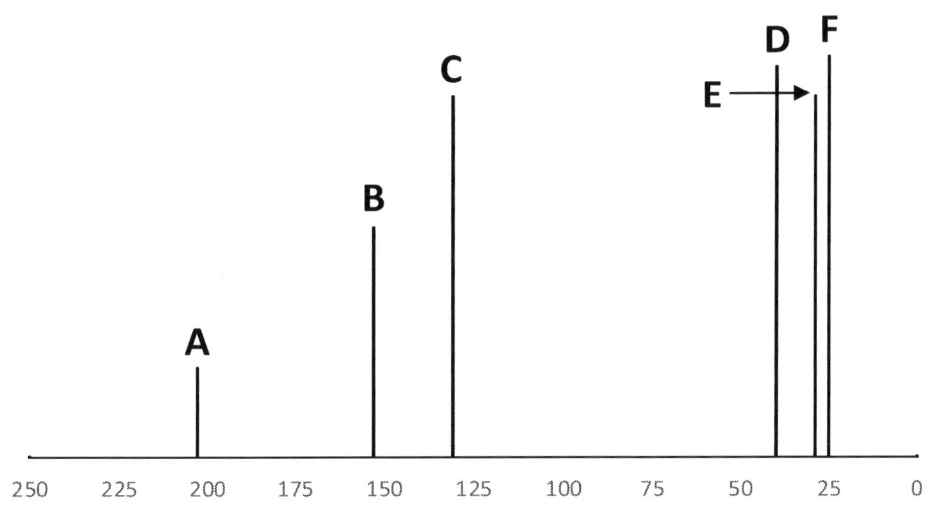

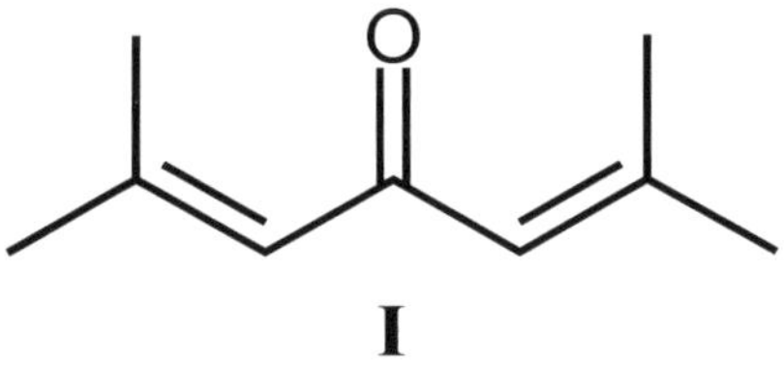

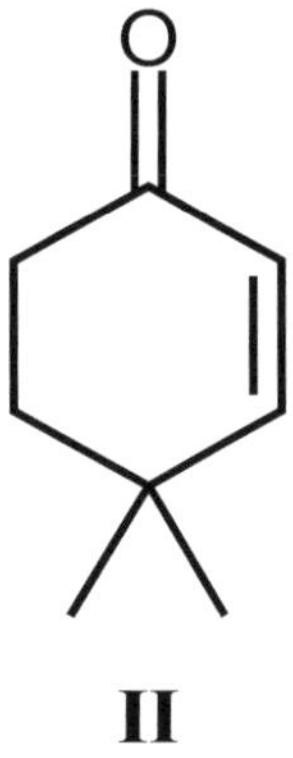

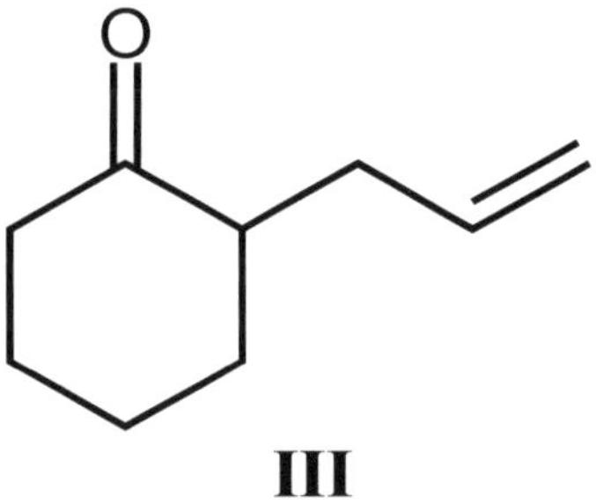

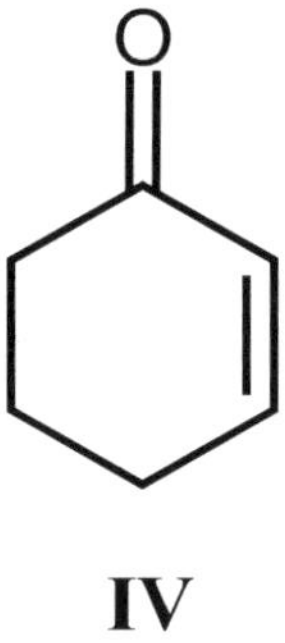

Chemical Shift (ppm)

I

II

III

IV

Carbon-13 NMR Answer 14.

**Peaks**

**Carbonyls:** 1

**C=C:** 2 (1 is borderline with **Carbonyls**)

**Sp/highly E.N.:** 0

**Alkyl:** 3 (1 is borderline with **Sp/highly E.N.**)

---

**Answer Choices:**

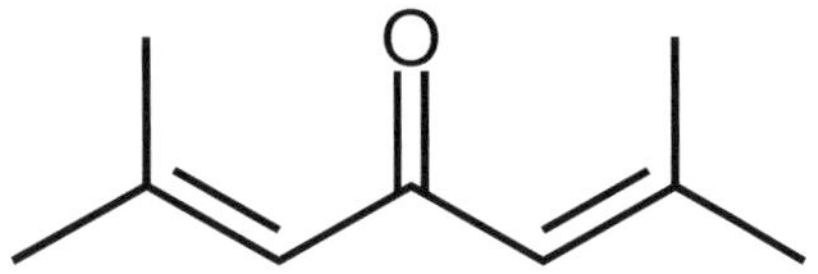

I

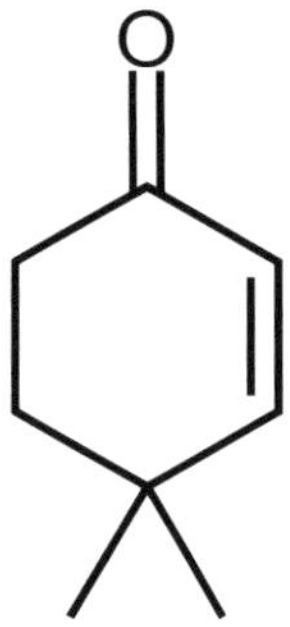

II

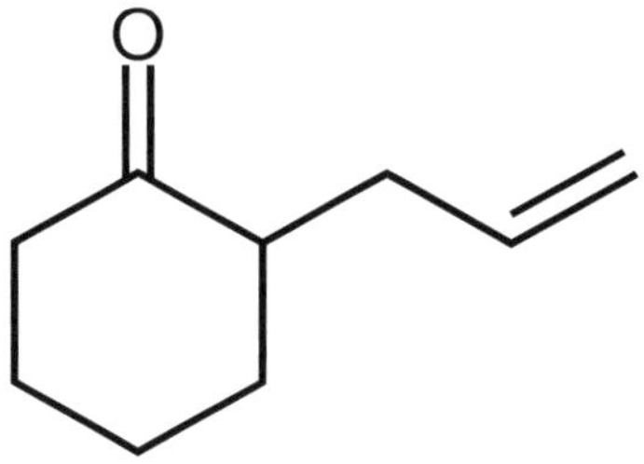

III

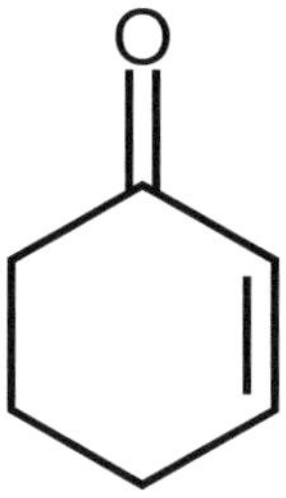

IV

**I** has a mirror plane and would thus exhibit only 2 peaks in the **Alkyl** region. The 2 methyl groups in **II** are equivalent, so this compound would exhibit 4 peaks in the **Alkyl** region. **III** would exhibit 6 peaks in the **Alkyl** region. **IV** is the only molecule consistent with the $^{13}$C NMR spectrum. We can easily assign peak **A** (only 1 carbonyl carbon in this molecule) and peaks **B-C** (only one alkene in this molecule), but we do not have enough information to assign peaks **D-F** (this is a non-trivial exercise even for advanced graduate students).

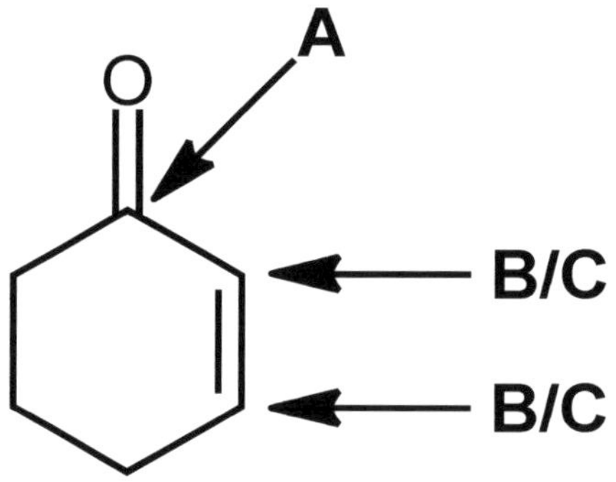

A

B/C

B/C

Carbon-13 NMR Question 15.

Which of the molecules **I-IV** would give rise to the $^{13}C$ NMR spectrum shown below?

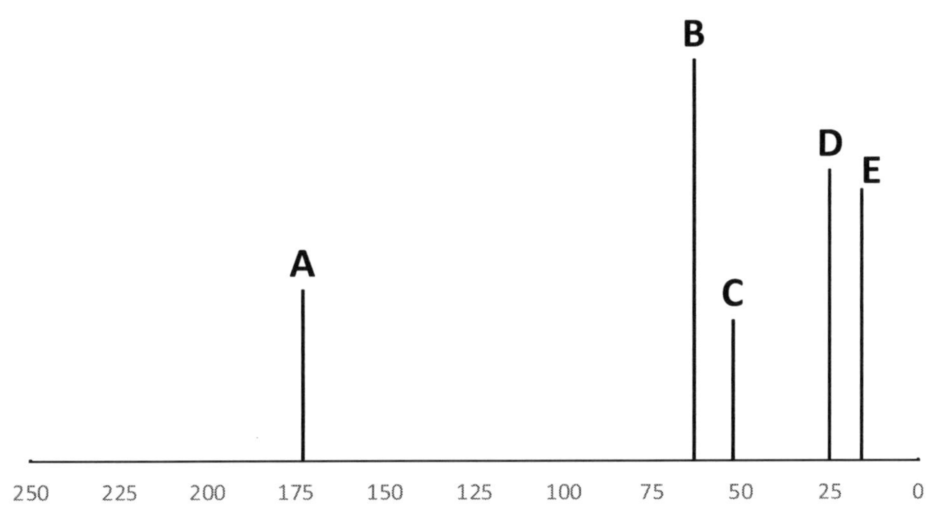


I II III IV

Carbon-13 NMR Answer 15.

**Peaks**

**Carbonyls:** 1

**C=C:** 0

**Sp/highly E.N.:** 2 (1 is borderline with **Alkyl**)

**Alkyl:** 2

---

**Answer Choices:**

I II III IV

**II** would exhibit 1 peak in the **Carbonyls** region, as well as 2 peaks in the **Sp/highly E.N.** region because the 2 methoxy carbons and 2 alkyne carbons are equivalent (imagine rotating the molecule 180º perpendicular to the plane), but it would not exhibit any peaks in the **Alkyl** region. **III** would exhibit 5 peaks in the **Alkyl** region, **IV** would exhibit 4 peaks in the **Alkyl** region, and neither molecule would exhibit any peaks in the **Carbonyls** region. **I** is the only molecule consistent with the $^{13}C$ NMR spectrum, and the peak assignments are shown below (a carbon between 2 carbonyl groups can get shifted into the **Sp/highly E.N.** region because it is beside **two** Cδ+ atoms!):

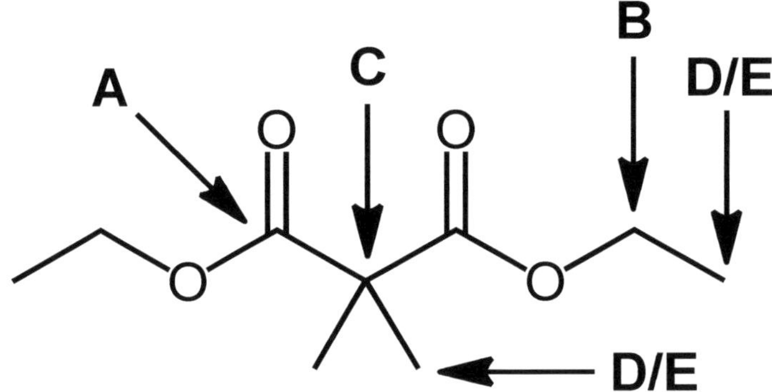

Carbon-13 NMR Question 16.

Which of the molecules **I-IV** would give rise to the $^{13}C$ NMR spectrum shown below?

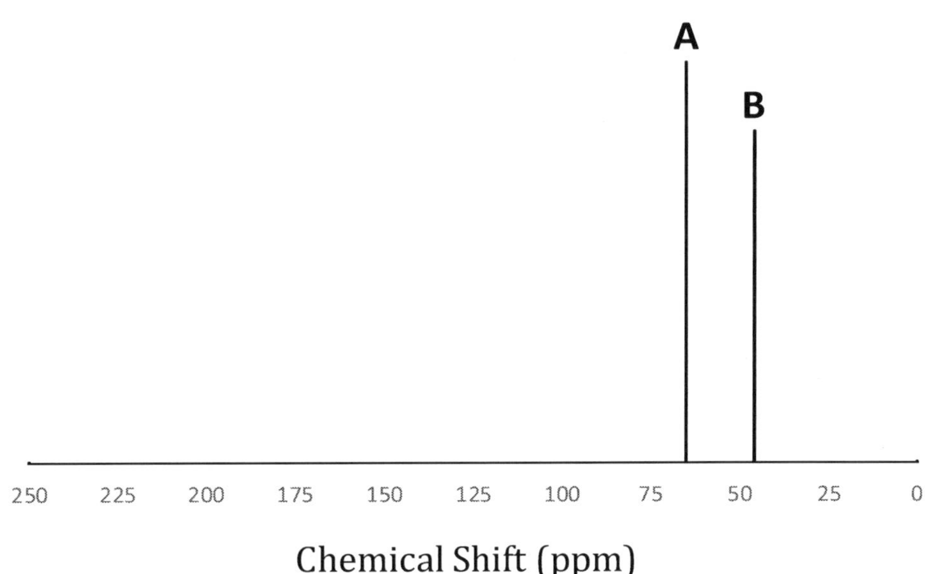


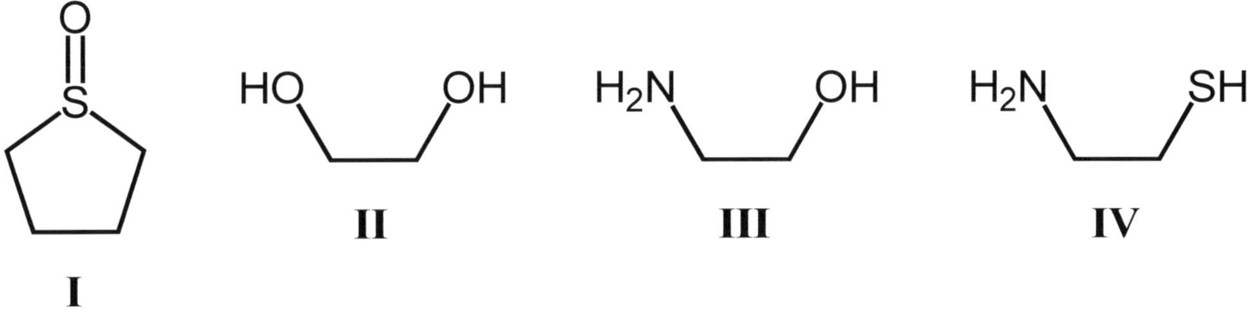

Carbon-13 NMR Answer 16.

**Peaks**

**Carbonyls:** 0

**C=C:** 0

**Sp/highly E.N.:** 2

**Alkyl:** 0

---

**Answer Choices:**

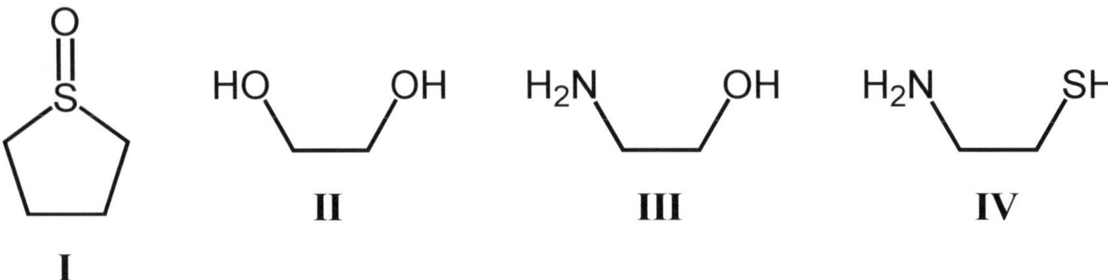


**I** would exhibit 1 peak in the **Alkyl** region, and not anywhere near borderline. The mirror plane in **II** makes the 2 carbons equivalent, thus it would exhibit only 1 peak total and it would be in the **Sp/highly E.N.** region. Sulfur is not a highly-electronegative element, therefore **IV** would exhibit 1 peak in the **Sp/highly E.N.** region and 1 peak in the **Alkyl** region. **III** is the only molecule consistent with this $^{13}C$ NMR spectrum, and the peak assignments are shown below (oxygen is more electronegative than nitrogen):

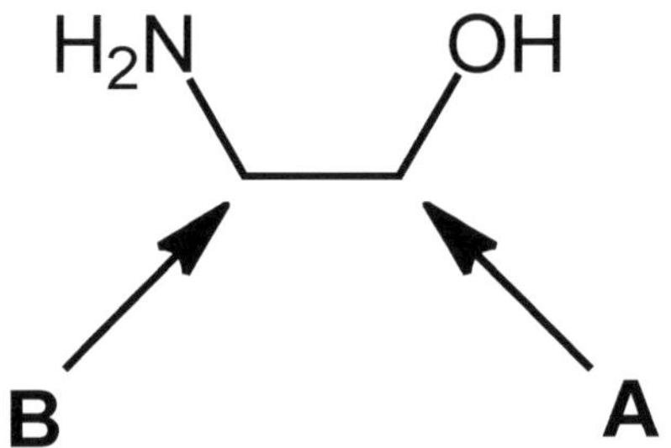

# INDEX

## 1



## A



## B



## C



## D



## E

## N



## O



## P



## R



## S

## T



## U



## V



## W



## X



## Y



## α

Made in the USA
Columbia, SC
30 December 2020

30063523R00187